Methods and Materials
of
Residential Construction

Methods and Materials
of
Residential Construction

Second Edition

ALONZO WASS

RESTON PUBLISHING COMPANY, INC., Reston, Virginia

A Prentice-Hall Company

Library of Congress Cataloging in Publication Data

Wass, Alonzo.
  Methods and materials of residential construction.

  Includes index.
  1. House construction. 2. Building materials.
I. Title.
TH4811.W36 1977   690'.8   76-39990
ISBN 0-87909-488-5

© 1977 by
Reston Publishing Company, Inc.
A Prentice-Hall Company
Reston, Virginia 22090

10   9   8   7   6   5   4   3

Printed in the United States of America

Contents

3 PRELIMINARY BUSINESS ORGANIZATION AND LAND GRADING   35

4 TYPICAL WALL SECTIONS, MATERIALS GUIDE LIST, TABLES, AND QUICK CALCULATIONS   62

5 PRELIMINARY BUILDING OPERATIONS   79

6   FOUNDATIONS, EXCAVATING, AND DEPRECIATION   86

7   FOUNDATIONS, PILES, CHIMNEYS, AND SLAB ON GRADE   98

8   FORMWORK AND CONCRETE   108

9   BASEMENT WALLS, COLUMNS, AND FIRST FLOOR ASSEMBLY   144

10   WALL FRAMING, ANCHORS, LUMBER STANDARDS, FASTENERS, AND SCAFFOLDING   165

11   POST AND BEAM CONSTRUCTION   195

12   MASONRY UNITS, GLASS BLOCKS, AND GLIDING DOORS   207

13   CEILINGS AND FLAT ROOFS   223

14   PITCH ROOFS, CHIMNEY FLASHINGS, AND ROOFING MATERIALS   236

15 WALL CLADDING, FIBER BOARDS, INSULATION, AND VAPOR BARRIERS — 269

16 STAIRS, FIREPLACES, AND INTERIOR TRIM — 293

17 PAINTING, DECORATING, FLOOR FINISHING, AND EXTRUDED METAL — 306

18 SUBTRADES — 313

19 LANDSCAPING, FINAL INSPECTION, NOTARIZATION OF DOCUMENTS, AND PROPERTY SALE 317

Preface

This book is designed for educators, students, builders of residences and apartment blocks, building superintendents, technicians, foremen, building subtradesmen, and building operatives.

For educators, the book offers a course of basic study, with review questions that should be supplemented with field trips and lectures by professionals with on-the-job local experience; it also offers a wide scope for those participating in continuing educational programs.

The book has been updated for its second edition, and the following material has been added: partnerships, corporations, bankruptcy, land grading, sewage disposal, architectural drawing practices and symbols, balloon and Western framing; also, more review questions have been added to stimulate student interest in local building by-laws and field trips as suggested in Article 1.7 Class Research.

A critically selected sampling of methods and materials for residential construction is presented in step-by-step order of chapters, with contemporaneous pictorial renderings of materials as they are used on the job.

An attempt has been made to alert the builder to be competitive, and to keep abreast with the latest technical advantages of methods and materials for residential construction, remembering that homes are becoming progressively more comfortable, colorful, hygienic, more aesthetically designed, and incorporate more and more up-to-date technical advances for easy and graceful living.

The first chapters discuss safety; job control; and an introduction to the critical path method. Many useful tables of weights, measures, and capacities are included, also methods of rapid calculations. Then follows: preliminary building operations, excavations, concrete formwork and tie loading, concrete, foundations, typical wall sections, and floor assemblies. These operations develop into discussions on wall framing, scaffolding, post and beam construction, masonry and glass units, curtain walls, ceilings, flat and pitch roofs, trusses, chimneys and chimney flashing. These presentations lead to discussions on wall cladding, exterior and interior trim, stairs, fireplaces, hardware, painting and decorating, driveways and landscaping.

The book would be incomplete without an introduction to methods and materials of some legal and other necessary paperwork; the text concludes with the final inspection of the property, its sale, and some legal aspects of notarization of documents.

The author acknowledges with sincere thanks the many contributors, who through their unstinting help have made this book possible. For information on new materials, the author recommends readers to consult "Sweet's Architectural Catalog File," which comprises twelve volumes (17,000 pages) and is represented by 1,500 leading manufacturers.

Again I extend my sincere thanks to Mrs. M. A. Olivier for her continued cheerfulness, efficiency and suggestions during the writing of the manuscript. Finally, acknowledgement is also made to Mr. Bryan H. Smith, educator, former builder, and friend for his many valuable suggestions, and to my wife Joan Rose Wass for her patience and firmness in having me lay down my writing when I should.

In conclusion for your consideration I would say, "He that builds *only* for profit, profits little."

Alonzo Wass

Job Safety, Workmen's Compensation, Insurance, Partnerships, Bankruptcy, and Class Research

In this chapter we shall discuss some aspects of job safety, workmen's compensation and insurance; afterwards, we shall present a discussion on partnerships, corporations, bankruptcy, and local and national construction associations. An awareness of all the foregoing is necessary to get in, be in, and stay in the challenging and rewarding business of residential construction. The chapter concludes with a suggested program for class research.

1.1 JOB SAFETY

Accidents (acts of God excepted) are nearly always preventable. In all human activities with which we are familiar, we tend to become casual; motor transport is a classic example. As soon as an accident occurs we hear said, "If only. . . ." **Remember that the more accident claims that are recorded against a builder, the more he has to pay for his building insurance coverage.**

The driving force for keeping a job site accident free lies with the superintendent. Some accident conscious superintendents call for a meeting every Monday at 8:00 a.m. for a pep talk on accident prevention. A friend of mine used to call a short meeting at irregular times when the job had been accident

free for a long time. He would address the men something like this:

I want to congratulate everyone for our long run without an accident; now, let's not become casual. When you get back to your work, I want you to clear the scaffolds or other work areas of all debris such as dried blobs of mortar or concrete, brick ends, broken pieces of concrete block, pieces of wood with nails sticking through them, short ends of piping and so on. See to it that all ladders extend at least 4′–0″ above the height of the scaffold, that all excavations on this particular job that are more than 4′–0″ in depth have shoring supporting their sides. Please be accident conscious yourselves, and see to it that no one else takes any risks by using machinery without guards, or grinding without safety glasses or working without a hard hat, or without earmuffs when necessary. Thank you for your care. Remember that keeping a job-site clean is almost a trade in itself. Good luck, good building, and thank you.

All posters, notices and warning signs around the job site should be changed at least once a week. The eye becomes accustomed to a pattern on a wall, but frequent startling and arresting new posters will attract the eye and thus become a weekly reminder to take care.

In the United States the average 10-year period of occupational injuries suffered was as follows:

     17,040 workers killed or totally disabled for life
     83,500 workers partially disabled for life
1,893,550 workers temporarily disabled.

The cost to the employers in the U.S.A. of occupational injuries during a single year is estimated by the U.S. Department of Labor at $3.5 billion, and the cost to the accident victims is $1.4 billion.

1.2 WORKMEN'S COMPENSATION AGENCIES

All workmen's compensation agencies publish general information, regulations, booklets, posters, films, notices, warning signs, signal cards, clues to accident protection, safety literature for inclusion in workmen's pay checks, first-aid service requirements, and methods of artificial respiration. You must become familiar with them!

Prominently display by each telephone the numbers to call in an emergency for such services as: doctors, ambulance, fire, police, power and gas companies, city engineer, sanitation department, building inspectors, and the workmen's compensation agency.

The supervisor or a senior assistant of the local workmen's compensation board would make an ideal guest speaker at one of your association meetings. He will also show you some very impressive films. The board issues some very striking posters which should be prominently displayed around your jobs. They have the right to come onto your project at any time to check that all safety precautions are being observed, and that your medical kit or first-aid center is adequate for the job. Minimum standard kits are required according to the number of employees.

The question often arises with both workmen and employers as to how serious an injury must be before a report is made and first-aid care is sought. *All accidents should be reported.* Many workmen's compensation boards will honor any claim up to three years providing that an entry has been made at the time of accident in the accident report book or the builder's diary. Assume that a person slips and hurts the lower part of his back. No one knows whether or not this will develop into a serious malignant growth. It would be very difficult for the individual to register a claim unless a record was made at the time of the mishap. In the case of a married man with children (unless he could register a successful claim), it would be a tragedy. **All accidents, however minor, should be reported at the time of accident; check with the local workmen's compensation board.**

The charges of the board for workmen's compensation is about two per cent of the wage bill for ordinary construction. Do not start to work on a building site until you have covered every workman by this insurance. If the subcontractor has not legally covered his own men, he and his men are in your employ and you will be held responsible. Check that each subcontractor has insurance for his own men.

Field First-Aid Rooms. The minimum requirements for any building operation *however small* is a first-aid kit, and the agency for workmen's compensation and employers' liability has information on the minimum first-aid requirements for any job, *however large.*

It is most important for you as an employer to be fully cognizant of your duties and responsibilities for accident prevention and for the administration of first-aid procedures to the sick and/or injured on your job site. You should get to know the officers of your insurance agency and have them give a talk to your staff.

First-aid notices (similar to the following) should be prominently displayed on the job near the first-aid room:

NOTICE

"When a qualified first aid attendant is engaged to render first-aid to injured workmen, he shall be in complete charge of all first-aid treatments required. No person shall interfere with him while he is thus discharging his proper first-aid duties, or while he is getting an injured workman to the nearest doctor or hospital to obtain treatment appropriate to the case.

Your cooperation is requested."
To be kept posted in all first-aid rooms.

Author's Note

See the following 'Employers Report of Injury or Industrial Disease.' *Be sure to make sufficient copies and be sure to answer every question, and sign it. About 30% of all official forms are returned because they have not been correctly filled out.*

The following material has been excerpted from the fourth edition of *Insurance For Contractors* by Walter T. Derk, published by Fred S. James & Company, 230 West Monroe, Chicago, Illinois, 60606. The

WORKMEN'S COMPENSATION BOARD OF BRITISH COLUMBIA

707 West 37th Avenue, Vancouver 13, B.C. Telephone 266-0211 Telex 04-507765

				Claim No.
Firm No.	Loc.	Class & Sub.	Coded by	

Employer's Report of Injury or Industrial Disease

FORM 7
REVISED 1969

SIGN HERE

Answer all questions, SIGN and mail to: Workmen's Compensation Board, 707 West 37th Avenue, Vancouver 13, B.C.

EMPLOYER'S NAME (please print)

WORKMAN'S LAST NAME (please print)
Mr.
Mrs.
Miss

Mailing Address

First Name(s)

CITY　　　　　ZONE

Full Address
NO.　　STREET/R.R　　CITY & ZONE

Location of Plant or Project

Date of Birth　　　　Social Insurance No.
MONTH　DAY　YEAR

Type of Business　　Employer's Telephone No.

Occupation　　　　Marital Status
MARRIED ☐　SINGLE ☐　OTHER

1. Date and time of injury　　OR Period of Exposure Resulting in Industrial Disease:
   19　, at　A.M./P.M.　FROM　19　TO　19

2. Injury was first reported to employer on
   19　, at　A.M./P.M.

IF EXPLANATIONS TO THE FOLLOWING QUESTIONS ARE REQUIRED, USE SPACE PROVIDED BELOW. USE SEPARATE NOTE OR LETTER IF NECESSARY.

3. Describe fully what happened to cause the injury and mention all contributing factors: description of machinery, weight and size of objects involved, etc.

OR In cases of industrial disease, describe fully how exposure occurred, mentioning any gases, vapours, dusts, chemicals, radiation, noise, source of infection or other causes.

8. Were workman's actions at time of injury for the purpose of your business? If NO, explain. ☐ YES ☐ NO

9. Were they part of workman's regular work? If NO, explain. ☐ YES ☐ NO

10. Are you satisfied injury occurred as stated by workman? If NO, explain. ☐ YES ☐ NO

11. Have witnesses been interviewed? ☐ YES ☐ NO
    If YES, do witnesses confirm workman's statement? ☐ YES ☐ NO

12. Did workman receive first aid? If YES, attach form 7A. ☐ YES ☐ NO

13. Is workman a relative of employer or a partner or principal in firm? If YES, specify. ☐ YES ☐ NO

14. Was any person not in your employ to blame for this injury? If YES, give details and name and address of such person. ☐ YES ☐ NO

State ALL injuries reported, indicating right or left if applicable.

4. City, town, or place where injury occurred.

5. How long has workman been employed by you?

15. Medical Insurance No. (MSA) or (BCMP)

6. How long at this particular job?

16. Name and address of physician or qualified practitioner.

7A. Was there any layoff beyond the day of injury? ☐ YES ☐ NO
7B. Was layoff for three working days or more? ☐ YES ☐ NO
    If YES, to part B, complete section below

1. Workman's gross earnings (enter one rate only) at time of injury
   per day $　　per week $　　per month $

7. Date and time he last worked.

2. If free room and/or meals were supplied in addition to above earnings indicate daily value. $

8. Enter normal working hours on day of lay off.
   From　　A.M./P.M., to　　A.M./P.M.

3. Do these earnings include rental of equipment? If YES, specify. ☐ YES ☐ NO

9. Enter normal working days and hours per day

Sun.	Mon.	Tue.	Wed.	Thu.	Fri.	Sat.

4. Enter particulars of any payment, allowance or benefit made or to be made for period of disability.

10. Wages paid on day of lay off　Normal days pay
    $　　　　$

5. Workman's gross earnings with you prior to injury
   3 mos. $　　12 mos. $

11. Is he working now? If YES, give date and time of return. ☐ YES ☐ NO

6. Duration of lay off, if any, from sickness during these periods.

12. If he worked after first lay off enter dates.
    From　19　to　19

Date　　TITLE:　　SIGNED BY

material is reproduced with the permission of the copyright holder.

Workmen's Compensation Reforms— What to Expect

As most readers are aware, there is now pending in Congress legislation to drastically "strengthen" state-regulated Workmen's Compensation Acts. Cited as the "National Workers' Compensation Standards Act of 1973" and identified as S.2008 (Williams-Javits) and H. R. 8771 (Perkins-Daniels), this legislation is expected to come to the fore during 1975. Contractors, as indeed all employers, should know what to expect.

Coming on the heels of the fifteen-member National Commission on State Workmen's Compensation Laws report to the President and Congress, this may be viewed as an attempt to take the Commission's recommendations for reform a giant step beyond even their sweeping changes.

Without attempting to editorialize on the relative merits or equity of the Commissions's very specific recommendations designed to upgrade state Workmen's Compensation benefits no later than July 1, 1975, we should be prepared to admit that changes were and continue to be necessary. The Commission reported that the present 50 state-administered Workmen's Compensation programs are, in general, neither adequate nor equitable.

Mike LaVelle in his Chicago Tribune Column, "Blue-collar views", points out that "In some states, an injured worker would be better off on welfare than on Workmen's Compensation." Certainly that situation is intolerable, but the Williams-Javits bill would provide radium therapy to a patient suffering from toothache.

The bill reads tough. Commencing January 1, 1975, unless the Workmen's Compensation Law of a state meets minimum standards set forth for each of three calendar years, the provisions of the Longshoremen's and Harbor Workers' Compensation Act shall apply within such state. The latter provides up to:

$210.54 weekly for temporary total disability;
$210.54 weekly death benefit without dollar or time limits;
$65,688 for loss of an arm (percentages thereof for partial loss of use);
$15,791 for loss of a thumb;
$60,636 for loss of a leg;
$42,108 for loss of hearing.

Compare those figures to benefits currently payable in your state and you will see that the threat is meaningful. And yet, the biggest impact of the pending bill is more subtle, hinging upon eradication of statutes of limitations so that a claim may never really be closed as paid. Indeed, it would invite reopening of cases now dormant, making more difficult the underwriters' task of adequate reserving.

In our opinion, the National Commission's estimate of additional costs to adopt those recommendations they labelled as being essential or to adopt all recommendations made are both light; that is, entirely inadequate to:

- Markedly increase weekly benefit levels;

- Establish *retroactive* benefit increase funds to counter the effects of inflation where the benefit level at the time of injury was less than current levels;

- Eliminate arbitrary limits on duration of time or total sum of weekly benefits payable;

- Fully cover injury, illness, or death resulting from both non-work-related and work-related causes where the latter part is "significant";

- Fully cover medical care and physical rehabilitation without any statutory limits as to dollar amounts or length of time payable;

- Provide vocational rehabilitation benefits unlimited by statutory time limitations;

- Give employees or their survivors the right to choose where their claim may be filed, in the state where injured, where hired, or where employment was principally localized;

- Permit workers to select their treating physicians;

- Limit lost-time benefit waiting periods to three days or less, retroactively picked up after no more than fourteen days of temporary total disability;

- Continue payment of death benefits to a widow or widower for life or until remarriage, continue payment to dependent children until age twenty-five if enrolled as a full-time student in any accredited school;

- Provide two years' benefits in a lump sum (presumably as a wedding present) to a widow or widower who remarries;

- Require loss prevention services to all employers, including smaller firms, to be monitored by a Workmen's Compensation agency established in each state;

- Finance such Workmen's Compensation agencies by assessment of employers via their insurance premiums or self-insured equivalent;

- Install more stringent accident reporting procedures applicable to employers in the event of lost time beyond one shift or workday;

- More sensitively reflect an employer's individual loss experience in his own insurance charges;

- Minimize approval of compromise and release settlements, particularly where the settlement offer is to terminate future medical and rehabilitation benefits.

The one hope we see for contractors in all of this is the likelihood of an opportunity to bid on erection of roughly fifty mini-pentagons to house the commissions and policing organizations required to effectively enforce these regulations and still more to come.

If readers concur that overthrow of state-regulated Workmen's Compensation benefits is too drastic a measure bound to have deep economic effect upon construction costs (which must ultimately be borne by the consumer, including injured workmen and their families), it is not too late to influence federal and state legislation. While individual effort can be meaningful, trade associations doubtless afford the best opportunity to influence the outcome of these ongoing deliberations. Your active participation is encouraged.

We commend the Council Of Construction Employers, Inc. for their statement of August 6, 1974, to the Senate Subcommittee on Labor as a realistic appraisal of S.2008. In admirably concise language, the Council scores effectively by pointing out that during 1973, the year after the Presidential Commission's report was made public, state legislatures enacted over 400 bills strengthening workers' compensation and bringing state programs closer to the goals set by the Commission. Still more have been passed in 1974, of course.

Representing fifteen national employer associations in the construction industry, with a combined membership of 66,000 contractors employing some 3,500,000 construction workers and totalling 5,000,000 persons engaged in key trades, the Council's summation must be regarded by Congress as a definitive statement on the issue and act accordingly.

The Presidential Commission reported no other existing program capable of matching the overall performance record of the state programs for effective delivery of benefits, encouragement of safety, and physical rehabilitation of disabled workers. Proponents of the Williams-Javits bill evidently need to be reminded of that finding. Now.

In the interim, be sure to ask your insurance representative to keep you closely advised of benefit/rate trends and developments, particularly as you bid fairly long-range work which might have to be performed under drastically increased Workmen's Compensation insurance overhead costs to come.

1.3 INSURANCES

Some forms of insurance such as workmen's compensation and automobile insurance are (by law) mandatory. Some others for builders are obligatory upon and after signing a building contract; these may include a performance bond, lien bonds, and so on. Still others such as fire and public liability are imperative for the builder.

The complete loss of a business or its survival may often be ascribed to adequate insurance coverage or a lack thereof.

New contractors pay much higher insurance charges than experienced ones. The contracting business is a high risk industry; insurance is also a higher risk for newcomers into the field, and building supply houses are also less prone to accept new accounts without reserve.

It is very important for the contractor to remember that if a member of the general public wanders onto a building site (even out of curiosity) and is injured, the contractor is liable for such injury. Open excavations subject to flooding have often been the cause of a child's death from drowning, and the owner is culpable. You should fence!

It is prudent to adequately insure, but unnecessary insurance is a charge against successfully bidding, profits, or both. It is recommended that the contractor build a good business association with reputable bonding and insurance companies so that they can work together with mutual confidence.

Insurance companies have formulae to assess their risk with each contractor. One method used is the loss ratio. This is the ratio of premiums paid by the contractor to the company and losses paid out by the insurance company to the contractor. Another method is the experience modification formula, by which the actual experience of the contractor over a period of time (say five years) is assessed, affording good reading. The contractor should discuss these points with his broker and ask for an annual report about himself. As with many insurances, it may be more profitable to the contractor to absorb small material losses so that he may maintain a good record with minimum future premium charges.

Workmen's compensation rates vary from state to state and in the provinces of Canada; they are also rated against the experience of the contractor.

Obviously it is only by progressing from simple, light construction to heavy construction that a contractor can build up his experience and reputation. Like everything else, success breeds success, and the more experience a contractor has with a minimum of claims, the cheaper the insurance.

Builder's Risk Insurance/Installation Floaters/Contractor's Equipment Floaters*

The special problems of providing adequate Property insurance during the course of construction are dealt with under Builder's Risk policy forms designed for that purpose. A good many variables

---

*Excerpted, see page 2.

tend to surround the subject, including not only the degree of coverage contemplated by several different policy forms, but often who is covered and even what the coverage is called. In a limited space we shall try to simplify matters a bit.

We shall concern ourselves here with the recommended Completed Value Builder's Risk policy and its Inland Marine counterpart, the Installation Floater. They may be written in the name of both the owner and the general contractor, either of them, or any of several variations; coverage may also apply to the work of subcontractors.

The prudent contractor, therefore, must review contract specifications to determine his obligation for this form of coverage.

He should guard against an outright duplication of coverage, however, since in many cases the owner or general contractor may have already purchased a policy protecting the entire venture.

Equipment and machinery, including electrical, plumbing, heating, and air-conditioning systems, should be insured as required while in transit from manufacturer to job site and during the period of actual installation or testing until fully released.

Coverage generally terminates when the insured's interest in the property terminates, when it is accepted, or when it is occupied by the owner. Covered building materials, such as bricks, steel, and lumber, are insured until actual installation into the real structure or until the insured's interest terminates, whichever happens first.

In some states, similar coverage is afforded the contractor under Special Builder's Risk Reporting Form, Contractor's Automatic Builder's Risk, or Contractor's Builder's Risk Completed Value Reporting Form policies. What is available in your area should be determined by your insurance representative, who should make known what options you have and determine which is recommended for you.

Even the perils covered by Builder's Risk and Installation Floater policies vary; they may insure against fire, lightning, the perils of Extended Coverage (windstorm, hail, explosion, riot, civil commotion, aircraft, vehicles, smoke), Vandalism and Malicious Mischief, or may be so-called "all-risk" floater policies designed to include still other risks. The latter are usually subject to variable deductible amounts, normally not applicable to the basic fire policy perils. (This is probably as good a time as any to stress that "all-risk" is a relative, not an absolute term; there are still exclusions applicable, although the coverage is decidedly broad.)

The Inland Marine policy form applicable to contractor's equipment, other than vehicles designed for use on public highways, is called an Equipment Floater because it applies to things of a mobile or "floating" nature. Almost anything movable can be insured, whether a large crane, power shovel, caterpillar tractor, lift truck, or small tool. Most large units carrying high values are specifically scheduled on such policies and a blanket amount takes care of smaller items. Coverage can be made to automatically apply to new or replacement equipment. Note that automatic coverage for leased equipment requires a specific policy statement or endorsement to that effect.

Coverage is largely tailored to suit the exposure. It may apply to named perils or be written "all-risk," or sometimes a combination of the two. To provide protection against meaningful losses at a reasonable premium level, your insurance representative should arrive at a suitable deductible amount. If the exposures can be clearly separated, there is no reason why differing deductibles cannot be selected.

Premium rates are largely negotiated but based upon prior loss history; in general, elimination of petty pilferage claims is preferred to ground-up coverage on the presumption that any insurance company is going to want more than a dollar of premium for every dollar of routine loss they are called upon to pay. The small ones are best chalked up as a business expense.

The task of proper arrangement of coverage, deductibles, and premiums demands imagination as well as the experience of a skilled professional.

Contract and Other Bonds

A necessary adjunct to administration of a contractor's insurance program is the performance of similar services with regard to his contract bond requirements. One complements the other because a close working knowledge of work in progress, projects being bid or completed helps to enhance the close relationship which should exist between the contractor and his insurance/bond counselor. Obviously, it is a relationship dependent upon confidence, in many respects parallel to a good banking connection.

We shall limit ourselves here to naming some of the bonds required of contractors and briefly stating what they do:

Bid Bonds — Given by the contractor to the owner, guaranteeing that if awarded the contract, he will accept it and furnish final Performance or Payment Bonds as required.

Performance Bonds — Given by the contractor to the owner, guaranteeing that he will complete the contract as specified.

Labor and Material
Payment Bonds —Given by the contractor to the owner, guaranteeing that he will pay all labor and material bills arising out of the contract.

Maintenance Bonds—Given by the contractor to the owner, guaranteeing to rectify defects in workmanship or materials for a specified time following completion. A one year maintenance bond is normally included in the performance bond without additional charge.

Completion Bonds —Given by the contractor to the owner and lending institution guaranteeing that the work will be completed and that funds will be provided for that purpose.

Supply Bonds —Given by the manufacturer or supply distributor to the owner guaranteeing that materials contracted for will be delivered as specified in the contract.

Subcontractor
Bonds —Given by the subcontractor to the contractor guaranteeing performance of his contract and payment of all labor and material bills.

Such bonds are required by statute for federal, state, and local government work and, of course, specified for a great deal of private construction. They are the best form of guarantee that construction will be finished as required and that all bills will be paid. Amounts of bond required may vary from 10% to 100% of the total price of the contract, but are normally 100%.

Any undue delay or outright failure to secure a required contract bond could cost the contractor the job, so performance of the bond agent is all-important to success of the contractor/counselor partnership. You can help by promptly supplying all financial information requested and, in general, keeping him posted about the status of your present and future work program. Those who are relatively new to the contracting business should strive to establish a strong working relationship with such an insurance/bond source.

Other bonds you may encounter include:

License or Permit Bonds —Given by the contractor/licensee to a public body, guaranteeing compliance with statutes or ordinances, sometimes hold-ing the public body harmless.

Sub-Division Bonds —Given by the developer to a public body, guaranteeing construction of all necessary improvements and utilities; similar to a completion bond.

Union Wage Bonds —Given by the contractor to a union, guaranteeing that the contractor will pay union scale wages to employees and remit to the union any welfare funds withheld.

Self-Insurers' Workmen's
Compensation Bonds —Given by a self-insured employer to the state, guaranteeing payment of statutory benefits to injured employees.

Others falling into the broad categories of Court Bonds and Fidelity Bonds will not be dwelled upon here; need for the former will be made known to you when and if the time comes, while the best method for protection against employee dishonesty will be brought to your attention by the professional charged with responsibility for your insurance/surety account.

Prompt Reports of Accident

Sprinkled throughout liability policies is a requirement that the insured give formal notice of claim to the insurance carrier as soon as practicable, an obvious requisite to efficient claim investigation and handling. By implication, this means promptly after the insured first has knowledge of a claim, although this could conceivably be a long time after the claimed occurrence itself.

Failure to do so may seriously inhibit your carrier's ability to investigate at all and further jeopardize their chances of arriving at a reasonable settlement. The company has a right to deny coverage under such circumstances.

It is always best in this respect to err on the side of reporting too much rather than too little. Notify your insurance representative at the first knowledge of any potential claim, leaving it to his judgment whether it is the opportune time to file a formal report, under which policies, etc. This will tend to convey a healthy feeling of loss control, professionalism in recognition of possible claims, and give your carrier a big jump on investigation while all the evidence is available, the facts fresh in everyone's mind.

Experience Rating

Contractors and others, when their premium level at normal manual rates reaches a practical level (very roughly $750 policy minimum), become eligible for experience rating; that is, calculation of individual credits or debits applied to manual rates, dependent upon the ratio of premiums to losses over a given number of years. These rating plans vary a good deal with the kind of coverage involved. In principle, they are similar, however; all are closely scrutinized by state insurance departments, intrastate and interstate rating authorities, and independent rating organizations.

Manual rates contemplate a certain average level of losses and, by mathematical formula, a comparison is made of actual losses reported over roughly three years compared with expected levels. Weights or credibility factors are allowed to minimize the effect of single catastrophe losses so that the small contractor who reports just one serious case does not pay an astronomical premium for eternity. In general, a frequency of claims will count more in experience rating than will severity, but this effect decreases as premium volume increases.

1.4 INDIVIDUAL OWNERSHIP, PARTNERSHIPS, LIMITED COMPANIES, AND CORPORATIONS

Individual Proprietorship

With individual ownership the total operation of the business is conducted by one person (frequently with the help of the family or employees). Legally, only one person is responsible for the indebtedness or profits of the business.

Some of the advantages and disadvantages of owning your own business are:

Advantages

(a) The profits and business expansion are yours.
(b) As your own boss you direct your own policy.
(c) There are no complications with others about responsibilities and rewards.
(d) You do not have to meet in committee to formulate policy.
(e) You set your own working days and hours.
(f) You may cut overhead expenses by operating the business from your own home.

Disadvantages

(a) Your liabilities include your *business* and *personal* assets.

(b) In case of business failure your personal and business assets are subject to seizure and sale.
(c) Assets that you may acquire from the time of your declaration of bankruptcy to the time of discharge are also subject to seizure.
(d) The liability of married men is restricted to their own personal property; that which is listed in the wife's name cannot be touched: *Married persons entering business in the Province of Quebec must go through legal proceedings to determine to what extent the property of the one is separate and distinct from the other.*
(e) The business is usually small in size and will have to compete with large organizations with specialists in every field.
(f) You cannot purchase materials in the same bulk with the same discounts available to larger companies.
(g) After your death, the administrators of your estate may carry on the business under letters of administration until the estate is settled and the business is handed over to your heirs. The business may suffer as a consequence, but you may purchase Key-Man Insurance.

If you wish to add "and Company" to your name or to use an entirely different name from your own, such a business name must be registered in the state or province in which you are conducting business. This is understandable since the public must always be protected by having access to the correct name of any organization with which any person wishes to do business or present claims.

Partnerships

In a partnership, at least two people pool their resources and abilities to conduct a particular business enterprise. A partnership is created by entering into a contract that has two main points: (a) it states the contribution to be made to the business by each partner; and (b) it specifies the manner in which earnings of the enterprise are to be shared by the partners. Note carefully that upon the demise of a partner, the partnership does not exist.

General Partnerships

In general partnerships, the members are jointly and severally (individually) liable for the debts of the entire business. That is to say, that if the creditors cannot get satisfaction from the business, they can enforce their claims against any partner(s).

It has been said that partnerships are the easiest things to get into and the most difficult from which to withdraw. Any partnership agreement should be given most serious thought. It should be drawn up and registered by an attorney who is acquainted with your type of business. *Consult an attorney before (not after) you enter into partnership.*

Limited Partnerships

In a limited partnership, one or more partners operate the business and one or more invest cash only. The former are *general partners* and are liable to creditors for both their business and private assets. The latter are *limited* to creditors claims, only to the amount of cash they have invested in the business; no claim may be made against their personal assets.

Individuals considering entering into partnership should itemize those points they wish to have included in the agreement. They should meet in committee once or twice and present a rough draft in legal terminology. The participants should again meet in committee and then present the final draft to the attorney for final drafting and notarization.

Partnership Agreements

A partnership agreement should contain a number of details. Among them are:

(a) The purpose of the agreement.

(b) The name and permanent address of the firm and of each partner.

(c) The commencement date of the partnership.

(d) The duration of the partnership.

(e) The manner in which the partnership is to be terminated.

(f) The name and address of the bankers.

(g) The name and address of the attorney.

(h) The procedure for continuing the agreement.

(i) The amount of capital to be invested by each partner, with a statement as to whether or not the amount or part of it is to be paid in property, and if in property, a full description thereof.

(j) The starting date of the fiscal year.

(k) The amount of interest, if any, that is to be paid on capital.

(l) The powers and duties of the partners.

(m) Whether or not the partners are to devote full, normal business hours to the partnership.

(n) Whether or not partners are allowed to enter into any other business that is in any way connected with the building industry.

(o) The manner in which profits or losses of the business are to be divided.

(p) The name of the person who will keep the books of the business to which all members will have access at any reasonable time.

(q) The percentage division of partnership assets at dissolution.

(r) The method of continuing the partnership in the event of the death or incompetence of any partner.

(s) In case of dispute, all differences in regard to the partnership affairs shall be referred to the arbitration of a single arbiter, if the parties agree upon one; otherwise to five arbitrators, two to be appointed by each party, and the fifth to be chosen by the four first named before they enter upon the business or arbitration. The award and determination of such arbitrators, or any three of such arbitrators, shall be binding upon the parties hereto and their respective executors, administrators, and assigns, always providing that the recommendation of the board shall not preclude any party the right of access to a court for a law ruling.

It is very important that you know that in the absence of express agreement the law provides that the profits shall be equally divided, regardless of the ratio of the partners' respective investments.
The foregoing list is not exhaustive.

Advantages of Partnerships

(a) Partners can specialize in their own fields.

(b) Each can direct workmen in his own area.

(c) There may be more collective working capital.

(d) The company may obtain better trade discounts.

(e) The company may operate on a larger scale than an individual enterprise.

(f) The company can maintain and keep operating more equipment.

(g) The company may more easily undertake projects in different places at the same time.

(h) Partners pay income tax on net profits. Net profits are those remaining after all operating expenses have been deducted, and are subject to personal income tax **whether or not the proceeds are taken out of, or remain in the business.**

Disadvantages of Partnerships

(a) The partnership is not a tangible entity for the person who enters it.

(b) Each *general partner* is individually responsible for the whole business.

(c) If creditors cannot get satisfaction from the business partnership, they may press the total claim against any partner having private assets.

(d) The dishonesty of a partner.

(e) The incompetence of a partner.

(f) The moral lapse of a partner.

(g) The neglect of the business by a partner.

(h) The repeated bad judgment of a partner.

(i) The partnership ceases to exist if a partner severs his connection with the business or if a partner dies, but this potential loss may be mitigated by taking out KEY MAN insurance.

Corporations

A corporation is an association of three or more persons that are able legally to act as one entity under a common name. A corporation continues to exist even though its members may change. It may engage in business or it may be a charitable enterprise or social religious association. It can own property, can be sued, and can sue. The extent of its activities are shown in a Charter that is given by federal, state, or provincial governments. (A charter is a document granting certain rights to a person, group of persons, or a corporation.)

1.5 BANKRUPTCY

Bankruptcy may be defined as the adjudication of a debtor's inability to pay his debts. Bankruptcy proceedings are of two kinds: voluntary and involuntary, that is, instituted either by the insolvent debtor or by his creditors. Two of the main purposes of these proceedings are to secure equitable division among creditors of the bankrupt's available assets, and to release or discharge the bankrupt from his obligations if he has complied with the law.

Bankruptcy Cases Filed and Pending. A bankruptcy case is a proceeding in a U.S. District Court under the National Bankruptcy Act. "Filed" means the commencement of a proceeding through the presentation of a petition to the clerk of the court; "pending" is a proceeding in which the administration has not been complete.

It is an offense for a person to continue to trade after he knows that he is insolvent. *As soon as he believes himself to be insolvent he should see an attorney for advice.* Sometimes the official receiver (person appointed by a court to manage property in controversy, especially in bankruptcy) may retrieve a situation in small building construction by managing the business until all properties are completed and sold.

The duties of a bankrupt include submitting to an examination, attending on creditors at the first meeting, fully disclosing all assets, and assisting in every way in the official administration of the estate. Punishable offenses include failing to comply with a bankrupt's duties, fraudulently disposing of property, concealing or falsifying books or documents relating to the business, refusing to answer fully and truthfully any question asked at an examination, and obtaining credit or property by false representation after or within twelve months preceding bankruptcy.

1.6 LOCAL AND NATIONAL CONSTRUCTION ASSOCIATIONS

These associations operate at the city level for local convenience, at the state or provincial level for area affairs, and at the federal capitals for affairs of national importance to affiliated members. Several such nonprofit associations are as follows:

Home Manufacturers Association, Bar Building, 910 17th Street, N.W., Washington, D.C. 20006

National Association of Home Builders of the United States,
1625 L Street, Washington, D.C. 20036

The Associated General Contractors of America, Inc.
1957 E Street, N.W., Washington, D.C. 20006

The National Association of Home Builders
The National Housing Center
1625 L Street, N.W.
Washington, D.C. 20036

Housing and Urban Development Association of Canada,
King Edward Sheraton Hotel, Toronto, Ontario

Canadian Construction Association,
Ottawa, Canada

Amalgamated Construction of British Columbia,
2675 Oak Street, Vancouver 9, B.C., Canada

Their objectives are the maintenance of a strict

code of ethics; they are the spokesmen for their members to the government for beneficial construction legislation. They aim to promote better relations between their members, owners, architects, and engineers; to establish high professional standards among contractors; to encourage methods of contracting, which will relieve the contractor of improper risks; to promote educational programs with members of the association and labor organizations; and to assure the public of the benefit of competitive contracting.

1.7 CLASS RESEARCH

For those offering courses and using this book, it is suggested that consideration be given to apportioning chapters among the students for research, and having them make an oral report to the class, and then giving a typescript paper of their findings to each student and to the instructor.

As an example, one or more students may be assigned to make a comprehensive study of the first chapter; job safety, workmen's compensation insurance, and other mandatory insurances for builders. They should list the name, address, and telephone number of each organization that they visit, and should return to class with sufficient copies of any acquired literature to give a copy to each student and one to the instructor. They should arrange to have a speaker from the workmen's compensation insurance company address the class and show films on safety; they should also arrange for other speakers from say an insurance company specializing in builders bonds and insurances; and a representative from one of the house builders associations, and also one from a successful builder. After the students have given a short oral report to the class, they should follow this with a question and answer period.

Another student committee should research the subjects discussed in chapter three. They should visit building projects, architects and engineers on large house building projects; relevant city departments, and building contractors. They should give a typescript paper of their findings (also a copy of any other acquired literature) to every member of the class, including the instructor. They should get a local building inspector to address the class, and should get a copy of a title deed to land, preferably one that includes some registered encumbrances.

It is suggested that this method be followed throughout all chapters, and that every student be involved. Students will not only be exposed to current methods, materials and philosophy of residen-

tial construction, they will also be attracted to specialize in particular aspects of construction that appeal to them. Many will be offered jobs.

At the end of the course, they will have been exposed to lectures by practicing professionals in the industry, have questioned each other and the instructor; in short they will have sufficient knowledge to enter the construction industry with confidence, competence, and to make a profit.

The class should have available to them a copy of the latest building code, and students should be made aware of the value of consulting "Sweet's Architectural Catalog File" which comprises twelve volumes, 17,000 pages, and is represented by 1,500 manufacturers of building materials. There is also a Canadian version of this catalog in which many papers are published in both English and French.

At the end of each chapter of this book are presented a number of pertinent review questions.

CHAPTER 1 REVIEW QUESTIONS

1. List ten methods of disseminating accident prevention literature in your area.

2. From what sources in your area would you seek the most recent information regarding accident prevention?

3. List ten important telephone numbers that should be prominently displayed at each telephone station on a building site.

4. Why is it important to an employee to report any accident in which he sustains any visible or invisible injury while employed on a building site, at the time of its occurrence?

5. Assuming that you were in charge of a small construction job (where there is no trained first-aid personnel) list in order all the steps you would take regarding the victim of a serious accident.

6. List three accident prevention devices for any construction job involving: (a) scaffolds; (b) stationary electrical equipment; (c) gas or diesel oil operated mobile equipment; (d) excavations; (e) electrical connections; (f) stacking of loose bricks, concrete units, or lumber on the job site.

7. Assuming that a member of the public who is neither employed by you, nor by any other person or organization engaged by you, wanders onto your building site and is injured; who is culpable; and what is the remedy for such an event?

8. Why do new builders have to pay higher insurance premiums than experienced ones?

9. What is the minimum first-aid requirement for any job however small?

10. Name a local or national housebuilding association in your area and briefly state its objectives, and the advantages to a builder of taking out membership.

11. List five advantages and five disadvantages of partnerships.

12. List five advantages and five disadvantages of being the sole owner of a building contractor's business.

13. Upon the demise of a partner in a business, what is the status of the surviving partner or partners?

14. If you order an expensive piece of equipment for installation on a project that you are building, at what point does the insurance coverage on such a piece of equipment become your responsibility?

15. State five advantages of class research.

Architectural Drawing Practices

This chapter is presented to enable the residential constructor to readily identify many symbols used in architectural drawing practices. These include: those used in frame construction; walls; windows; electrical; plumbing; heating; door and room schedules; a miscellany of other symbols, and a drawing sheet index and drawing references.

The following material (sections 2.1 through 2.13) have been excerpted from publication 33-GP-7 on Architectural Drawing Practices, published by the Canadian Government Specifications Board, and are reproduced here through the courtesy of the Board. Copies of 33-GP-7 may be purchased from:

Canadian Government Specifications Board
88 Metcalfe Street
Ottawa, Ontario
K1A OS5

2.1 FOREWORD

Architectural working drawings have the important functions of (a) recording clearly the clients' requirements so that cost estimating and bidding are facilitated, (b) forming a part of the contract between the client and builder and (c) providing instruction to the builder for the purpose of construction. To fulfill these functions most efficiently the drawings must be complete, accurate and concise. The use of uniform practices in producing the drawings is a valuable aid toward achieving this goal.

This standard is presented as a guide toward such uniformity and toward clarity and simplicity rather than artistry in architectural drawing. It is intended primarily for use within Canadian government agencies, but is available for use generally and, indeed, its general and widespread use is encouraged.

The standard is not exhaustive in its treatment of architectural drawing practices, particularly as regards symbols for electrical, plumbing and heating applications, but deals with what are considered to be the most basic and commonly encountered elements of architectural drawing.

2.2 LINE CONVENTIONS

Three line thickness are recommended—thick, medium and thin.

Visible Object Lines

Medium

Hidden Lines

Length of Dashes may be Increased Slightly for a Longer Line (Medium)

Center Lines

Thin

Reference or Datum Lines

Thin

Section Lines

Thick and Used with Reference Symbol

Suggested Reference Symbols — Letter Designates Section — Numeral Indicates Sheet Number

Dimension & Extension Lines

Thin

2.3 SYMBOLS FOR WALLS & PARTITIONS

Solid Wall

Option (1) Where the wall material symbols are not continuous but show only material and construction changes

Option (2) Where wall material symbols are not used but wall construction details are shown in a wall type schedule or detail

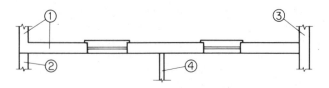

Especially applicable to small-scale drawing

Option (3) Where wall material symbols are continuous throughout the work

Window Walls

Wall of sheet glass or other transparent or translucent sheets in wood or metal frame, e.g., store window

Wall of glass block or other transparent or translucent masonry

A window built into a transparent or translucent masonry wall

Curtain Walls

Panel type curtain wall, without windows

Panel type curtain wall, with windows (see page 17 for elevation)

Partitions

FRAME PARTITION—When showing frame partition in new work, material symbols are not shown

Finished Both Sides   Finished One Side Only

GLAZED PARTITION—Indicates the installation of any transparent or translucent sheet material, either flat, corrugated or textured in any way

EXISTING PARTITION—When showing existing partitions (regardless of construction) in alteration work, material symbols are not shown

PARTITION TO BE REMOVED (in alteration work)—Shown in broken lines

NEW PARTITION (in alteration work)—Material symbols are used to distinguish from existing partitions

Frame   Concrete   Brick

Examples of New Partitions in Alteration Work

NEW OPENINGS IN EXISTING WORK—Material symbols are shown or shading is used for short distances on each side of the new openings

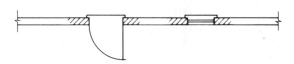

TOILET PARTITIONS—The same partition symbol is to be used for all types of toilet partition construction, at scales of ¼" = 1'−0" & ⅛" = 1'−0"

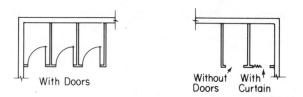

With Doors   Without Doors   With Curtain

MOVABLE PARTITIONS—A partition that can be removed intact and relocated. (A movable partition is never a bearing or structural partition)

Solid   Glazed

WIRE PARTITIONS—Where wire mesh material is used to separate areas within a building

2.4 SYMBOLS FOR DOORS

When used in conjunction with a door schedule, door symbols include the appropriate door schedule reference. See examples 'A' & 'B' below. It is optional when using this method to give the door size as shown in example 'B'

Single Swing Doors

Use option 1 except where space & clarity dictate the use of option 2

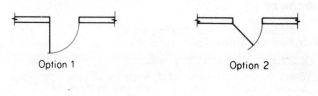

Option 1   Option 2

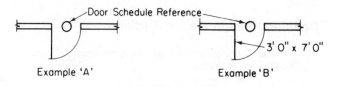

Door Schedule Reference

3' 0" x 7' 0"

Example 'A'   Example 'B'

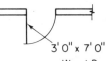

The Size and Type of Door are Given When a Door Schedule is not Included

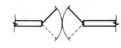

3' 0" x 7' 0" x 1¾"
Wood Door

Double Acting Doors

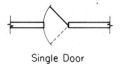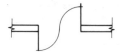

Single Door   Pair of Doors

In and Out Doors

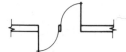

In and Out Combination Without Mullion   In and Out Combination With Mullion

Folding Doors and Partitions

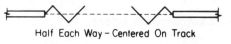

Half Each Way − Centered On Track

All One Way − Centered On Track

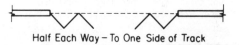

Half Each Way − To One Side of Track

All One Way − To One Side of Track

Without Pocket   With Pocket

Shower Curtain, Draperies, Etc.

Sliding Doors

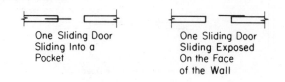

One Sliding Door Sliding Into a Pocket   One Sliding Door Sliding Exposed On the Face of the Wall

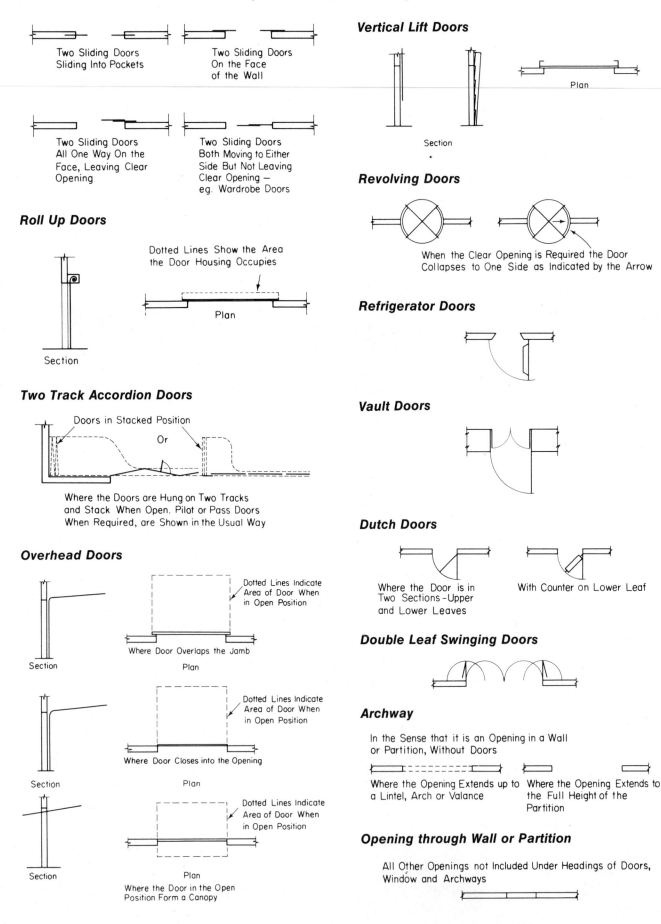

Two Sliding Doors Sliding Into Pockets

Two Sliding Doors On the Face of the Wall

Two Sliding Doors All One Way On the Face, Leaving Clear Opening

Two Sliding Doors Both Moving to Either Side But Not Leaving Clear Opening — eg. Wardrobe Doors

Roll Up Doors

Dotted Lines Show the Area the Door Housing Occupies

Plan

Section

Two Track Accordion Doors

Doors in Stacked Position

Or

Where the Doors are Hung on Two Tracks and Stack When Open. Pilot or Pass Doors When Required, are Shown in the Usual Way

Overhead Doors

Section

Dotted Lines Indicate Area of Door When in Open Position

Where Door Overlaps the Jamb

Plan

Section

Dotted Lines Indicate Area of Door When in Open Position

Where Door Closes into the Opening

Plan

Section

Dotted Lines Indicate Area of Door When in Open Position

Plan

Where the Door in the Open Position Form a Canopy

Vertical Lift Doors

Plan

Section

Revolving Doors

When the Clear Opening is Required the Door Collapses to One Side as Indicated by the Arrow

Refrigerator Doors

Vault Doors

Dutch Doors

Where the Door is in Two Sections - Upper and Lower Leaves

With Counter on Lower Leaf

Double Leaf Swinging Doors

Archway

In the Sense that it is an Opening in a Wall or Partition, Without Doors

Where the Opening Extends up to a Lintel, Arch or Valance

Where the Opening Extends to the Full Height of the Partition

Opening through Wall or Partition

All Other Openings not Included Under Headings of Doors, Window and Archways

2.5 SYMBOLS FOR WINDOWS

Windows in Plan for All Types of Sash

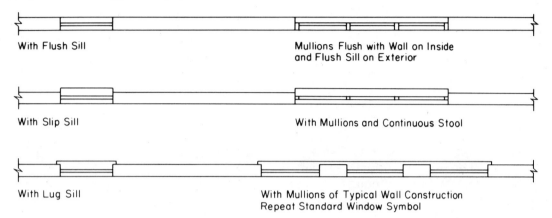

With Flush Sill

Mullions Flush with Wall on Inside and Flush Sill on Exterior

With Slip Sill

With Mullions and Continuous Stool

With Lug Sill

With Mullions of Typical Wall Construction Repeat Standard Window Symbol

Windows in Elevation

These symbols apply regardless of the material used in the manufacture of the sash, i.e., wood, steel, aluminum, etc.

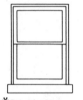
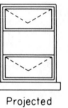
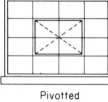

*Double Hung　　　Projected　　　Pivotted　　　Casement

*　Abbreviation D.H. if Thought Necessary for Clarity

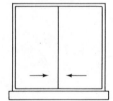
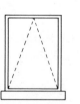
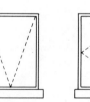
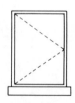

Horizontial Sliding　　Top Hinged　　Bottom Hinged　　Left Side Hinged　　Right Side Hinged

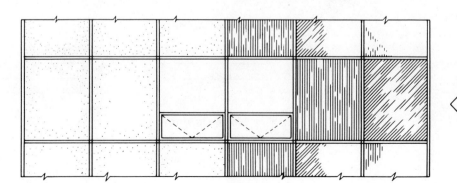

Note: Panels that are Not Windows are Darkened Either by Shading (Right Side) or by Stippling (Left Side)

Window in Panel Type Curtain Wall

2.6 MISCELLANEOUS SYMBOLS

Ramp

In some instances it will be necessary to indicate the gradient

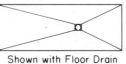

Surface Slope

Usually used to indicate floor conditions. Can also apply to roofs or other sloping surfaces in plan. Arrows are sometimes required to indicate direction of slope.

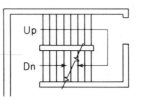

Shown with Floor Drain

Stairs

It is desirable to indicate the number of risers from floor to floor

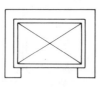

Elevators

Dumbwaiters

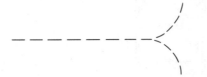

Overhead Track

Cranes

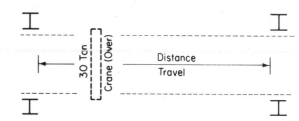

Escalators

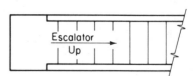

Access Doors

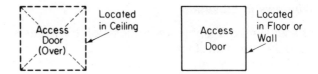

Skylights

Dormers

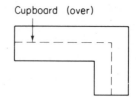

Cupboards

Medicine Cabinets

Inset   Surface

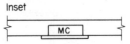   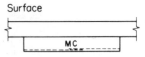

Fire Hose Cabinets

Inset   Surface

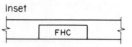   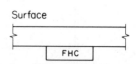

Lockers

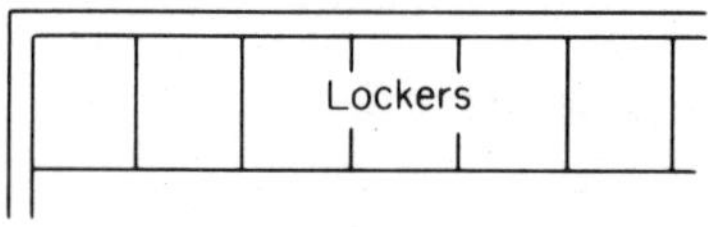

Lockers

2.7 ELECTRICAL SYMBOLS

CEIL   WALL

○	−○	Outlet ○_{PC} Ceiling Outlet with Pull Chain
Ⓕ	−Ⓕ	Fan Outlet
Ⓧ	−Ⓧ	Exit Light Outlet
Ⓒ	Ⓒ	Clock Outlet. Specify Voltage

Duplex Convenience Outlet

1 = Single, 3 = Triplex, etc.

Range Outlet

Switch and Convenience Outlet

Motor Outlet

Special Purpose Outlet

Fluorescent Fixture. Letter Designates Type

Fluorescent Fixture with Outlet Box

S   Single Pole Switch

S₂   Double Pole Switch
S₃   Three Way Switch
S₄   Four Way Switch

S_P   Switch and Pilot Lamp

Outside Telephone

Interconnecting Telephone

Lighting Panel

Power Panel

Home Run to Panel Board. Indicate No of Circuits by No. of Arrows

Overhead
Underfloor

Cable Designation

Dia $\frac{3}{4}$ — 5 — 12

No. of Wires
Gauge of Wire

2.8 PLUMBING SYMBOLS

Soil and Waste

Soil Waste or Leader Below Grade

Soil and Waste Underground

Vent

Cold Water

Hot Water

All Other Special Lines. Insert Initials as Required and Shown On Legend

---

FL / DR
or
RF / DR   Floor or Roof Drain

RWC ○   Rainwater Conductor

Soil ○   Soil Stack

Wall Type / Floor Type   Urinals

Water Closet Tank Type

Water Closet Flush Valve Type

Recessed Bath   Free Standing Bath

Sink

SS   Service or Slop Sink

Drain Board

Shower

HB   Hose Bib

Compa

Compressed Air Outlet

Riser   Valve   Elbow   Tee

Fire Hose Cabinet

2.9 HEATING SYMBOLS

High−Pressure Steam

Medium−Pressure Steam

Low−Pressure Steam

High−Pressure Return

Medium−Pressure Return

Low−Pressure Return

Air Relief Line

Boiler Blow Off

Hot Water Heating Supply

Hot Water Heating Return

Gate Valve

Riser   Elbow   Tee

Rad or Convector

Finned Tube

Baseboard Convector

12 x 20   Duct { 1st. Dim. Applies to Side Shown / 2nd. Dim. to Side Not Shown

Duct Section (Supply)

Duct Section (Exhaust or Return)

Ceiling Register or Grille

Duct and Direction of Flow

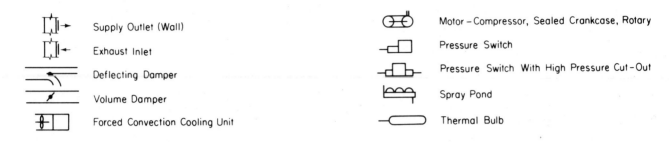

	Supply Outlet (Wall)		Motor-Compressor, Sealed Crankcase, Rotary
	Exhaust Inlet		Pressure Switch
	Deflecting Damper		Pressure Switch With High Pressure Cut-Out
	Volume Damper		Spray Pond
	Forced Convection Cooling Unit		Thermal Bulb

2.10 SYMBOLS FOR MATERIALS

| | ¼" SCALE AND SMALLER | DETAILS | |
	Plan and Section	Plan and Section	Elevations
Cut Stone Masonry			
Brick Masonry			Spacing to Represent Actual Coursing
Artificial Stone Masonry	Line Spacing Twice as Wide as for Brick		Stipple Density One-half of that for Cut Stone
Marble	Usually too Fine to Hatch		
Concrete			Stipple Density One-third of that for Cut Stone
Concrete Block Cinder Block etc.	Line Spacing 3 Times as Wide as for Brick		Spacing to Represent Actual Coursing
Clay Tile Masonry			Spacing to Represent Actual Coursing

¼" SCALE AND SMALLER	DETAILS	
Plan and Section	Plan and Section	Elevations
Ceramic Tile — Usually too Fine to Hatch		Show Actual Pattern
Terrazzo — Usually too Fine to Hatch		Show Actual Pattern of Strips and Stipple
Cinder or Slag Fill			—
Fire Brick			
Gypsum Masonry			—
Plywood	—		
Glass Block			
Natural Stone i.e. Rip Rap. Field Stone, Etc.,			
Sand, Fill Plaster and Cement			

	$\frac{1}{4}''$ SCALE AND SMALLER	DETAILS	
	Plan and Section	Plan and Section	Elevations
Glass	Too Fine to Hatch		GL Include GL where Req'd for Clarity
Wood Framing	For New Work For Alteration Work		No Symbol Required
Wood (Finished)	Usually too Fine to Hatch		—
Insulation			—
Structural Steel			—
Bronze Brass Copper and Associated Alloys	Same as for Detail when Area to be Hatched is Large Enough		—
Earth			—
Rock			—
Gravel Fill			—

2.11 DIMENSIONS

(See modular dimensioning in Appendix A.) Dimensions up to 12 inches are given in inches; dimensions of 12 inches and over are given in feet and inches.

Linear Dimensions

Used to Indicate the Distance Between Two Points

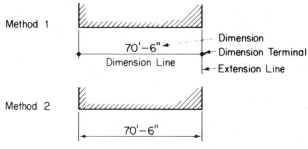

Method 1

> 70'-6"
> Dimension Line

Dimension
Dimension Terminal
Extension Line

Method 2

> 70'-6"

The Use of Any Other Dimension Terminal Than The Arrowhead or The Dot is Not Recommended

Radius Dimensions

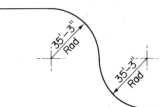

35'-3" Rad
35'-3" Rad

Used to Indicate the Distance from the Center to the Arc of a Circle

Diameter Dimensions

Pipe and Tube sizes are Nominal Inside Diameter Unless Indicated Otherwise

O·D· After a Dimension Indicates Outside Diameter

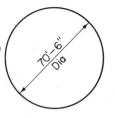

70'-6" Dia

2.12 FRAME CONSTRUCTION DIMENSIONING

Basement Plan

Walls and Footings

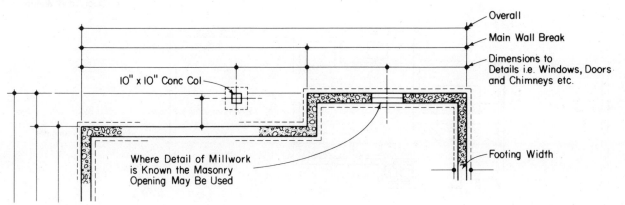

10" x 10" Conc Col

Where Detail of Millwork is Known the Masonry Opening May Be Used

Overall
Main Wall Break
Dimensions to Details i.e. Windows, Doors and Chimneys etc.

Footing Width

Partitions and Columns

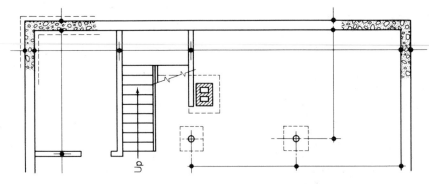

Chimney and Interior Footings

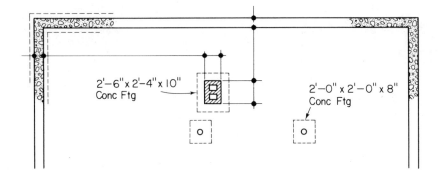

Floor Plan

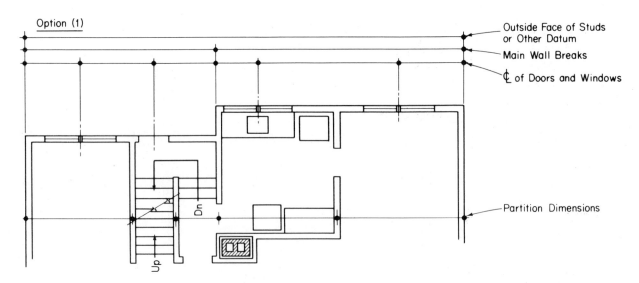

Option (2)

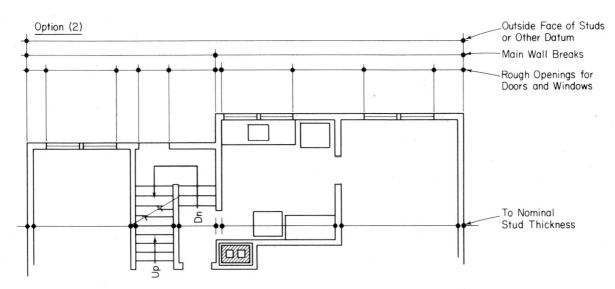

Outside Face of Studs or Other Datum

Main Wall Breaks

Rough Openings for Doors and Windows

To Nominal Stud Thickness

Dn

Up

Elevation

Option 1   Use where Window Millwork is Optional

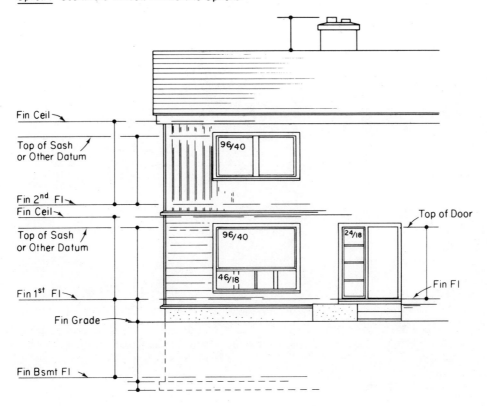

Fin Ceil

Top of Sash or Other Datum

Fin 2nd Fl

Fin Ceil

Top of Sash or Other Datum

Fin 1st Fl

Fin Grade

Fin Bsmt Fl

Top of Door

Fin Fl

96/40

96/40

46/18

24/18

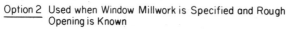

Option 2 Used when Window Millwork is Specified and Rough
Opening is Known

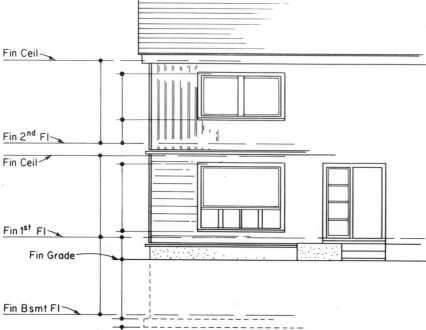

Fin Ceil

Fin 2ⁿᵈ Fl

Fin Ceil

Fin 1ˢᵗ Fl

Fin Grade

Fin Bsmt Fl

Steel Frame Construction Dimensioning — Plan

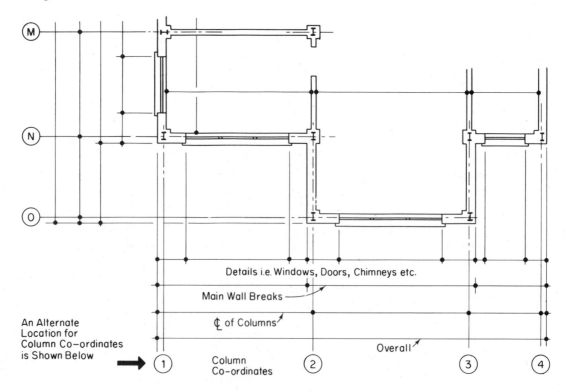

M

N

O

Details i.e. Windows, Doors, Chimneys etc.

Main Wall Breaks

$\mathbb{C}$ of Columns

An Alternate
Location for
Column Co-ordinates
is Shown Below

Overall

1   Column
Co-ordinates   2   3   4

Reinforced Concrete Construction
Dimensioning — Plan

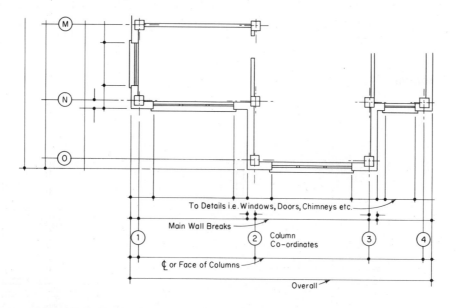

To Details i.e. Windows, Doors, Chimneys etc.

Main Wall Breaks

Column Co-ordinates

₵ or Face of Columns

Overall

Steel Frame or Reinforced Concrete
Construction Dimensioning — Elevation

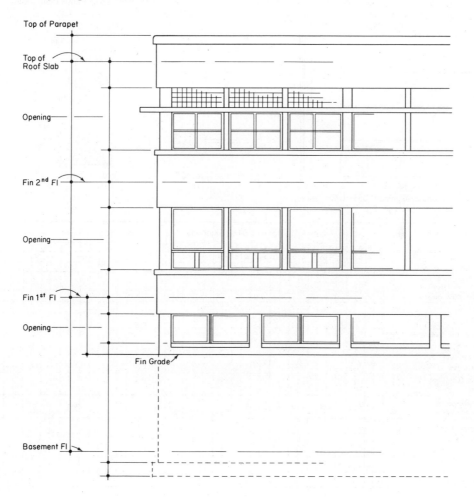

Top of Parapet

Top of Roof Slab

Opening

Fin 2nd Fl

Opening

Fin 1st Fl

Opening

Fin Grade

Basement Fl

2.13 DOOR AND ROOM FINISH SCHEDULES

It is recognized that no strict rules can or should be proposed concerning the exact form and arrangement of door and room finish schedules. The examples shown on the following pages illustrate door and room finish schedules that are in common use in some areas of architectural drawing, and they illustrate the kind of information that should be provided by such schedules.

Door & Hardware Schedule

Example 1

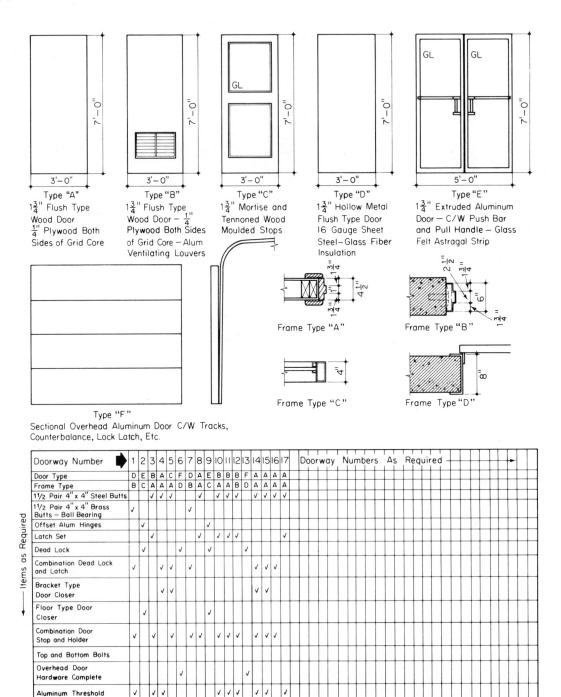

Type "A"
1¾" Flush Type
Wood Door
¼" Plywood Both
Sides of Grid Core

Type "B"
1¾" Flush Type
Wood Door – ¼"
Plywood Both Sides
of Grid Core – Alum
Ventilating Louvers

Type "C"
1¾" Mortise and
Tennoned Wood
Moulded Stops

Type "D"
1¾" Hollow Metal
Flush Type Door
16 Gauge Sheet
Steel–Glass Fiber
Insulation

Type "E"
1¾" Extruded Aluminum
Door – C/W Push Bar
and Pull Handle – Glass
Felt Astragal Strip

Type "F"
Sectional Overhead Aluminum Door C/W Tracks,
Counterbalance, Lock Latch, Etc.

Frame Type "A"

Frame Type "B"

Frame Type "C"

Frame Type "D"

Doorway Number ➡	1	2	3	4	5	6	7	8	9	10	11	12	13	14	15	16	17	Doorway Numbers As Required ➡
Door Type	D	E	B	A	C	F	D	A	E	B	B	B	F	A	A	A	A	
Frame Type	B	C	A	A	A	D	B	A	C	A	A	B	D	A	A	A	A	
1½ Pair 4" x 4" Steel Butts			✓	✓	✓		✓		✓	✓	✓		✓	✓	✓	✓	✓	
1½ Pair 4" x 4" Brass Butts – Ball Bearing	✓					✓												
Offset Alum Hinges		✓					✓											
Latch Set			✓				✓		✓	✓	✓				✓			
Dead Lock	✓				✓		✓					✓						
Combination Dead Lock and Latch	✓			✓	✓		✓					✓	✓	✓				
Bracket Type Door Closer				✓	✓								✓	✓				
Floor Type Door Closer		✓						✓										
Combination Door Stop and Holder	✓		✓		✓		✓	✓		✓	✓	✓		✓	✓	✓		
Top and Bottom Bolts																		
Overhead Door Hardware Complete						✓							✓					
Aluminum Threshold	✓		✓	✓				✓	✓	✓		✓	✓		✓			
Push and Pull Hardware		✓					✓											

Items as Required

Example 2

DOOR SCHEDULE

Floor	Number of Door	Revisions	Location of Door — Outside Room	Location of Door — Inside Room	Door Swing	Door Type	Door Width	Door Height	Door Thickness	Door Material	Door Finish	Frame Type	Frame Thickness	Frame Material	Hardware	Remarks
First Floor	2		Exterior	Vestibule 1–01	PD	D–1	3'-0"	7'-0"	$1\frac{1}{2}$"	Al		F–1		Al		See Dwg #14
	2	*A	Vestibule 1–01	Foyer 1–02	PD	D–2	3'-0"	7'-0"	$1\frac{3}{4}$"	Wood	Laq.	F–2	14GA	Met		
			Foyer 1–02	Gen. Office 1–04	RH	D–4	2'-8"	6'-8"	$1\frac{3}{4}$"	Wood	Laq.	F–3	14GA	Met		
			Foyer 1–02	Corridor 1–03	RHR	D–3	3'-6"	6'-8"	$1\frac{1}{2}$"	Met		F–5	14GA	Met		
			Corridor 1–03	Public Lav. 1–07	LH	D–6	2'-6"	6'-8"	$1\frac{3}{4}$"	Wood	Laq.	F–3	14GA	Met		
			Coffee Shop 1–10	Kitchen 1–12	DA	D–5	3'-0"	6'-8"	$1\frac{1}{2}$"	Al		F–7		Al		

Swings
PD – Pair of Doors
DA – Double–Acting
RH – Right Hand
LH – Left Hand
RHR – Right Hand Reverse
LHR – Left Hand Reverse

Revisions

	Date	Revisions
A	15·2·61	Swing Changed to LH From RH

*Revision Table Added Only When Necessary

29

Room Finish Schedule

Example 1

Room		Floor			Base		Walls				Ceiling		Notes
							Dado		Field		Field		
No.	Name	Vinyl Tile	Terrazzo	Lino Tile	Rubber	Terrazo	Matl	HT	Paint	Block	Acoustic	Painted	
1	Dining Rm	●			●				●		●		
2	Kitchen Vest.		●		●				●			●	
3	Kitchen		●		●				●			●	
4	Snack Bar	●			●				●		●		
5	Office			●	●				●		●		
6	Baggage Rm	●			●					●		●	
7	Concessions		●		●				●		●		
8	Concessions		●		●				●		●		
9	Concessions		●		●				●		●		
10	Waiting Rm		●		●				●		●		
11	Airport Attend.			●	●				●			●	

Extend Downward as Required by No. of Rooms

The Schedule is Usually Extended in Width to Include More Detailed Subdivisions Under Floors, Walls, etc.

Room Number should be Indicated on Plans Thus [1]

Room Finish Schedule

Example 2

Leave Extra Spaces in Each Column for Additions

No.	Name	Floor					Base			Dado		Wall			Ceiling			Accessories		Remarks
		Concrete	Terrazzo	Quarry Tile	Vinyl Tile 0.080"	Broadloom by Owner	Terrazzo	Quarry Tile	Vinyl	Vinyl Fabric 4'–6"	Ceramic Tile 7'–0"	Unfinished	Plaster	Walnut	Plaster	Gypsum Board	Acoustic Tile	Counter Cupboards	Lockers	
1	Vestibule			•				•		•			•		•					Inset Door Mat
2	Display Area		•				•						•				•			See Dwg 7 for Terrazzo
3	Information Office				•				•				•				•	•		
4	Conference Room	•				•			•					•			•			
5	Cloak Room				•		•						•			•			•	
6	Projects Office				•				•				•				•			
7	Women's Lav.		•								•		•		•					

31

Example 3

Schedule of Finishes

Item Number	Rooms and Areas	Floor				Base		Walls						Ceiling				Trim			
		Border		Field				Dado		Field		Cornice		Border		Field		Door		Window	
		Material	Finish	Material	Finish	Material	Finish	Material	Finish	Material	Finish	Material	Finish	Material	Finish	Material	Finish	Material	Finish	Material	Finish
	Column Symbol	a	b	c	d	e	f	g	h	i	j	k	l	m	n	o	p	q	r	s	t
1.	Entrance Vestibule, Lobby, Waiting Rooms and Corridors	To	Pd	To	Pd	To	Pd	Pl	VP1	Pl	PT1	–	–	–	–	At	PT1	St	PT2	Al	–
2.	Below Grade, Offices, Laboratory, Exam Room, Canteen, Locker Room, Pharmacy, Pack Stores, C.R., Rest Rooms	To	Pd	Va	Wx	To	Pd	Pl	VP1	Pl	PT1	–	–	–	–	At	PT1	St	PT2	Al	–
3.	Above Grade, Wards	To	Pd	Hv	Wx	To	Pd	Pl	VP2	Pl	PT1	–	–	Pl	PT1	Pl	PT1	St	PT2	Al	–
4.	Utility Rooms, Janitor Rooms	To	Pd	To	Pd	To	Pd	Pl	VP1	Pl	PT1	–	–	–	–	At	PT1	St	PT2	Al	–
5.	Toilets and Bath Rooms	Ct	–	Ct	–	Ct	–	Gt	–	Kp	PT1	–	–	–	–	At	PT1	St	PT2	Al	–
6.	Auditorium	Wd	Vd	Wd	Vd	Wd	Vd	Pl	VP1	Pl	PT1	–	–	–	–	At	PT1	St	PT2	Al	–
7.	Dining Room	To	Pd	Hv	Wx	To	Pd	Pl	VP1	Pl	PT1	–	–	–	–	At	PT1	St	PT2	Al	–
8.	Arts and Crafts	To	Pd	Hv	Wx	To	Pd	Pl	VP1	Pl	PT1	–	–	–	–	At	PT1	St	PT2	Al	–
9.	Stairways	To	Pd	To	Pd	To	Pd	–	–	Tg	–	–	–	–	–	–	PT1	St	PT2	Al	–
10.	Serveries	Qt	–	Qt	–	Qt	–	–	–	Gt	–	–	–	–	–	At	PT1	St	PT2	Al	–
11.	Overhead Passage	To	Pd	Hv	Wx	To	Pd	Pl	VP1	Pl	PT1	–	–	–	–	At	PT1	St	PT2	Al	–

Key

Symbol	Materials and Finishes
At	Acoustic Tile
Cp	Cement Plaster
Ct	Ceramic Floor Tile
Gt	Glazed Wall Tile
Hv	Homogeneous Vinyl
Kp	Keenes Cement Plaster
Pl	Gypsum Plaster
PT1	Paint-Gloss
PT2	Paint-Enamel
Pd	Polished
Qt	Quarry Tile
St	Steel
Tg	Glazed Structural Terra Cotta
To	Terrazzo
Vd	Varnished
Va	Vinyl Asbestos
VP1	Vinyl Plastic 0.020"
	Clear Vinyl
VP2	Vinyl Plastic 0.012"
Wd	Wood Hardwood
Wx	Wax

2.14 DRAWING SHEET INDEX

Drawings usually fall under four main headings; that is, architectural, identified by a prefix capital letter A; structural, identified by a prefix capital letter S; mechanical, identified by a prefix capital letter M; and electrical, identified by a prefix capital letter E. Each series of drawings is allotted a block of numbers such as the A series, which may range from numbers A1 and anywhere up to and including A100, S201 to S300, and so on. For very large buildings the allotted series of drawing numbers may go into the thousands; for example, A1 to A1000, S2001 to S3000, and so on.

Title and revision blocks are positioned in the lower right-hand corner of the drawing sheets and normally should provide the following information: the name and location of the project, the name of the agency responsible for the drawings; the title of the drawing; the number of the drawing including the prefix identifying letter; drawing scales; the initials of the designer, checker, draftsman, and supervisor, together with the appropriate dates. The revision block should be adjacent to the title block, either directly above it or to its left, and should be read from the bottom upwards. *It is very important that latest revisions be carefully noted by the estimators when taking off of quantities.*

Upon receiving the material from the architect, the contractor will first examine the index sheet which will list the title of each drawing. He will then check that all indexed drawing sheets are included in the roll. Again, he would make sure that all revised (and dated) copies are included. Similarly, he will check the specifications to see what addenda are included.

Following is a typical index for a roll of drawings:

ARCHITECTURAL DRAWINGS
No.   INDEX

1	Index
2	Plot and Site Plan
3	Topographical
4	Soil Analysis
5	Services and Utilities
6	Basement Plan
7	First Floor Plan
8	Second Floor Plan
9	Roof Plan
10	North and South Elevations
11	East and West Elevations
12	Sections C/6 D/6 E/6
100	Sections $\frac{\text{G-H-J-K-L-M}}{100}$
101	Sections $\frac{\text{N-O-P-R-S-T}}{101}$
102	South and East Entrances
103	Column Sections, Window Details
104	Column Sections, Window Details
105	Stair Details
106	Outside Doors and Details
107	Isometric View of Columns
108	Office Partitions
109	Room Schedule
110	Room Schedule
111	Roof Dome Details

Carefully note that all the above drawing sheet numbers will be prefixed with the letter A for architectural.

STRUCTURAL DETAILS
No.   INDEX

201	Pile Layout
202	Footing Plan
203	Plan of Grade Beams
204	First Floor, Framing and Details
205	Second Floor, Framing and Details
206	Roof Framing and Details
207	Details
208	Column Layout (Precast)
209	Beams and Slabs
210	Details of Loading Ramp
211	Receiving Platform Details

Carefully note that all the above drawing sheets will be prefixed with the letter S for structural.

MECHANICAL DRAWINGS
No.   INDEX

300	Site Plan
301	Footing Plan, Drainage, Exhaust System
302	Basement Plumbing
303	First Floor Plumbing
304	Second Floor Plumbing
305	Roof Plan
306	Steam Trench
307	Basement Floor Heating
308	First Floor Radiation
309	Second Floor Radiation
310	Exhaust System
311	Mechanical Room, Monoxide Exhaust Paint Shop, Ventilation and Exhaust
312	Plumbing Details
313	Service Piping Layout
314	Boiler Room, Diagram and Layout

Carefully note that all the above drawing sheets will be prefixed with the letter M for mechanical.

ELECTRICAL DRAWINGS
No.   INDEX

Carefully note that all the above drawing sheets will be prefixed with the letter E for electrical.

There is a considerable difference in the renditions of drawings by architects and engineers. Some are difficult to read and understand; others are exceptionally well-done with lots of easy references that enable the contractor's staff to flick the drawings over from one sheet to another by easily identifiable symbols.

2.15 DRAWING REFERENCE SYMBOLS

Be aware that at this time of writing there is no one set of symbols used throughout the profession. It is suggested that you examine carefully new sets of drawings to establish in your mind what type of symbols are used; this will certainly save time during the estimating and building process.

Drawing reference symbols are sometimes called flags; two types are as follows:

The letter designates the section, that is, A for architectural, S for structural, and so on. The numeral indicates the sheet where the detail appears. Such symbols are placed on small-scale drawings to direct the attention of the reader to a larger, detailed scale drawing appearing on another sheet. As an example, many such symbols appear on the column drawings for large buildings and would probably show the detail construction at each intersection with other framing members.

CHAPTER 2 REVIEW QUESTIONS

1. Draw the line conventions for the following: (a) a hidden line; (b) a dimension line; (c) a visible object line.

2. Draw the symbols used for the following: (a) earth; (b) gravel fill; (c) concrete; (d) wood framing; (e) insulation.

3. Draw a neat freehand plan sketch of a wide "U" staircase — one that leads to a half space landing and returns in itself.

4. Draw a neat freehand plan sketch of a single swing doorway showing the Schedule Reference, Type "C".

5. Describe in one hundred words or less: (a) a door and hardware schedule; (b) a room finish schedule.

6. Explain the use of drawing reference symbols (sometimes called flags).

Preliminary Building Organization and Land Grading

In this chapter we will discuss a builder's license; making a district survey for a building lot; design professionals; overhead expenses; land grading; the critical path method; title deeds; progress reports; change orders; time cards; progress schedules; payments to subcontractors; inspections; and progress certificates.

3.1 BUILDER'S LICENSE AND THE BUILDING CODE

In many areas before a person starts to operate as a builder, it is necessary that he obtain a builder's license from city authorities. There is a different fee for builders living and contracting within the city limits and for builders living outside of city limits and working in the city. There is also a difference in fees for subcontractors. The purchase of a builder's license is not an indication that the purchaser is a good builder. His work has to stand up to progressive inspection by the city building inspectors during all phases of construction.

The builder's license will enable the holder to make building material purchases at normal trade discounts. In business there are several different discounts which apply, depending upon the number of days within which payment for delivered goods must be made. The differences are considerable, and you should make very careful inquiries from the merchants with whom you may deal. A license holder may also, upon nomination, be accepted into the local and national construction associations.

The contractor and all subcontractors shall comply with all local, state, provincial, or national government rules, regulations, and ordinances. They will prepare and file all necessary documents or information, pay for and obtain all licenses, permits, and certificates of inspection as may be specified or required.

Wherever municipal bylaws or state or provincial legislation require higher standards than those set forth in the drawings and specifications, such higher standards shall govern.

Note: *Assume there is some peculiar subsoil condition in a certain area. It is fair to assume that the local authority, knowing the conditions, may require special precautions to be taken which are more stringent than those set forth in the specifications. In all cases, the highest building standards shall prevail.*

3.2 MAKING A DISTRICT SURVEY FOR A BUILDING LOT[1]

The object of speculation building is to erect the right type of house in the right place to suit the purse and aesthetic taste of a certain class of people. *Remember that it costs just as much to build an expensive house on a poor lot as it does to build the same house on an appropriate and sales-appealing good lot.*

Here are some of the points that you should carefully consider before building residential units:

(a) The price range of the units you wish to build.

(b) Where the houses should be built.

(c) The cost of the land in the area you wish to build.

(d) The probable taxes on the completed home. (You may get an assessment from the local authority.)

(e) The topographical features of the land. Will it have to be cut and filled (landscaped) or is it reclaimed land? (If you are purchasing the land from the local authority, they may know its history.)

(f) What about drainage? Is the area subject to extremes of rain, snow, freezing temperatures, or wind?

(g) Will any trees or existing buildings have to be removed and at what cost to you?

(h) What kind of water is available? Is it city, rural, spring, or well water, or will you have to drill a well? What about the sewage system? Will you have to put in a field yourself? How do these things affect the end price of your product?

(i) What about power, natural gas, telephone, and transportation?

(j) How far is it to existing or projected schools, shopping centers, libraries, parks, hospitals, and churches?

(k) What other residential construction is being built in the area, and what is the price range?

(l) Where will people work? Are there any projected industries near this area and what are the zoning laws? (Contact the Department of Industrial Development for Trade and Commerce.)

Land Purchase

The land in North America is owned either by the governments or by private parties. The title to all parcels of land is recorded in the land registry office of the area in which the land is located.

Assume that you want to purchase a small parcel of

vacant land within the city limits and that you do not know the owner. Take the following steps:

(a) Obtain a map of the area from the local authority and identify the land and the address of the nearest adjoining property.

(b) Inquire at the city taxation department for the name of the person paying the taxes—the owner.

(c) Obtain from the taxation department the official description of the property. This bears no relation to the postal address. It may be something like: Lot 7 of Block 14 City of _____ AP 2816.

(d) Proceed to the city planning department and check for zoning restrictions on the property.

(e) Using the official description of the property, go to the land registry office (or have your agent go for you) and search the title.

(f) Write to the owners of the property asking if they wish to sell the land. They may live miles away, possibly even abroad.

Land Registry Office

In every state and province there is a land registry office where all the land in that particular registration direct is registered according to ownership on a document called a title. All the land in North America is registered by individual parties or by the government.

It is important that you make yourself familiar with the system of land registry and the searching of titles in your district; there are differences between some states.

Searching the Title. For a very nominal fee (about one dollar) any citizen, or his lawyer or agent, may scrutinize any title to any land registered in any particular office. This is called "searching the title." The prospective purchaser of a parcel of land may wish to search the title with a view to having any encumbrance removed from it. If there are no encumbrances registered on the title, the property is said to have a "clear title."

Some of the encumbrances registered on the title may be as follows:

(a) A reservation by any party or the government for the mineral rights on the property.

(b) A first, second, third, or more mortgages registered against the property.

(c) *Mechanics Lien.* This is a statement of claim registered against the property for work alleged to have been done on the property for which the mechanic has not been paid. Such liens must be registered by the individual who did the work within a certain number of days of completion

[1] Alonzo Wass, *Construction Management and Contracting* (Englewood Cliffs, N.J.: Prentice-Hall, Inc., 1972), pp. 3–4.

of the work. You should check in your area. Why not arrange a field trip by your local association to visit one of these registry offices and get all the information firsthand?

The mechanics lien may have to be proved in a court of law. It is of prime importance that a very careful record be kept of the date and time that the last work was done, and this should be witnessed.

(d) *Easements.* An easement is the right of one party to enjoy some privilege on the land of another. Easements are often registered for such purposes as sewer, gas, power and telephone lines, and the owners of these have the right to enter the property for the maintenance and servicing of such lines. Sometimes temporary easements may have to be taken out by a contractor building in a metropolitan area where he wants the use of the adjoining property (if vacant) for temporary storage of his materials and machines. There are, of course, many other types of easements.

(e) *Power of Attorney.* This entry on a title shows that the title owner has given another party the authority to act for him in any dealings in connection with the land specified. These powers are very sweeping and are usually given to highly reputable law firms, but the power of attorney may be given to almost anyone over the age of twenty-one. This instrument is useful where the owner is living at a great distance from his holding.

(f) *Writ or Judgment.* This is a recorded statement on the title deeds showing that, at a certain time and place, a judgment was handed down by a court of law to the effect that the party in whose name the title stands was indebted at that time to another party for some services or goods that were not paid for. These services or goods could even be indebtedness to multiple stores, garages, and so on.

(g) *Caveat.* This instrument is a *"notice to beware"* that the party who has filed the caveat has an interest in this property; this interest must be taken into account before any change in ownership is made.

3.3 THE ARCHITECT, ENGINEER, AND THE BUILDING CONTRACTOR

On large projects, the architect and engineer are employed by the owner and are responsible for interpreting into physical being the structure that the owner wants, at the price he is willing to pay. The architect conceives of the type of building, considering its functional use and aesthetic value; the engineer designs the structural drawings, together with

their details, which make the erection of the building feasible.

Wherever there is the slightest doubt about the boundaries of a building site, it is imperative to get the services of a registered surveyor. The boundary pegs must be visible to the owner and the contractor; they should be specially tagged and witnessed by a third party. It is imperative that the residence be correctly located on the correct lot. Even while you are reading this, some newly built houses are either being removed from an incorrectly located lot or are being properly relocated on their correct lots. All buildings must be positioned on their lots to conform to the local building code's minimum allowable distance from the road or from the other boundaries of the property.

After inspecting the site, the contractor may stockpile all the topsoil; then cut and fill to the grade prescribed. He then assembles all the temporary buildings such as office, toilet facilities, workshops, store-sheds, first-aid room (or as a minimum, a first-aid kit), and tool shed. (He checks with the insurance company about who is responsible for the safety of workmen's equipment overnight and during holidays.) The contractor will make such necessary things as fences, barricades, hoardings, ramps, roads. Also he will get permission from the local authority to erect a signboard advertising himself as the contractor and listing the names of all his subcontractors. He must obtain a permit for the temporary closing of streets or the breaking of the public sidewalk. He will make application to the authority for the installation of temporary utilities such as water, power, heat, telephone. He must also get permission for blasting, for night work and other unusual work. He may also have to get permission from the union to work overtime.

All drawings should be individually glued to thin pieces of fiber board. They should be varnished to protect them from the weather and to prevent people from making notes or drawings on them. The varnish may be written over with a chinagraph pencil and later washed off.

3.4 GUIDE LIST OF OVERHEAD EXPENSES

There are two types of overhead expenses, i.e., those that are constant, irrespective of the amount of building being done at any one time and those that are chargeable specifically to each individual job. An example of the former is as follows: A contractor completed four separate jobs in one year which were valued as follows: (a) $116,000; (b) $146,000; (c)

$36,000; and (d) $52,000 respectively. He had therefore completed $350,000 worth of construction in one year. The percentage proportion of general overhead expenses chargeable to each of the jobs would be as follows:

$$\text{(a)} \quad \frac{116,000 \times 100}{350,000} = 33\%;$$

$$\text{(b)} \quad \frac{146,000 \times 100}{350,000} = 42\%;$$

$$\text{(c)} \quad \frac{36,000 \times 100}{350,000} = 10\%; \text{ and}$$

$$\text{(d)} \quad \frac{52,000 \times 100}{350,000} = 15\%$$

These percentages of the total cost of all general expenses would be apportioned to each of the four jobs respectively. Note the following general overhead expenses:

General Office: Rent or interest on invested capital; depreciation of office building and contents; fuel; light; telephone; stationery; postal services; business machines; furnishings; fire and public liability insurance; property tax; heat; telex; office supplies; business machines; office furnishings

Staff Salaries: Executives; accountant; estimators; draftsmen; stenographers; clerks; janitors; staff traveling expenses

Advertising: Radio; television; magazines and journals; daily newspapers; club association dues; billboards

Literature: Trade magazines; trade journals; trade reports; company library (new publications); association dues

Tools and Equipment: Purchases; maintenance; depreciation. *Note: There is a fixed table of allowances; see Inland Revenue Department for details.*

Legal Retention Fees: Attorneys, barristers, and lawyers

Professional Services: Architects; engineers; surveyors; certified public accountant; auditors

The overhead expenses totally chargeable to each individual job are as follows:

Salaries: Superintendent; job sponsor (usually an assistant to the superintendent); foreman, carpenter, mason, and others; timekeeper and storeman; watchman; cleanup man

Temporary Buildings: Field office; toolshed; store sheds; workshops; toilet facilities; fencing; hoardings; job signs; roads and trackage; ramps; platforms; temporary doors and stairs; barricades; protecting new and adjoining properties; job billboards for advertising

Legal: Public liability insurance; workmen's compensation; fire insurance; bonds—performance, and so on; attorneys and lawyers (barristers and solicitors); social security benefits; accountants and auditors; real estate fees

Building Permits: Town; out of town; special permits for roads; sewer connections; excavating sidewalks and roads

Utilities: Temporary light and power; poles and meters; temporary heat—electric, gas, or salamanders; temporary water meter and taxes; digging of well, haulage of water by tanker; temporary telephone

Testing: Soil compaction; concrete; decibel readings

Progress Reports: Written; photographs; progress diary. (See page 47.)

Out of Town Jobs: Traveling; premium rates for staff and workmen; room and board; freight and trucking of own or hired materials and equipment

Professional Services: Survey; engineers; accountant; attorney; engineers/architects

Protection of Work: Curing, covering, water-spraying, or keeping concrete ice-free; protecting finished work such as wood floors and staircases

Attending on Other Trades: Cut away and make good; breaking and repairing sidewalks and streets

Plant: Own equipment; concrete mixers; wheelbarrows, buggies, hoses, cutoff saws; hired equipment; pumps; maintenance and haulage of equipment; company transport

Completion Dates: Penalties or rewards

Final Cleanup: Replace and clean all damaged glass; janitor service

Computer Services: Estimate annual expenses

Landscape and Lawn: Shrubs

Notarization Costs

3.5 AN INTRODUCTION TO LAND GRADING

A Glossary of Land Grading Terms:

Cut
— Land that has to be cut (usually with heavy machinery) to a lower level (grade) and the cut earth (burrow) either to be hauled from the site and **legally dumped,** or compacted in a low lying area of the same parcel of land.

Parcel of Land
— Any area of land registered as one unit such as a building lot. It could be several acres which may later be divided (and registered) into smaller parcels.

Fill
— Land that has to be brought to a higher level (grade) by filling with earth from cut areas, or by hauling loose fill from other parcels of land to the job site.

Grade
— The existing grade (level) of land. For a proposed new building, the future grade of the sidewalk and/or the center of the proposed new road may be obtained from the city engineers department.

Grade Line
— A predetermined line indicating the proposed elevation of the ground for an area such as: a parking lot or the area around a new building and so on.

Grade Striker
— The title given by some authorities to the person whose duty it is to place **official stakes** on the city property adjoining a new building project and indicating the depth of the sewer below the existing grade and the future grade of the sidewalk and the center of the road.

Test Holes
— Holes drilled or dug on a proposed building site to determine the nature of the subsoil.

Top Soil
— Organic earth which will sustain vegetation.

Soil
— Subsoil or the inert earth beneath the top soil such as cobbles, boulders, gravel, fine sand, soil and clay.

Interstices
— The voids between pieces of cut rock (or grains of sugar and so on). The larger the volumes of the pieces of rock, the greater the collective volume of voids between the lumps.

Grid
— Baselines intersecting at right angles as used on survey maps. *(See Fig. 3.8.)*

Station
— A point from which an observation is taken as at the corners of a parcel of land. *(See Fig. 3.1.)*

Interpolate
— To determine an intermediate (average) term between two stations, such as 112.7' and 105.1'. The intermediate reading is the difference between the elevations of the two stations divided by two; thus: 112.7' − 105.1' divided by two is 3.8' average. Notice that survey readings are given in feet and decimal fractions of feet.

Swell Percentage of Cut Earth

The actual amount of swell for any given cut is determined by soil tests or by studying the soil reports given in the specifications. Soil means earth. Topsoil is expensive and is estimated separately for either buying or selling.

Type of Earth Cut	Swell Percentage
Sand and gravel	6 to 20%
Loam and clay	20 to 25%
Rock	30 to 50%

Bench Mark or Data

The grade to which land has to be cut and/or filled is taken from a bench mark having this symbol: ⊼ The center of the horizontal bar is the bench-mark level or datum line. For small jobs this symbol is usually cut onto a stout stake which is erected near the perimeter of the site and fenced around for protection during operations. The center of the bar is identified as 100.0'; for example, 112.2' equals 12.2' of cut and 89.2' means the land is 100.0' − 89.2' and requires 10.8' of fill.

Many bridges and large public buildings bear a bench mark showing their height above sea level; other levels may be taken from these to determine sewer levels, street drainage, and so on. A manhole cover in the center of a road could be taken as an initial reference. Internationally, the sea level is recognized as a datum.

Decimal Equivalents of Inches in Feet

In this chapter all land dimensions are given in feet and decimals of a foot:

Inches	Feet	Inches	Feet
1	0.083	7	0.583
2	0.1667	8	0.667
3	0.25	9	0.75
4	0.333	10	0.833
5	0.417	11	0.917
6	0.5	12	1.0

Remember that when estimating quantities, computations are not worked out to mathematical exactitudes.

Estimating Cut

Problem 1: Estimate the number of cu yds of cut to be removed from a square parcel of land 50.0' × 50.0' as shown in Fig. 3.1. The land is to be graded to 100.0' as shown on the benchmark. As an example station (a) is shown as 107.1' which is 7.1' higher than the benchmark.

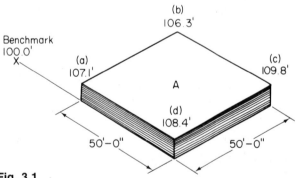

Fig. 3.1

Step 1: From the given benchmark of 100.0', estimate the average height of the four stations (a) (b) (c) and (d)

$$\frac{7.1 + 6.3 + 9.8 + 8.4}{4} = 7.9 \text{ average height}$$

Step 2: Multiply the area of the land in sq ft by the average height of the stations in ft: 2500 × 7.9 = 19,750 cu ft. Since excavating is reckoned in cu yds, the estimate would be 731 cu yds.

Using graph paper (or draw to scale) a section of the land on line (d) (c). See Fig. 3.2.

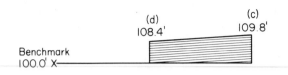

Fig. 3.2

Problem: Example 2. From the given bench mark, estimate the number of cu yds of earth to be cut and hauled from two separate parcels of land dimensioned as shown in Fig. 3.3.

Note that in the previous example, *A* has already been estimated as 731 cu yds, as in Step 2, Example 1.

The average height of the stations in lot *B* is 7.0' and the estimated cu yds is 648. The total cut of lots *A* and *B* is 731 + 648 = 1379 cu yds. The total number of stations taken into consideration for the two lots is eight, four for each lot.

Remember that estimating *is* estimating and does not necessarily require mathematical exactitude.

Examine Fig. 3.4, where both lots are placed end-on-end. It will be seen that the adjoining lot stations —that is 6.3 and 9.8—were taken twice; once for parcel *A* and once for parcel *B*. Since eight stations must be taken into account, set out the work as follows:

Station Readings	Times Taken	Totals
7.1	1	7.1
6.3	2	12.6
4.4	1	4.4
7.5	1	7.5
9.8	2	19.6
8.4	1	8.4
	8	59.6

Average cut is $\frac{59.6}{8} = 7.45$ and $\frac{7.45 \times 5000}{27} = 1379$ cu yds. The figure 5,000 is the area of the two lots together.

From the given bench mark, using six station elevations only, make a calculation of the cut required for the land shown in Fig. 3.4. Subtract the result of calculations using six elevations from the result of calculations using eight elevations. Satisfy yourself that it is more accurate to take into account eight elevations than six. Remember this is a very small parcel of land on an even plane.

Be sure to always check the number of elevations taken into account.

Rule: Take the number of station elevations in adjoining grids which have the same shape and perimeter and multiply by the number of grids. This equals the number of elevations to be taken into account.

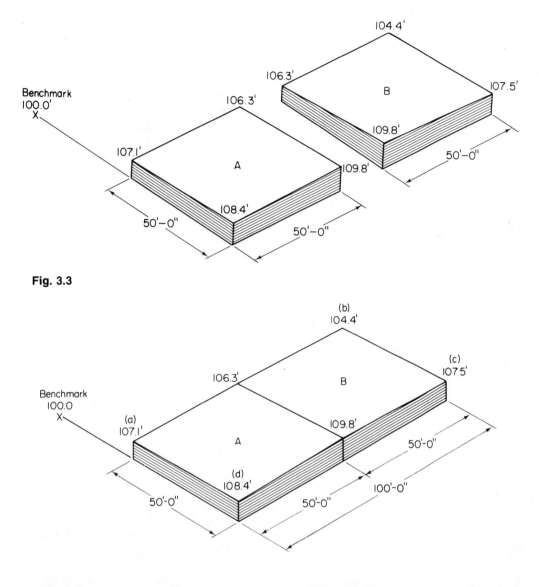

Fig. 3.3

Fig. 3.4

Problem: Example 3. From the given bench mark, estimate the number of cu yds of cut on a parcel of land with the dimensions and stations as shown in Fig. 3.5. Read the rule again. Note that the area of Fig. 3.5 is the same as in Fig. 3.4. The station elevations are the same, the bench mark is the same, *but the grids are not the same dimensions.* Take each grid estimate separately, add them together, and then subtract the differences of cuts for Figs. 3.4 and 3.5.

Make a scaled sectional drawing on line *d–c*, Fig. 3.4. Impose on this section a section in colored ink of line *d–c* of Fig. 3.5 (see Fig. 3.6).

Satisfy yourself on the validity of the rule.

Problem. Estimate the number of cu yds of earth to be cut from a parcel of land with the dimensions, stations, and bench mark as shown in Fig. 3.7.

Re-read the rule and set out your work as shown in Example 2 on page 40. How many times are you going to take station reading 110.7?

Problem: Example 4.

(a) Estimate the number of cu yds of earth to be cut from a piece of land with the dimensions, stations, and bench mark as shown in Fig. 3.8. Re-read the rule. There are fifteen equal area grids and four station elevations to each grid. There are 60 stations to be reckoned, thus:

Stations

(i) The four outside-corner stations are reckoned once: $4 \times 1 = \quad 4$

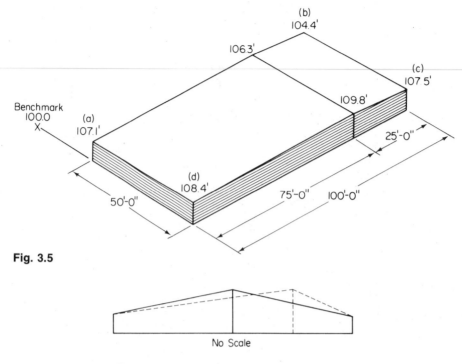

Fig. 3.5

No Scale

Fig. 3.6

(ii) The outside adjoining
     equal area and perimeter
     grid stations are reckoned
     twice: $12 \times 2 =$ 24

(iii) The internal adjoining
      equal area and perimeter
      grid stations are reckoned
      four times: $8 \times 4 =$ 32
                                                  ___
                                                  60

For this problem, set out your work on a cut and fill
sheet similar to the one shown on page 44.

(b) Allowing for a swell of 15 per cent for cut earth,
    how many 8-cu-yd truckloads of burrow (cut
    earth) will have to be hauled from the site?

3.6 AN INTRODUCTION TO THE CRITICAL PATH METHOD

Part of a dictionary definition of the word "critical" is "quick to find fault." The CPM was devised
for the construction industry for assessing the minimum time required to erect a structure and for planning the critical consecutive jobs to be done to accomplish this end. See the following Critical-Path
Flow Chart in Fig. 3.9 on page 45.

Observe how the arrowed flow indicates the progress of the units of work which follow each other,
such as: preliminary operations, hoarding, fencing,
delivery of lumber (these activities should be going
on at the same time); then land grading, *(see page 45)*
and so on. Note that the glazing and painting may be
going on from the readiness of the second floor until
the job is finished. Those items are critical that have
to be completed before other items may be started;
and when such items are given a time limit for completion, we are *quick to find fault* if they fall behind
schedule.

The term **project** denotes the whole project. The
term **activity** denotes any identifiable job that takes
time to complete within the project. Every activity
occurs before, after, or at the same time as another
activity or activities. Study the chart and note that
when the formwork is being made and erected, it is
an identifiable activity; it occurs after the layout, but
before the activity of placing concrete footings. It
also occurs at the same time that reinforcing is being
set in the formwork; furthermore, while the forms
are being made, excavating may proceed. The critical
activity in the above operations is formwork because
other activities cannot proceed until it is finished.
This is an activity that must be carefully watched so
that it may be completed in the time schedule.

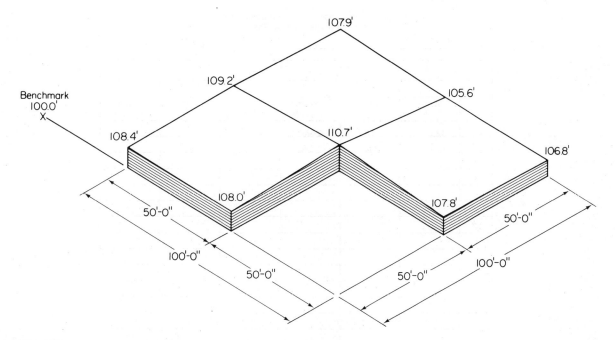

Fig. 3.7

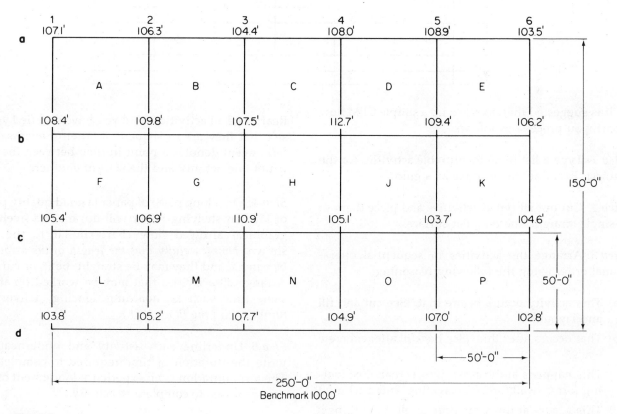

Fig. 3.8

John Doe Construction Company Ltd
Cut and Fill Estimating Sheet

Job __Apartment Block at............................__   Date __9 September__

1	2	3	4	5	6	7
	Cut				Fill	
Grid	Area Sq Ft	Average Cut	Volume Cu Yds	Area	Average Fill	Volume Cu Yds
a	2500	17.5	1620			
B	2401	8.2	730	99	1.67	6
C						
D						
E						
F						
G						
H						
I						
K						
	TOTAL CUT				TOTAL FILL	
	Note: This is an Example of Cut and Fill Sheet Only.					

It is suggested that, to make up a simple CPM flow chart, you proceed as follows:

Step 1: Type a list of all identifiable activities for the project. Examine the example as a guide.

Step 2: Cut out all typed activities and place them in a single straight line on a flat surface.

Step 3: Arrange the activities in sequential operational order using the following reasoning:

(a) This activity occurs before that. (See cut and fill and layout.)

(b) That occurs after this. (See backfill after concrete walls.)

(c) This happens at the same time as that. (See footing forms, wall forms, excavating, and ditching.)

(d) This occurs at the same time as all these things. (See glazing and plumbing, which occur during lots of other activities.)

Rearrange all activities until you have satisfied yourself that they are in sequential order of events. The term **event** denotes a point in time between the finish of one activity and the start of another.

Step 4: On a long piece of paper (a good quality piece of kitchen shelving paper will do) make a freehand graphical arrowed flow chart similar to the one shown. *Note carefully that the length of the arrows is immaterial and they may be straight, bent, or curved.* Arrange all activities that may be worked on at the same time such as: hoarding, fencing, delivery of lumber. *(See the flow chart.)*

Step 5: Underline each activity and underneath it write the duration of time required to complete it. The term **duration** is an experienced judgment of the time required to complete an activity.

Step 6: When you are completely satisfied that all the activities (together with their durations) are in correct

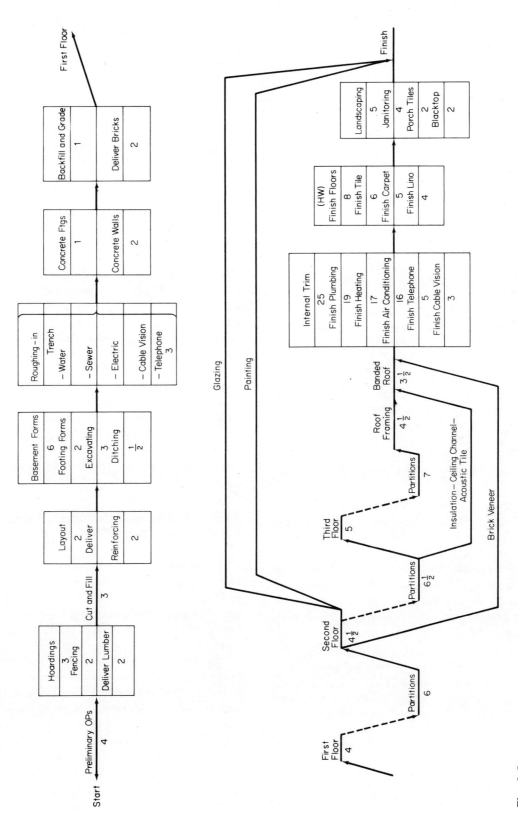

Fig. 3-9

order, make a final CPM chart. *Be aware that the making of a* CPM *chart* takes time and patience and can only be attempted against the background of your experience in the field.

You will notice the dotted lines on the chart between first, second, and third floors, and the following partitions. These lines indicate that some of the carpenters will leave the floor framing to start working on the partitions before the whole floor assembly is completed. You will find the experience of making a CPM chart interesting and rewarding, and it will fix in your own mind the best order of operations; you will also be better able to communicate with others about the job.

Using the CPM Chart

Step 1: To find the total duration of the project, add together the number of days required for all activities that appear on the top arrowed paths. See the following CPM Analysis Chart at column (a).

Step 2: Add up all the days required for all activities appearing on the arrowed path, but including the second-row arrows instead of the first—column (b).

Step 3: Add up all the days required for activities appearing on the arrowed path, but including the third-row arrows instead of the first—column (c).

Step 4: Add up all the days required for all activities appearing on the arrowed path, but including the fourth-row arrows instead of the first—column (d).

Step 5: Compare the total times for each arrowed path in columns (a), (b), (c), and (d). *The longest path is the critical path.* Anything that can speed any of the activities on this path will shorten the total time required for the project. Assuming that by using more men and equipment, or by working overtime, the longest critical path was shortened to fewer days than the second-longest path, then the next-longest path would become critical, and an endeavor would be made to speed it. This is how we examine the CPM chart for critical events. Where more men, machines, or equipment are used to shorten a critical path, it reflects itself in direct costs; but indirect costs will go down by completing the project in a shorter period of time.

Computing CPM Time

Time is measured by a five day week, excluding holidays. If a building takes 75 working days to com-

plete, this is $75 \div 5 = 15$ weeks plus public holidays. This could mean as much as 100 calendar days in all.

To calculate on the CPM chart when brickwork will commence we make what is termed a **forward pass.** This is done by adding together all the preceding time activities and relating this total to a calendar date (allowing for public holidays), which yields the date for the bricklaying to commence. The subtrades must know when they are to arrive on the job, and they must know quite definitely when they must have finished their part. A delay in any subtrade can very seriously affect the whole course of the progress of the project. *You will notice that on the* CPM *chart the delivery of bricks is shown at the time of backfill.* It is important to show delivery times for important items, including mechanical items.

To calculate on the CPM chart when the carpet layer should commence work, we make what is termed a **backward pass.** This is done by adding together all the timed activities (including that of the carpet layer), starting with his allotted duration time until the finish of the project, and subtracting this number of days from the total number of working days for the whole project. Remember to transcribe this total of working days into calendar dates.

For housing units and small jobs, you may make a CPM chart yourself, but for large, complex jobs you may, *according to the terms in the specifications,* have to make a computerized CPM chart. There are specialists in this field who will set up a program and revise it each month if required. The rental service for a computer will range from $90 to $450 per hr., but most jobs can be processed in less than 1 hr.

3.7 TITLE DEEDS, EASEMENT, AND MECHANICS LIENS

A **title deed** for land is a legal document on which is recorded the legal description of a parcel (piece) of land and the registered owners thereof. A title is also a certificate constituting legal evidence of ownership of a thing other than land. An **easement** is the right of one party to enjoy some privilege on the land of another. Easements are often registered (after an agreed compensation to the owner or owners of a parcel of land for such things as sewer, gas, power, and telephone lines); the owners have the right to enter the property for the maintenance of such lines.

Sometimes a contractor while building in a metropolitan area may negotiate with the owner of a vacant lot for a temporary easement to enable him to use it for his field office, storage sheds, and to park his equipment so as to be near his job. Thus the con-

Analysis Chart CPM

Item	(a) Activity		(b) Activity		(c) Activity		(d) Activity	
1	Preliminary ops	4	Preliminary ops	4	Preliminary ops	4	Preliminary ops	4
2	Deliver lumber	2	Deliver lumber	2	Deliver lumber	2	Deliver lumber	2
3	Hoardings	3	Fencing	2	Fencing	2	Fencing	2
4	Cut and fill	3	Cut and fill	3	Cut and fill	3	Cut and fill	3
5	Layout	2	Layout	2	Layout	2	Layout	2
6	Deliver reinforcing	2	Deliver rein	2	Deliver rein	2	Deliver rein	2
7	Basement forms	6	Footing forms	2	Excavating	3	Ditching	½
8	Roughing-in	3	Roughing-in	3	Roughing-in	3	Roughing-in	3
9	Concrete ftgs	1	Concrete walls	2	Concrete walls	2	Concrete walls	2
10	Backfill	1	Backfill	1	Backfill	1	Backfill	1
11	Deliver bricks	3	Deliver bricks	3	Deliver bricks	3	Deliver bricks	3
12	First floor	4	First floor	4	First floor	4	First floor	4
13	Partitions	6	Partitions	6	Partitions	6	Partitions	6
14	Second floor	4½	Second floor	4½	Second floor	4½	Second floor	4½
15	Partitions	6½	Partitions	6½	Partitions	6½	Partitions	6½
16	Third floor	5	Third floor	5	Third floor	5	Third floor	5
17	Partitions	7	Partitions	7	Partitions	7	Partitions	7
18	Roof framing	4½	Roof framing	4½	Roof framing	4½	Roof framing	4½
19	Bonded roof	3½	Bonded roof	3½	Bonded roof	3½	Bonded roof	3½
20	Interior trim	25	Finish plumbing	19	Finish heating	17	Finish telephone	5
21	Finish floors	8	Finish tile	6	Finish carpet	5	Finish lino	4
22	Landscaping	5	Janitoring	4	Porch tiles	2	Blacktop	2
	Days:	109	Days:	96	Days:	92	Days:	76½

Note: Items 12–16 have similar activities for each floor level; the extra time is taken up by the raising of material and placing at progressively greater heights.

tractor for a negotiated fee (and after the easement has been legally registered) will have the right, under law, to use such property for the agreed purposes for the legally agreed length of time. Inquire at your local land office for the location of the land titles office of your area and *visit it!*

A **Mechanics Lien** is a statement of claim officially registered on the title deed of a property for work alleged to have been done on such property, or materials supplied to such property, for which the mechanic or supplier has not been paid. Such liens must be registered by the individuals who did the work or the individuals who supplied the materials *within a certain number of days of completion of such work or such delivery.* The mechanics lien may have to be proved in a court of law. It is of prime importance that a very careful record be kept of the date and time that the last work or delivery was completed, *and this should be witnessed and recorded.* Some specifications state that the contractor shall hand over the property free from all liens and encumbrances. *Read your specifications!*

3.8 FIELD DIARY: PROGRESS REPORTS AND PHOTOGRAPHS

Many specifications state that a daily progress diary shall be maintained by the contractor from start to finish of the project; such records shall show the daily progress of the work, daily weather conditions, and such activities as excavation work, concrete placing and finishing, removal of forms, subtrades progress. It is recommended that the contractor pay a good price for the diary; he will then be uninclined to make frivolous and badly written notations in it. *This record may become an important exhibit in case of arbitration or a law suit.* The daily progress report must be reconciled with the progress schedule to see if there are any delays in the work, and if so, how they may be rectified—possibly by employing more men, working overtime if approved by the union, or using more machines. See the following daily progress report. *(See page 48.)*

Progress reports and glossy print photographs on large projects are required at specific intervals by the

DAILY PROGRESS REPORT

THE MANSFIELD ENGINEERING & CONSTRUCTION CO. LTD.

JOB No.

FOREMAN'S SIGNATURE   DATE

JOB NAME   AND LOCATION

WEATHER CONDITIONS   LOW TEMP.   HIGH TEMP.

Men On Job	No.	Total Hours Worked	Sub Trades On Job
Carpenters			
Laborers			
Others			

WORK ACCOMPLISHED AND REMARKS

CHANGE ORDERS OR EXTRAS AS REQUESTED OR APPROVED

EQUIPMENT ON JOB	EQUIPMENT ON RENTAL

Superintendent's Signature

Daily progress report.

CHANGE ORDER

DAILY EXTRA WORK ORDER

Send To Head Office Daily

Contract No. _____   Date _____

Job _____   Authorized by _____

Charge to _____   Extra Order No. _____

Detail Supervision, Materials, Equipment Hire or Own				
List Numbers of Men by Trades	$	cts	$	cts
Fringe Benefits				
$				
Materials :				
Hired Equipment :				
$				
Overhead Expenses % Commission %				
Own Equipment :				
TOTAL $				

Work done or charged to subcontractors.   To be approved

Authorized by _____   Approved by _____

Reported by _____   Superintendent _____

Typical change order form.

architect/engineer. These give a written and visual evaluation of the progress of the job on the dates stated.

A small speculative builder may be spared these requirements; but where a contractor is erecting a number of residences, town houses, apartment blocks, or condominiums, the architect/engineer, surveyor, and financial investors require reports and photographs to evaluate the progress of the work in order to program their own activities against them.

It is important for the contractor to complete his projects on time. On large jobs there is usually a penalty (liquidated damages) for each day of delay on the handing over of a building, and sometimes there is a reward for each day of earlier delivery than the date specified.

3.9 CHANGE ORDERS FOR ALTERATIONS AND ADDITIONS

It is quite common for the owner to require some alterations from the original drawings and specifications during actual construction.

No extra work of any description should be started without written instructions and negotiations of cost. Tradesmen dislike tearing down newly completed work; it is not good for their morale. Also, the time taken to make extensive changes may delay the contractor from completing the total project on schedule and thus delay his starting date for another contract. See the specimen of a typical change order form on page 49.

3.10 TIME CARDS

The following specimen weekly (or daily) time card is an excellent device for keeping a check on the length of time required for a man, men, or machines to complete a unit of work. This card should be made out by the foreman in a very legible hand or by lettering. The men should be told that time checks are necessary so that the costing department and estimators may keep their records up-to-date for bidding on new jobs. The timing for specific units of work will alter with the introduction of new methods, materials, or machines. Men will respond to situations

The Mansfield Engineering & Construction Co. Ltd.
Weekly Time Card

Job *Rosemont School*
Name *B. Ross*   Week Ending *5 May*

Date	Job Unit No.	Discription of Work Done	Hours	
30 April	505	First Floor joists	8	00
1 May	505	First Floor joists	5	30
1 May	506	✓ ✓ ✓ Bridging	3	30
2 May	507	First Floor Sub Floor	8	00
3 May	507	First Floor Sub Floor	7	15
3 May	508	✓ ✓ Plate layout	1	00
4 May	509	Wall Framing First Floor	8	00
		Total Hours	41	15

Forman *L. Brown*
Processed Cost A/c Dpt. Date *2 June*

Time card.

if they are given good reasons. At first the men's response to a time card might be, "Don't they think we are doing enough work already without keeping a check on us?" When they realize that it is only by knowing how long it takes to complete a certain unit of work that future successful bids may be made, they will realize their future prospects for work depend on accurate estimates. These time cards are analyzed by the costing department to establish actual unit costs of work completed against the original estimates.[1]

Note the different job unit numbers that are given by the original estimator as shown on the specimen weekly time card. When completed, the cards are returned from the job site to the estimating department for analysis, and if the total number of hours *actually* taken to complete a unit of work is greater or lesser than the estimate, a correction must be made on future estimates for similar unit numbered operations.

3.11 PROGRESS SCHEDULES

Two types of progress schedules are shown. The first is a typical specimen that may be called for by the architect (for a small job) about one week before building operations begin. It is prepared and submitted by the contractor. The second is for multiple dwelling units.

From the first one the architect would be able to see, at any time during the building operation, the actual progress of the work against the assessed time. He would be in a position to advise his client of the financial obligations to the contractor at the agreed progress payment scheduled times.

The contractor would make great use of his copy of the progress schedule in the following ways:

(a) He will mark on the schedule in a different color the actual progress of the work done under the progress of the work estimated.

(b) He will be able to see at a glance if more men and/or machines are necessary on the job to maintain the progress scheduled. It must be remembered that a large contractor will have several jobs in progress at any one time, with a separate progress schedule for each. He may see the advisability of moving men or equipment from one job to another, alternatively recruiting or laying men off.

(c) He will send extracts of the schedule to subtrades so that they will know when they are expected on the job *and quite definitely when they are to be off the job*. If a subtrade falls behind on his schedule, this could throw the whole schedule and completion date off, with possible penalties for late delivery.

(d) He will know when he may expect progress payments for completed work. Such payments may be expected on most jobs, monthly, upon issuance of an architect's certificate to the owner, who may then instruct the bank to release to the contractor the amount of cash enfaced on the certificate. The progress schedule percentage column shows a breakdown of the cost per main item.

(e) Some contracts stipulate the amount of daily liquidated damages assessed the contractor for late delivery of a structure. Some contractors may feel more competent to control costs than time. For a time study see the critical-path method (CPM) in section 3.6.

Housing Project Daily Progress Report

The source of the second daily progress report was devised (with minor exceptions) and used by my friend and colleague, Mr. H. B. Smith, formerly of the Southern Alberta Institute of Technology. I extend my sincere thanks for permission to reprint it here. *(See page 53.)*

There is no hard-and-fast rule on how a progress schedule should be drawn up. However, some kind of progress schedule must be made to keep close control over the job.

---

[1] For a full treatment of estimating, see Alonzo Wass, *Building Construction Estimating,* 2nd ed. (Englewood Cliffs, N.J.: Prentice-Hall, Inc., 1970).

THE MANSFIELD ENGINEERING AND CONSTRUCTION CO. LTD.

PROGRESS SCHEDULE

Project 18 Suite Apartment   Date: April 24
for Mr. J. Doe

Starting Deadline: May 1   Scheduled Completion: September 3

	%	May	June	July	Aug	Sept
Preliminaries	3/4	—				
Survey	1/2	—				
Cut and fill	2	—				
Fencing	1/2	—				
Excavating	2 1/2	—				
Ditching	1/2	—				
Rough plumbing (subtrade)	5	——				
Concrete footings	3 1/2		—			
Concrete walls	5 1/2		—			
Backfill	1/4		—			
Rough carpentry, basement	4		—			
First-floor assembly	3		—			
Electrician (subtrade)	4		——			
Heating engineer (subtrade)	4 3/4			——		
Plumbing (subtrade)	8			—		
Insulation and soundproofing	1/2			—		
Brickwork, first floor	2			—		
Second-floor assembly	4			—		
Rough carpentry, second floor	4			—		
Brickwork, second floor	2 1/2			——		
Third-floor assembly	5				—	
Rough carpentry, third floor	4 1/2				——	
Brickwork, third floor	3				—	
Rough carpentry, third floor	5				——	
Ceiling, third floor	3				—	
Flat roof assembly	4					—
Bonded roof (subtrade)	2 1/2					—
Glazing (subtrade)	4 1/2					—
Interior finish	7					—
Painting and decorating (subtrade)	3 1/2					—
Finish floor (subtrade)	2					—
Sidewalks	1 1/2					—
Landscaping	2 1/2					—
Final Clean-up	1/4					—
Final Handover	100					—

THE MANSFIELD ENGINEERING AND CONSTRUCTION CO LTD.

DAILY PROGRESS REPORT

Project No. _____   Date _____   Sheet 1 of 2

Starting Deadline _____   Scheduled Completion _____

JOB NUMBER:	1	2	3	4	5	6	7	8	9	10	11	12	13	14	15	16	17	18	19	20	21	22	23	24
1 Permit																								
2 Layout																								
3 Utilities — gas																								
4 — water																								
5 — sewer																								
6 — telephone																								
7 — cable television																								
8 Excavation																								
9 City and mortgage inspection no. 1																								
10 Formwork for footings																								
11 Concrete footings																								
12 Posts, forms, beams, and floor joists																								
13 Concrete walls																								
14 Floor sheathing																								
15 City and mortgage inspection no. 2																								
16 Framing walls																								
17 Masonry unit walls																								
18 Chimney																								
19 Framing roof																								
20 Walls and roof sheathing																								
21 Roofing complete																								
22 City and mortgage inspection no. 3																								
23 Rough plumbing																								
24 Rough wiring																								
25 Rough heating																								
26 Rough telephone																								
27 No 1 progress payment applied for																								
28 No 1 progress payment received																								
29 Exterior trim — windows/doors																								
30 Insulation																								
31 City and mortgage inspection no. 4																								

THE MANSFIELD ENGINEERING AND CONSTRUCTION CO LTD.

DAILY PROGRESS REPORT

Project No. _____   Date _____   Sheet 2 of 2

Starting Deadline _____   Scheduled Completion _____

JOB NUMBER:	1	2	3	4	5	6	7	8	9	10	11	12	13	14	15	16	17	18	19	20	21	22	23	24
32 Gyprocking / lathing																								
33 Basement floor																								
34 Taping/plastering																								
35 Sheet metal / heating																								
36 City and mortgage inspection no. 5																								
37 No 2 progress payment applied for																								
38 No 2 progress payment received																								
39 Exterior wall finish																								
40 Meter applications																								
41 Hardwood floors laid																								
42 Interior finish																								
43 Painting																								
44 Floor sanding																								
45 Linoleum																								
46 Sidewalk and steps																								
47 Backfill																								
48 Finish — plumbing																								
49 — wiring																								
50 — heating																								
51 Lawn seeded																								
52 No 3 progress payment applied for																								
53 No 3 progress payment received																								
54 City and mortgage inspection no. 6																								
55 Sold proposal																								
56 Approval of sale																								
57 Purchaser's mortgage complete																								
58 Down payment received																								
59 Final inspection and holdback																								
60 Occupied																								
61 Garage																								

Notice that on the first line the job numbers are shown. In this particular example, they would refer to each separate house as numbered on the contractor's projected overall plan. They are in no way associated with the address of the houses.

Building permits and applications for utility services on this type of project may be taken out at the rate of ten or twelve at a time (entered under each house number). The date would be written when each application was granted.

The progressive inspections are stressed because these are very important dates. Builders' or mortgage progressive loans are not issued before inspections are completed.

The first item on "Sheet 2 of 2" shows "Progress payment applied for" and, underneath, is shown "Progress payment received." Item 60 shows the final disposal of the property. You must make your own progress schedules.

3.12 PAYMENTS TO SUBCONTRACTORS

Payments to the subcontractors may be made progressively as they complete portions of their work. As an example: the plumbers, electricians, and heating engineers, all have a certain amount of work to do before the contractor can cover their work with wall or ceiling, cladding. Remember, too, that such subcontractors' work must be approved (and a certificate issued) by the building inspector before such work is covered; otherwise, it may have to be opened up for inspection. At each major stage of completion of their work, the subcontractors are entitled to a percentage payment of their total contract. Assume that the plumbing contract for a residence is $2,600, and the first portion of the work that has been completed (and passed inspection) is in the sum of $1,500. The plumber should then be paid that amount, less a holdback of 10% ($150), leaving a payment to be made of $1,350. The holdback (sometimes called retainage) is retained in case of default by the plumber in completing his subcontract. He may have left the district, become bankrupt, met with an accident, or even have died before the remainder of his work could be completed.

Remember that once the framing for a house is completed, the financing for all the subtrades is imminent; the contractor's financial arrangements should be regulated to meet this situation. Remember that the bulk of housebuilding expenses are for subtrades.

Corrections after the handover of residential units should be made by the contractor up to two or three months. Such items may include the easing of doors, drawers, or the making good of minor plaster cracks due to settlement. But there is a time limit (say, three months approximately) after which the contractor should make a charge for his services. It is important for the contractor to build good homes and good public relations.

A holdback of final payment to the contractor is usually recommended to the purchaser by his attorney. This amount may vary according to circumstances in order to cover contingencies. As an example: assume that a residence was completed in every respect, in the late fall, except for the placing of the concrete paths and driveway. If the climate of the area is severe, the purchaser may live in the house and the concrete may not be placed until the following March. In such a case, the final payment would be withheld until the work was complete. If the contractor for any reason was unable to do the work himself, the attorney would have retained sufficient funds to have the work completed by someone else. I know of one young builder who (without a written contract) started to build a house on the property of the owner. The builder stockpiled the topsoil, excavated and placed the sewer lines, built the concrete basement, completed the laying of the subfloor, and supplied all the materials and labor. Then, the owner met with a fatal accident, and his land and everything on it went into the estate of the deceased. It took many months before the will was probated, and the funds released to the builder. **Be careful and see your attorney before, not after, you sign any important document.**

3.13 INSPECTIONS AND PROGRESS CERTIFICATES

To ensure that the builder complies with all local and/or national building codes and standards, the authority having local jurisdiction will have its own inspectors examine the quality of the builder's work at critical times. One inspector may specialize in the framing of the building, another in plumbing, still another in the electrical work and so on. It is very important for the builder to notify the building inspectors' department when each phase of the work will be ready for its inspection. It is illegal to cover up some work such as plumbing lines, electrical lines, and insulation until a certificate of approval has been given by the building inspector. If earth has been filled over plumbing lines, or if wall boards have been placed covering electrical wiring or any other runs such as for heating, the inspector can enforce the opening up of all such closures for inspection. Be careful!

The Architect, if one is engaged, or his own inspector (sometimes called the clerk of the works) will also inspect the building during construction to ensure that all the materials and workmanship are being implemented as directed according to the plans and specifications. To ensure that all safety precautions are being taken, a representative from the workmen's compensation agency may also inspect the site at any time. He has the authority to suspend operations, if necessary, in certain areas until the minimum safety precautions are implemented.

The architect issues progress payment certificates to the contractor at stipulated times upon satisfactory completion of different phases of the construction of the building.

An inspection certificate is a recommendation to the owner, by the architect, that the contractor is now entitled to a progress payment in the amount shown on the certificate. The owner may then instruct the bank to release to the contractor the amount stated on the certificate. *Some specifications call for evidence that all materials and labor used in the erection of the work certified as satisfactorily completed have indeed been paid for by the contractor before the certificate will be issued.* It is usual for the architect to withhold 10 to 15 per cent of the value of the work performed. This represents the contractor's profit and maybe more. When the building is completed, the architect then makes a final holdback for the period of time stated in the contract.

The following progress payment schedule shows the manner of payments for an $80,000 residence with a 10 per cent holdback at six stages of progress. The last 10 per cent of the final payment is held back for a further six months until the contractor has made the final check and made good all necessary adjustments due to settling of the building or for any other legitimate reason.

At the end of September the total value of work completed was $80,000 (Column 2). The total value of certificates issued was $72,000 (Column 3) and the total value of holdback payments was $8,000 (Column 4). Let us examine the position with the issue of the July payment certificate of $10,080 (Column 3).

Step 1: The value of work completed as of July was $51,200 (Column 2).

The Architect's Progress Certificates

	PROGRESS PAYMENTS		
Date	Value of Work Completed (cumulative)	Value of Certificates Issued (individual)	Value of 10% Holdback (cumulative)
1	2	3	4
April	$ 9,600	$ 8,640	$ 960
May	24,000	12,960	2,400
June	40,000	14,400	4,000
July	51,200	10,080	5,120
August	72,000	18,720	7,200
September	80,000	7,200	8,000
		Total $72,000	

Step 2: The value of all previous certificates issued was:

April $ 8,640
May 12,960
June 14,400 (see Column 3)
$36,000

Step 3: The value of the 10 per cent holdback as of July was 10 per cent of $51,200 = $5,120 (Column 4).

Step 4: The value of the July certificate was equal to the *total value* of all work completed as of July, which was $51,200 as in Step 1, less the cumulative value of certificates already issued—$36,000 as in Step 2—and also less the new holdback on the July work completed, amounting to $51,200 as at Step 3.

Step 5: The value of the July certificate was:

$51,200 - (36,000 + 5,120) = $10,080.

Problem

Pattern your work after the example shown and analyze the position with the payment certificate of $18,720 at the end of August.

3.14 SEWAGE DISPOSAL

The greatest contributing factor towards the maintenance of good public health has been modern plumbing and safe sewage disposal. In many areas adjoining cities, septic tanks are installed until such time that they can be connected to city sewers. For summer cottages and farms they are imperative. The builder should get a copy of the local building by-

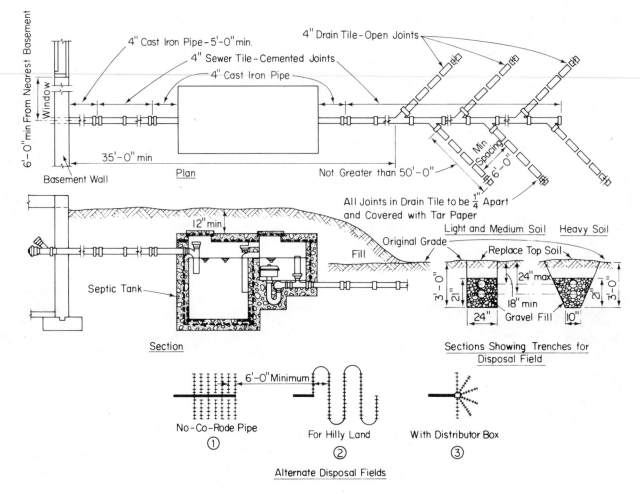

Fig. 3.10. Typical family-size sewage disposal system.

laws, and he should talk with the district Sanitary Engineer; **this is important.**

The source of the following material is "Canada Lafarge Ltd."

House Sewer Drain Line

The septic tank is connected to the house drain with a line of pipe called the house sewer *(See Fig. 3.10).* It is usually built of 6 in. bell and spigot sewer pipe. All joints should be thoroughly filled with a mortar composed of 1 part portland cement, 3 parts mortar sand, and enough water to make a plastic mix. When properly made, such joints will normally keep out roots. Sometimes, as a further precaution against root penetration, a mortar band 1 in. thick and 3 in. wide is made around the joint *(Fig. 3.11).*

Double Chamber Septic Tank

Septic tanks with a separate chamber which discharges intermittently are required for large house-

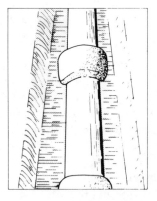

Fig. 3.11. Mortar band around joints in house sewers provides protection against root penetration.

holds of more than 12 people and are required in parts of Canada where severe freezing weather endangers operation of the normal single-chamber tank. This type of system consists of a main chamber from which the outlet pipe leads to a smaller connected tank which is fitted with a siphon. This tank

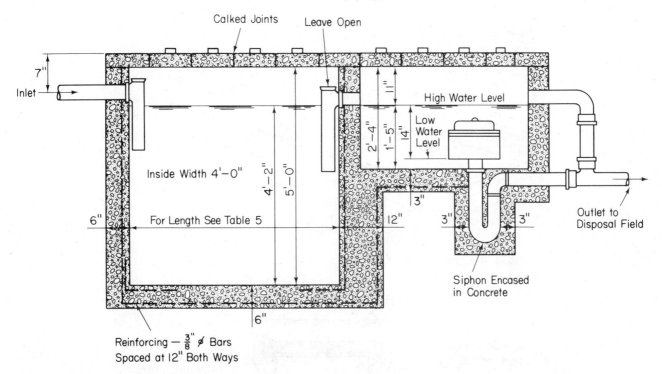

Fig. 3.12. Double chamber septic tank with siphon.

automatically discharges when it becomes full, and thus distributes an even load to all disposal tile at a higher temperature than effluent reaching the field of a single-chamber tank. *(See Fig. 3.12.)*

Inlet and Outlet Tile to Septic Tank

Fig. 3.13 shows a method of building an inlet or outlet pipe to the septic tank. These joints must be positively watertight.

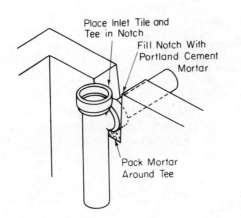

Fig. 3.13. Detail showing how inlet and outlet tile are mortared into notch at end of septic tank.

Concrete Cover Slab

The concrete cover slab (Fig. 3.14) is made 3½" thick and 12" wide and long enough to reach across the tank. Each slab is reinforced with ⅜" round bars spaced 3" apart and placed about 1" above the btm.

Distribution Box

The distribution box is a small tank which distributes the liquids from the septic tank to disposal lines. This box helps to equalize the flow into the disposal lines and permits the inspection of the sewage liquids. See Fig. 3.15.

Layout for Disposal Field

The disposal field consists of two or more lines of drain tile laid with open joints in specially prepared trenches. The sewage liquid flows from the distribution box into the tile lines, where it seeps out through the open joints into the gravel fill. Here it is worked on by bacteria, which complete the disposal process. See Fig. 3.16.

Disposal Tile Line

Fig. 3.17 shows a cross section and a longitudinal section of disposal tile lines. The whole of the sewage disposal system should be laid down with absolute thoroughness.

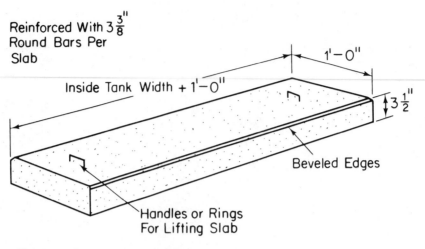

Fig. 3.14. Concrete cover slab.

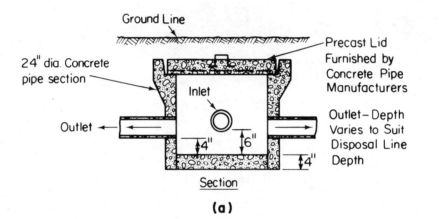

(a)

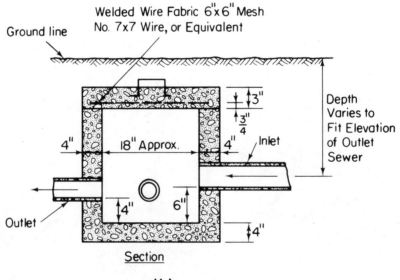

(b)

Fig. 3.15. Section through distribution box with three outlets (top), with four or more outlets (bottom).

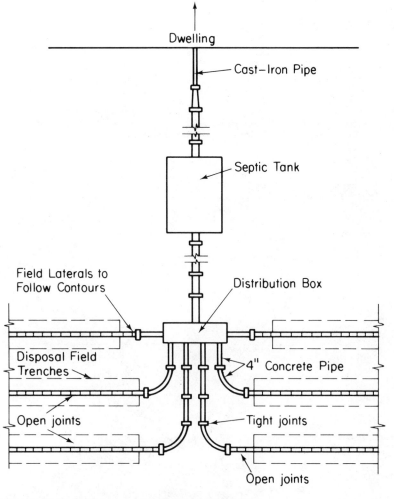

Fig. 3.16. Layout for disposal field on hillside.

The foregoing drawings are presented for serious study. If we were limited to making only one contribution to newly emerging countries, I would unhesitatingly say, "This is it."

Assignment:

Go to your nearest local Health Board or Government agency, obtain all the material available on drinking water and sewage disposal, and file it in your library.

CHAPTER 3 REVIEW QUESTIONS

1. List six temporary buildings that may be seen on a building site.
2. Give the name, address, and telephone number of the building licensing authority in your district; state the cost of the license to build; the cost of the builder's license, and the necessary qualifications of the applicant for a builder's license.
3. How many building inspections are made by the authority having jurisdiction in your area for the erection of a house?
4. State whether or not building construction safety inspections are made by a member of the workmen's compensation insurance company in your district; also state whether or not inspections are made by any of the companies having a financial interest in the erection of residential projects. If you personally loaned someone $20,000.00 to help them to build a house, would you inspect its erection?
5. State in not more than one hundred words the function of an architect's progress certificate.
6. Define a retainage (or hold back payment) by the owner to a builder.

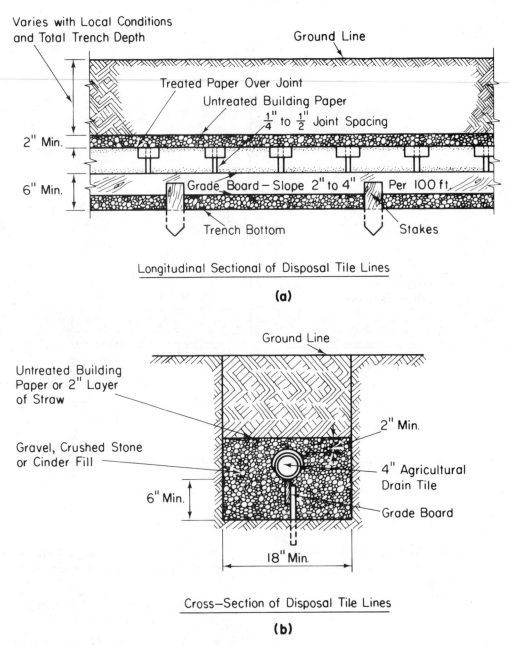

Longitudinal Sectional of Disposal Tile Lines

(a)

Cross—Section of Disposal Tile Lines

(b)

Fig. 3.17. Sections of disposal tile lines.

7. List twelve general overhead expenses.

8. List twelve individual job overhead expenses.

9. Define a progress schedule and state its function.

10. In what way does a critical path method of job control differ from a simple progress schedule?

11. What is the function of a workman's time card?

12. Define a change order; a daily progress report from the job site; and the use of progress photographs.

13. Define the following terms: (a) a bench mark; (b) a parcel of land; (c) grade; (d) interstices; (e) station; (f) cut.

14. The parcel of land described on page 61 is to be cut and levelled to a benchmark of 100.0′. Allow for a percentage of swell for cut earth as being 22%. Estimate the number of truckloads of earth that will have to be hauled away from the site. *(Answer—480—500 loads.)*

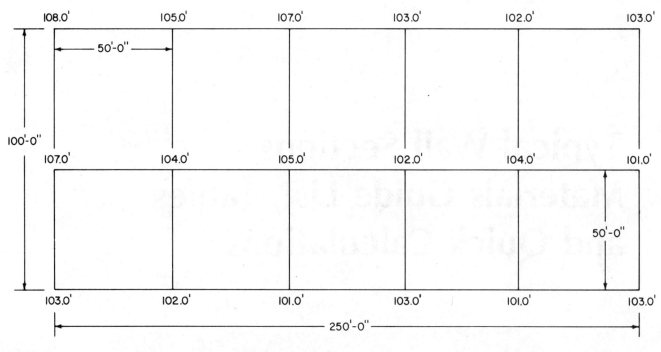

Fig. 3.18.

Typical Wall Sections, Materials Guide List, Tables and Quick Calculations

In this chapter we will examine a number of typical wall sections for residential construction as used in America and Canada. Then follows a list of materials used in house building. The chapter concludes with a number of useful tables: methods of rapid calculations, miscellaneous capacities, measures, and weights of materials used by house builders.

4.1 WALL SECTIONS

Figure 4.1 illustrates a typical North American traditional house wall section. To refresh our memories, all parts of this drawing are named, and it is recommended that all components of any typical wall section be memorized.

4.2 PLATFORM OR WESTERN FRAMING

Figure 4.2 illustrates a view of a platform or what is known as western framing. The first floor framing originates with the sill plate, over which are placed the rangers (rim and header joists); then, follows the subfloor like a platform; the walls have a top plate and a cap plate. The operation is then repeated for the second floor. In some jurisdictions, not more than four stories may be built in this manner, but

check locally for the bylaws. An advantage of this type of construction is the ease with which the 8'–0" walls may be raised from the floor (platform) into place. A great disadvantage lies in the fact that many longitudinal members are subject to shrinkage. Add all those members together to really appreciate the total thickness subject to shrinkage. Lengthways of the grain the shrinkage is negligible; but where platform framing is used for a four-story apartment that must, by law, have a concrete fire wall, the difference in shrinkage between the lumber and the concrete could cause uneven floors, especially in the third and fourth floors. See also Figs. 10.9 and 10.10.

4.3 BALLOON FRAMING

Figure 4.3 illustrates a view of balloon framing. The 18'–0" (plus or minus) wall studs originate on top of the sill plate. The first floor joists are spiked to the feet of the wall plates and the floor joists and extend to the wall plates and the cap plates at the ceiling level of the second floor. Note carefully that the only shrinkage possible with this construction for the two stories is in the sill plate, and the wall plate and cap plate of the second floor at the ceiling height level. A disadvantage with this type of construction is in the fabricating and raising of the walls.

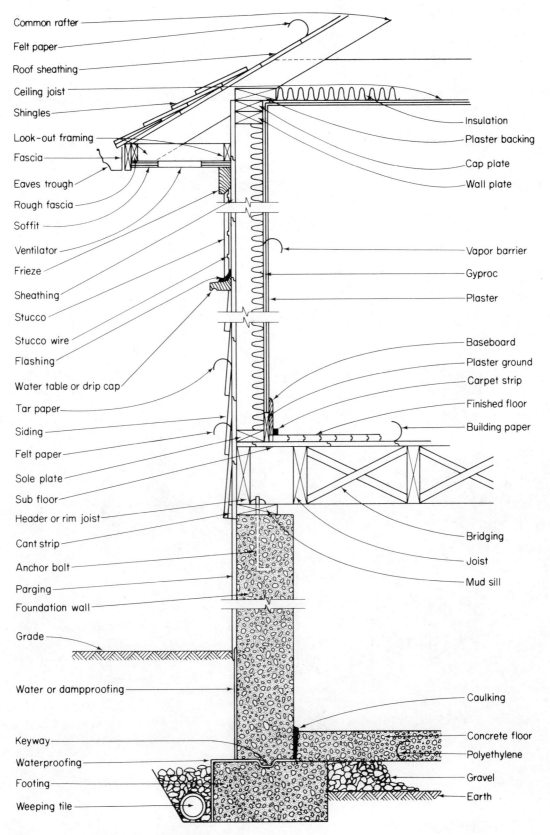

Common rafter
Felt paper
Roof sheathing
Ceiling joist
Shingles
Look-out framing
Fascia
Eaves trough
Rough fascia
Soffit
Ventilator
Frieze
Sheathing
Stucco
Stucco wire
Flashing
Water table or drip cap
Tar paper
Siding
Felt paper
Sole plate
Sub floor
Header or rim joist
Cant strip
Anchor bolt
Parging
Foundation wall
Grade
Water or dampproofing
Keyway
Waterproofing
Footing
Weeping tile

Insulation
Plaster backing
Cap plate
Wall plate

Vapor barrier
Gyproc
Plaster

Baseboard
Plaster ground
Carpet strip
Finished floor
Building paper

Bridging
Joist
Mud sill

Caulking
Concrete floor
Polyethylene
Gravel
Earth

Fig. 4.1. A typical wall section.

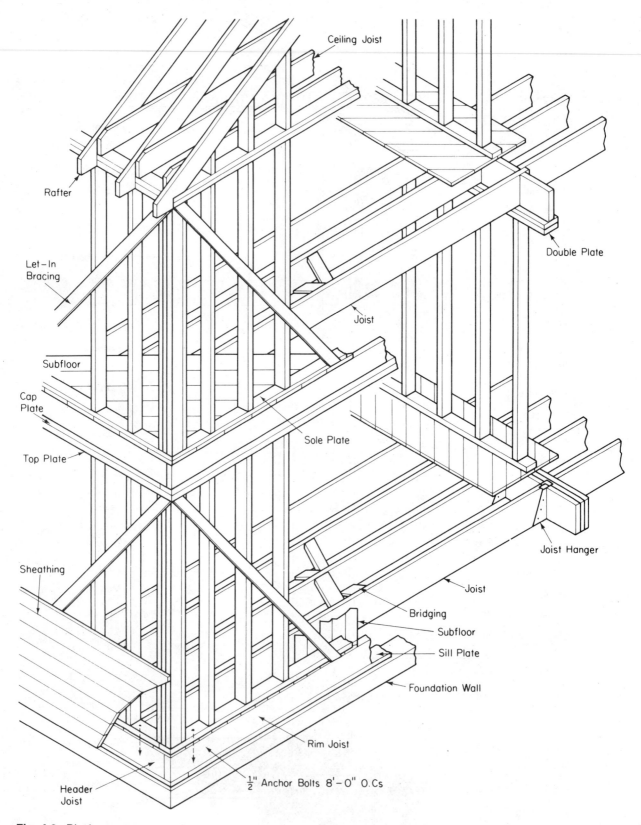

Fig. 4.2. Platform or western framing.

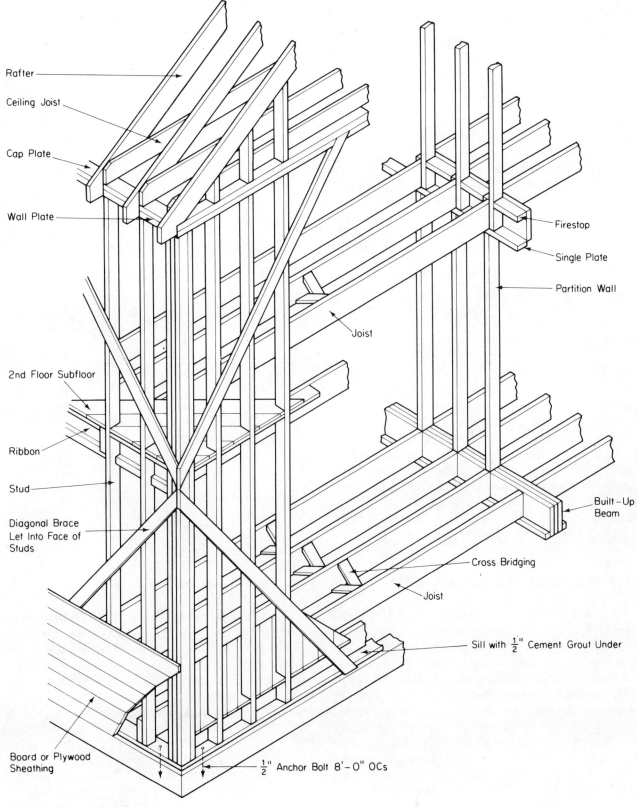

Rafter

Ceiling Joist

Cap Plate

Wall Plate

Firestop

Single Plate

Partition Wall

Joist

2nd Floor Subfloor

Ribbon

Stud

Diagonal Brace
Let Into Face of
Studs

Built−Up
Beam

Cross Bridging

Joist

Sill with $\frac{1}{2}$" Cement Grout Under

Board or Plywood
Sheathing

$\frac{1}{2}$" Anchor Bolt 8'−0" OCs

Fig. 4.3. Balloon framing.

Note how the $\frac{3}{4}'' \times 6''$ diagonal bracings are let into the outside faces of the studs. To support the second floor joists, a $\frac{3}{4}'' \times 6''$ ribbon is let into the inside face of the studs. The floor joists to the second floor rest upon the ribbon and are spiked to the studs. These joists also form the ceiling joists of the first floor.

4.4 BRICK VENEER 4"

Figure 4.4 illustrates the component parts of a 4" brick veneer finish to a wood-frame building.

An advantage of brick veneer construction is that the maintenance costs for brickwork are negligible. A disadvantage is that it requires the services of an additional subtradesman during construction. This type of building affords excellent insulation against either heat or cold penetrating or leaving the building according to the climatic conditions. It is excellent construction for all regions of North America except for the very far north.

4.5 THIN BRICK VENEER OR CERAMIC TILE

Figure 4.5 illustrates the component parts of a thin brick or ceramic tile veneer finish to a wood-frame building.

This is a somewhat similar treatment as for the 4" brick, except that the thin brick or ceramic tile is grouted onto a mortar base which is supported by stucco wire. The tiles may also be of longer dimensions than those of the standard brick. Once installed, such a veneer is virtually maintenance free, whereas fiber boards and many other wood siding finishes require frequent maintenance.

4.6 BRICK VENEER WITH 1" AIR SPACE

Figure 4.6 illustrates a brick veneer wall, tied to the sheathing of a wood-frame structure with corrugated metal ties spanning a 1" space. With this type of construction the 1" dead air space between the wood sheathing and the brick veneer acts as an insulator. Note carefully that the corrugated metal ties are placed at 4'0" OCs both horizontally and vertically. To keep the 1" air space free of mortar droppings, the mason uses a 1" board suspended at each end with a cord; this he places between the sheathing and the inside of his brickwork. At every 24" in height and before he places another row of corrugated metal strips he then raises the board, cleans it, and replaces it again.

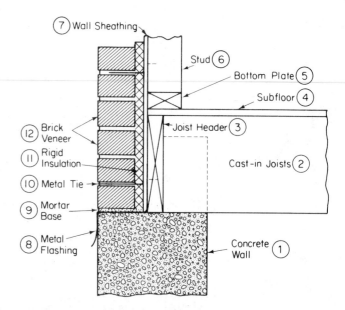

Fig. 4.4. Four-inch brick veneer.

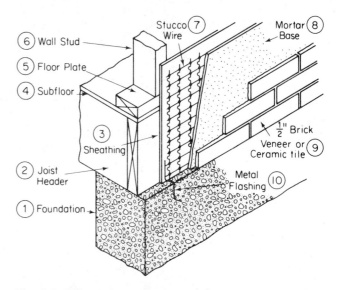

Fig. 4.5. Thin brick or ceramic veneer.

A weep hole should be left at 4'0" OCs on the first course of bricks. *A weep hole has no mortar in the vertical joints; this is done so that moisture may weep through these openings.*

4.7 TYPICAL WALL SECTION WITH BRICK VENEER

Figure 4.7 illustrates a typical wall section with brick veneer up to the eaves. Read the drawing together with the text and learn all the names of the parts as follows: *concrete foundation; metal flashing; floor joist header; sheathing; building paper; air space;*

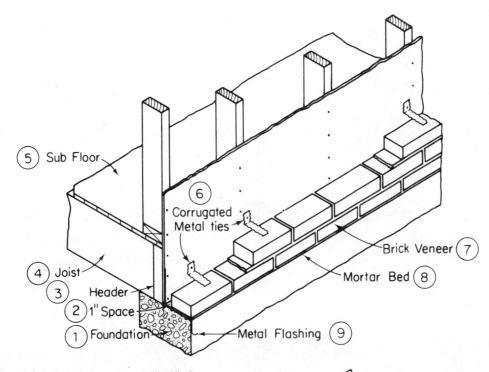

Fig. 4.6. Brick veneer with 1" air space.

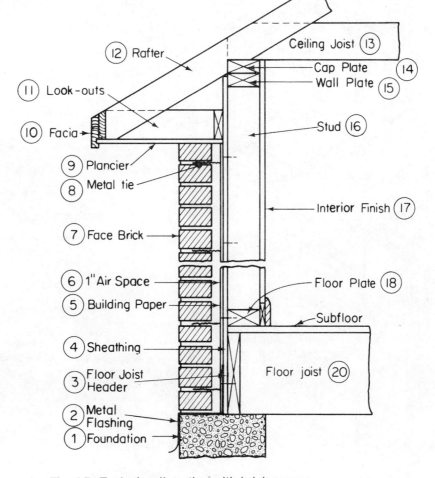

Fig. 4.7. Typical wall section with brick veneer.

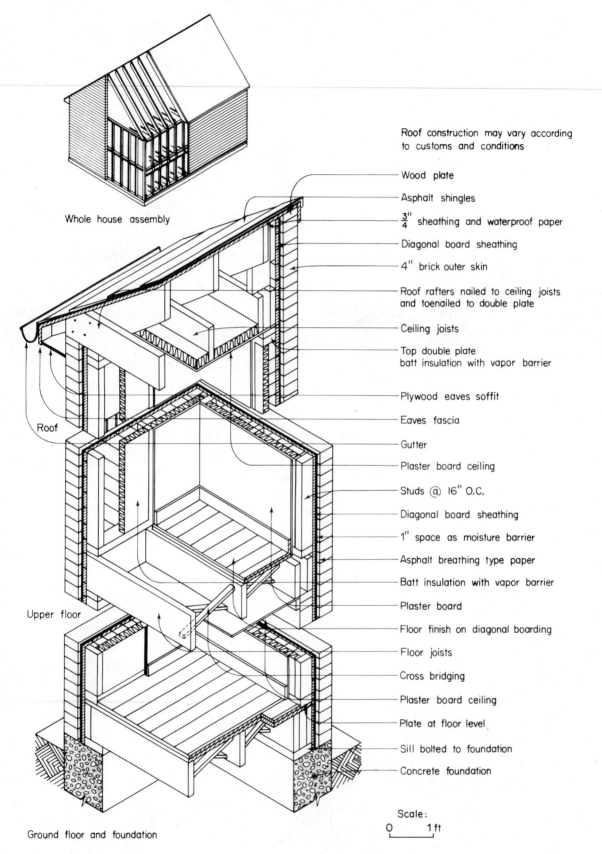

Whole house assembly

Roof

Upper floor

Ground floor and foundation

Roof construction may vary according to customs and conditions

Wood plate

Asphalt shingles

$\frac{3''}{4}$ sheathing and waterproof paper

Diagonal board sheathing

4" brick outer skin

Roof rafters nailed to ceiling joists and toenailed to double plate

Ceiling joists

Top double plate
batt insulation with vapor barrier

Plywood eaves soffit

Eaves fascia

Gutter

Plaster board ceiling

Studs @ 16" O.C.

Diagonal board sheathing

1" space as moisture barrier

Asphalt breathing type paper

Batt insulation with vapor barrier

Plaster board

Floor finish on diagonal boarding

Floor joists

Cross bridging

Plaster board ceiling

Plate at floor level

Sill bolted to foundation

Concrete foundation

Scale:
0   1 ft

Fig. 4.8. Brick veneered frame.

face brick; metal tie; plancier; facia board; look-outs; rafter; ceiling joist; cap plate; wall plate; stud; interior finish; floor plate; subfloor; and *floor joist.*

In addition there would be some form of insulation between the studs and a vapor barrier between the studs and the interior finish.

4.8 BRICK VENEERED FRAME

The exploded view of brick veneered construction in Fig. 4.8 is published through the courtesy of the Central Mortgage and Housing Corporation, Ottawa. This is a clearly defined drawing with the added advantage of named parts.

4.9 A GUIDE LIST FOR RESIDENTIAL CONSTRUCTION

The guide list below (which is not exhaustive) covers, in order of building, the operations and materials required in residential construction. The list should be continually revised with new materials added and redundant matter excised. The list is offered as a check so that when originally estimating, and later when costing, nothing is overlooked. Costing means the comparing of actual job costs of every unit of completed work with its original estimate, and adjusting the pricing of such units for future estimating. It is only in this way that good pricings may be made for other competitive work. When contracting, anything forgotten on the original successful bid is a four-way loss, i.e., the cost of the material, the labor to install it, the time, and the profit. Refer to this guide in conjunction with the typical wall sections and other building methods throughout the book. See Chapter 14 for flashing in walls and roofs.

Preliminary Building Operations: Lot purchase; transfer of title; surveyor's certificate; architect/engineer's fee; locating survey pegs; location of building on site, bench mark; legal costs; real estate fees; city taxes; plans and specifications; building permits; billboard and sign permit; road closure permit; application to excavate street for installing sewer, water, gas, electric conduit, telephone, and paving; city water permit; cost of alternate water supply.

Temporary Amenities: Buildings on site; mobile office; toilet facilities; power pole and meter; light; heat; sewer application; application to erect a billboard for advertising

Builders Loan Charges: Bank, real estate, insurance company; credit union; other

Insurance: Bonds; public liability; workmen's compensation; fire; vehicle; equipment; bonds; fringe benefits

Permanent Utilities: Heat; light; power; telephone; cable vision; sewer; water

Advertising: Newspapers; trade papers; billboards; radio and TV; realtors

Supervision: Superintendence; job runner (junior estimator); foreman; transportation costs

Equipment: Machinery (own or hired); ladders; scaffold; wheelbarrows, hoses, picks, shovels, wrecking bars; saw horses

Topographical: Clearing the site; demolitions; soil tests; establishing a bench mark; grading; lot layout; batter boards; see Fig. 6.5 on page 90.

Excavations: Remove and stockpile topsoil; cut, fill and grade; mass excavations — basement, septic tank, soak-away area; swimming pool

Extras: *You must get signed agreements for all extra work done over and above that which is called for on the original drawings and specifications.* **This is most important.**

Trenches: Sewer, water, power, telephone, cable vision; backfill and landscaping

Concrete and Framework: Estimate concrete and formwork together; concrete slab on grade floor heating; footings, house and garage; reinforcing; pier pads; walls (internal and external) wood or masonry units; floors; steps; sidewalks and driveways; septic tank; concrete test specimens; swimming pool

Driveways: Black-top; gravel stone flagging; concrete; other

Drain and Weeping Tile: Straight runs; elbows, tees, inspection chambers; fabric; application for connection to city drain

Waterproofing and Damp-Proofing: Special concrete; pargeting; emulsion

Progress Reports: Written; photographs

Underpinning for First Floor Joists: Wood posts; wood partitions; masonry walls; lally posts; masonry columns; tele-posts; steel columns; termite shield

Beams: "I" beam; wood, solid; wood, laminated; glu-lam; reinforced concrete

Rough Floor Assembly: Joists; joist hangers; headers (rangers); tail joists; bridging, wood or metal; subfloor; floor deadening material; building paper; reinforced concrete; chimney, stair and linen chute openings

Walls: Plates; studs, wood or metal; expanding metal; girths or girts; metal strapping for plumbing; concrete block and reinforcing; plumbing wall; cripples to doors and windows; headers for doors and windows, exterior and interior; milk chute; fuel chute

Ceiling Joists: Blocking at wall plates; backing; bridging, wood or metal; metal hangers; hanging beams; ceiling access door

Roof Framing: Gable studs; barge boards; barge board mouldings; roof bracing members; purlins; collar ties and bracing; rafters, common—hip—jack—valley; trusses; vents; dormer window framing; hanging beam or strong-back; wall and roof flashing

Carport: Framing; posts; fire protection, footings, ceiling (see local fire regulations)

Garage: (See local fire regulations)

Roof Eaves: Look-out framing; soffit and soffit ventilators; fascia board—rough and finished; eaves troughs and downpipe

Roofing Materials: Sheathing; flashing; building paper; shingles, wood or wood shakes; concrete tile; asbestos; masonry; asphalt tile; asphalt roll roofing; underlay for roofing; corrugated metal; plastic; slate

Chimney: Concrete reinforced footing; bricks—common, decorative fireclay; mortar fireclay; flue lining; flat concrete arch; metal for arch; damper; clean-out; fuel chute; ash dump; parging—cement, lime and sand; flue lining; chimney pots; chimney cap; flashing—roof to chimney; insulation—chimney to wood through floors and roof (see local fire regulations); hearth; fireplace surrounds

Cladding Exterior: Sheathing, board, plywood, other; building paper; siding; stucco and stucco wire; brick, veneer and wood framing; brick, solid; brick and cinder block; stone; decorative masonry units

Stairs: Wood, metal, concrete; stringers; undercarriage; treads; nosings; risers; wedges; newel posts; handrail and brackets; concrete forms and ties; stair finish—resilient material, carpet

Cladding Interior: Insulation—floor, walls and ceilings; vapor barriers—floors, walls and ceilings; grounds for doors and windows; lath and plaster; drywall and taping; decorative finish; expanding metal; corner beads

Window Units: Wood, metal and storm; basement, main floor, dormer; flywire screens; flashing to all openings; headers, wood, metal or concrete lintels

Outside Trim: Gable ends; eaves; frieze; fascia and soffit; drip cap flashing; drip cap; ventilators

Door Frames: Inside and outside; garage

Doors: Front, back, storm and screen; room; closet; linen cabinets; clothes and storage; ceiling access door

Inside Trim: Baseboard; window and door trim; window stool-apron-stop; carpet strip; closet rods and shelves; valance; decorative beams

Bathroom: Medicine cabinet; vanity; towel bars; soap and grab bar; tissue paper holder; shower-rail and curtain; wall and floor finish

Glass Block: Front entrance and patio; decorative light to stairs

Decorative Concrete Units: Bricks and metal ties; stone

Tiling: Living room; kitchen; bathroom; den; hall and other

Hardware: Form wire or ties; screws and fastening devices; door locks, hinges, checks and stops; towel rails; cabinet tracks, hinges, catches and pulls; coat hangers; handrail brackets; weather stripping; letter-drop chute; clothesline posts; glue and sandpaper; nails—common and finish, double headed nails for formwork, roofing, galvanized, copper, aluminum or shingle, wall board

Attending on other Trades: Cutting away and making good after plumbers, electricians, heating engineers, etc. Some allowance must be made

Cabinets: Meters; storage; ironing boards; den; medicine; books; kitchen

Special Fixtures: Chandeliers; dishwasher; oven and range; refrigerator; deep freeze; garbage-disposal; hood and fan; planter boxes; mirrors; ceiling fans; ceiling decorative light fixtures; shower doors; vacuum cleaning outlets; cable vision and telephone outlets; installation of vacuum cleaning service lines; door chimes

Subtrades: Excavator; concrete; plumber; chimney and fireplace; plasterer and dry-wall; electrician; heating engineer; hardwood floor layer and finisher; linoleum, tile and carpet layer; tinsmith; air conditioning; telephone and cable vision painter and decorator; sidewalk and cement finishers; landscaping; swimming pool; seeding and planting; fencing, wood, metal, decorative, brick, stone or other; final clean up—windows, doors, floors, vacuum and polish; clean up of surrounds; black-top, gravel, concrete or other

Overheads: Cancel all temporary utilities; proportion of general office expenses to each job; individual job overheads; superintendent, job runner and foreman salaries; sinking fund appropriations; builders loan fees; mortgage fees; federal taxes; remove all temporary buildings

Handover of Property: Professional fees; attorney; fixed fees for registrations; notarization fees, real estate; accountancy; accrued interest charges; maintenance clause; sign release and occupation certificate

Profit: You may be surprised to know how many bids are submitted where an allowance for profit has been forgotten. **Take care!**

4.10 TABLE OF DECIMAL EQUIVALENTS

$\frac{1}{16}$	=	.0625
$\frac{1}{8}$	=	.1250
$\frac{3}{16}$	=	.1875
$\frac{1}{4}$	=	.2500
$\frac{5}{16}$	=	.3125
$\frac{3}{8}$	=	.3750
$\frac{7}{16}$	=	.4375
$\frac{1}{2}$	=	.5000
$\frac{9}{16}$	=	.5625
$\frac{5}{8}$	=	.6250
$\frac{11}{16}$	=	.6875
$\frac{3}{4}$	=	.7500
$\frac{13}{16}$	=	.8125
$\frac{7}{8}$	=	.8750
$\frac{15}{16}$	=	.9375

4.11 DECIMAL EQUIVALENTS OF INCHES IN FEET

Inches	Feet	Inches	Feet
1	0.083	7	0.583
2	0.1667	8	0.667
3	0.25	9	0.75
4	0.333	10	0.833
5	0.417	11	0.917
6	0.5	12	1.0

4.12 FRACTIONAL EQUIVALENTS IN PERCENTAGES AND DECIMALS

$\frac{1}{8}$	=	$12\frac{1}{2}\%$	=	.125
$\frac{1}{4}$	=	25%	=	.25
$\frac{3}{8}$	=	$37\frac{1}{2}\%$	=	.375
$\frac{1}{2}$	=	50%	=	.5
$\frac{5}{8}$	=	$62\frac{1}{2}\%$	=	.625
$\frac{3}{4}$	=	75%	=	.75
$\frac{7}{8}$	=	$87\frac{1}{2}\%$	=	.875
$\frac{1}{5}$	=	20%	=	.2
$\frac{2}{5}$	=	40%	=	.4
$\frac{3}{5}$	=	60%	=	.6
$\frac{4}{5}$	=	80%	=	.8
$\frac{1}{6}$	=	$16\frac{2}{3}\%$	=	.163+
$\frac{1}{3}$	=	$33\frac{1}{3}\%$	=	.33+
$\frac{2}{3}$	=	$66\frac{2}{3}\%$	=	.66+
$\frac{5}{6}$	=	$83\frac{1}{3}\%$	=	.833+

4.13 MORTGAGE PERCENTAGES AND YIELDS PER ANNUM

A man purchases a house with an $18,000 mortgage at 6 per cent per annum. What are the interest charges for one year?

Step 1: An interest charge of 6 per cent per annum is equal to a rate of $60 per $1000 per annum.

Step 2: An $18,000 mortgage at 6 per cent per annum is equal to $18 \times 60 = \$1080$ per annum.

Complete the following table by mental process:

Loan	Rate per cent per annum	Charges per $1000 per annum	Total Charges for Total Loan per annum
$14,000	$4\frac{1}{2}\%$	$45	$630
$16,000	5		
$18,000	$5\frac{1}{2}$		
$20,000	6		
$22,000	$6\frac{1}{2}$		
$25,000	7		

Mark 5 points per correct answer: possible 50 points.

4.14 TO CHECK MULTIPLICATIONS

Example

Multiply 1379 by 3741 and check the answer by casting out nine's. *It is important to completely read this paper and then study it.*

Step 1: Multiply:

		Check (sum of digits in each row)	Product	Digital remainder
(a)	1379	2	$2 \times 6 = 12$	3
(b)	3741	6		
	1379			
	5516.			
	9653..			
	4137...			
(c)	5158839	The answer line adds up to 39, which divided by 9 leaves 3→		3

Step 2: Add together crosswise the figures in line (a) of Step 1: $1 + 3 + 7 + 9 = 20$ which when divided by nine (cast out 9's) yields a remainder of 2. This figure 2 is shown under the *Check* column in Step 1.

Step 3: Repeat the operation and write the remainder in the *Check* column as in Step 2 for line (b).

Step 4: Multiply the digital remainders of lines (a) and (b): $2 \times 6 = 12$ (see Step 1, *Product* column). Now cast out 9's from the product of 12 leaving a final digital remainder of 3 in the *Digital remainder* column as in Step 1.

Step 5: Add all the figures (crosswise) in the answer line (c), Step 1. Cast out the 9's, which leaves an *answer-line remainder of 3.*

Step 6: If the last *Digital remainder of lines (a) and (b) and the answer-line remainder* (see Step 1) are equal, the multiplication is correct. In this case both the remainders are 3.

Problems

Multiply and check:

3584×753 2478×4853 $78,901 \times 234$

To multiply by $\frac{3}{4}$: Calculate one-half the number and add one-half of *that* number; thus: $\frac{3}{4}$ of 4672 = $(\frac{1}{2} \times 4672) + \frac{1}{2}(\frac{1}{2} \times 4672) = 2336 + 1168 = 3504$. Try finding $\frac{3}{4}$ of the following:

56	49	473.08	1777.68
64	53	654.032	2413.44
76	97	998.72	8194.54

4.15 HOW TO MAKE RAPID MENTAL CALCULATIONS OF EQUAL NUMBERS

Example

The following shows how to mentally multiply *two numbers of equal value* together, such as 16×16.

Step 1: Change one number to a number ending in zero; thus $16 + 4 = 20$.

Step 2: Deduct from the other number the amount that was added to the first number; thus $16 - 4 = 12$.

Step 3: Multiply the altered numbers together $12 \times 20 = 240$.

Step 4: Add the square of the figure which was added and subtracted; thus $4 \times 4 = 16$. Add 16 to 240 = 256.

Exercise

$13 \times 13 = 10 \times 16 = 160 + (3 \times 3) = 169$

14×14	25×25	37×37
15×15	26×26	38×38
16×16	27×27	39×39
17×17	28×28	41×41
18×18	29×29	42×42
19×19	31×31	54×54
21×21	32×32	63×63
22×22	33×33	78×78
23×23	34×34	81×81
24×24	36×36	95×95

Now check this work by the long method. Check your time. Practice every day while you are shaving. Allow yourself ten points for each correct answer, and see how many points you can make out of each column of 100.

A man spends 5 minutes shaving every day for 40 years. Estimate mentally how many 8-hour working days this represents.

Try this: Multiply 17×16.

Step 1: Find 16×16 mentally.

Step 2: Add 1 × 16 to the result of Step 1.

4.16 HOW TO MAKE QUICK MENTAL CALCULATIONS

Quick Mental Calculations

To multiply a number ending in a zero with another number such as 120 × 27:

Step 1: The answer must end in a zero, so write a zero (0).

Step 2: Discard the zero from 120 leaving 12.

Step 3: Multiply 12 × 27 = 324 which must now be placed before the zero: 3240.

Multiply mentally the following:

110 × 34	441 × 30	773 × 50
90 × 28	60 × 431	20 × 5562
80 × 56	756 × 90	9784 × 70
40 × 115	544 × 60	543 × 80
30 × 179	80 × 987	60 × 5436

Rapid Check.

To multiply two numbers where one or a combination of them ends in two zeros such as 120 × 270:

Step 1: The answer must end in two zeros, so write them thus: 00.

Step 2: Discard the zeros from the original numbers, thus 12 × 27 = 324 which must now be placed before the two zeros thus: 32,400.

Multiply by mental process the following:

240 × 120	2300 × 90	690 × 700
80 × 980	80 × 4500	8723 × 600
466 × 700	700 × 4500	54,700 × 400
401 × 600	5000 × 400	90 × 6574
850 × 110	600 × 7890	120 × 9672
692 × 500	4678 × 800	2769 × 1100

Check your answers freehand.

4.17 TO FIND THE APPROXIMATE AREA OF A CIRCLE

Step 1: Examine Fig. 4.9, where the area of the 12″ × 12″ square is 144 sq. in.

Step 2: For all practical estimating purposes, the area of a circle is πR^2 where π is $3\frac{1}{7}$ (approximately) and R equals the radius.

Step 3: The area of the circle shown is $3\frac{1}{7} \times 6 \times 6 = 113$ sq. in.

Step 4: The approximate area of the circle is:

(a) the area of three squares 6″ × 6″ = 108 sq. in.
(b) add $\frac{1}{7}$ of one square 6″ × 6″ = $\frac{36}{7}$ (= approx. 5) + 108 = 113 sq. in.

For a rough estimate, the area of any circle is a little more than three times the square of the radius. Thus the area of a circle 30′-0″ × 30′-0″ is a little more than 3 × 15 × 15 — say, 700 sq. ft. The area just could not be in the seventies nor seven-thousands of sq. ft. If you do use a slide rule, remember this: you must learn to estimate.

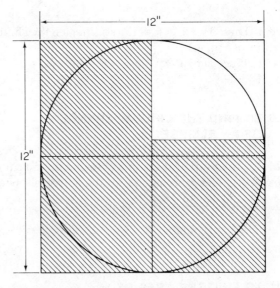

Fig. 4.9. The area of the shaded portion equals the area of the circle.

4.18 TO FIND THE CIRCUMFERENCE OF A CIRCLE

Multiply D (diameter by $3\frac{1}{7}$. The dimension C (circumference) of the circle shown in Fig. 4.9 is 12 × $3\frac{1}{7}$ ≃ 38. (The symbol ≃ means "approximately equals.") *Your thinking should be: 3 × 12 plus what?*

4.19 TO FIND THE AREA OF AN ELLIPSE

Step 1: Examine Fig. 4.10, where the area of the rectangle containing an ellipse is 12 × 16 = 192 sq. in.

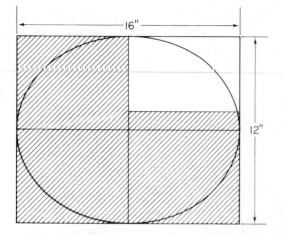

Fig. 4.10. An ellipse.

Step 2: The area of an ellipse is $ab\pi$, where π is $3\frac{1}{7}$. Note that a is half the major axis and b is half the minor axis.

Step 3: Then $3\frac{1}{7} \times 8 \times 6 \simeq 151$ sq. in. (see the shaded portion of Fig. 4.10).
Your thinking should be: $3 \times 8 \times 6 = 144$ plus what?

4.20 TO FIND THE CIRCUMFERENCE OF AN ELLIPSE

Multiply one-half the sum of the major and minor axes by $3\frac{1}{7}$. The dimension C of the ellipse shown in Fig. 4.10 is $\frac{12 + 16}{2} \times 3\frac{1}{7} \simeq 44$ in.

Your thinking should be: $3 \times 14 = 42$ plus what?

4.21 TO FIND THE AREA OF PLANE FIGURES

(a) Hexagon

Square the short diameter and multiply by 0.866.

Your thinking should be: $12 \times 12 = 144$ sq. in. less what?

(b) Octagon

Square the short diameter and multiply by 0.707.

Your thinking should be: $12 \times 12 = 144$ sq. in. less what?

(c) Parallelogram

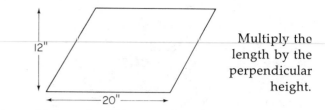

Multiply the length by the perpendicular height.

Your thinking should be: an oblong $20 \times 12 = 240$ sq. in.

(d) Trapezoid

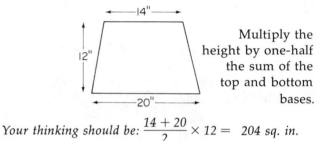

Multiply the height by one-half the sum of the top and bottom bases.

Your thinking should be: $\frac{14 + 20}{2} \times 12 = 204$ *sq. in.*

4.22 TO FIND THE VOLUME OF A SPHERE

Step 1: The approximate volume V of a sphere is equal to $\frac{2}{3}V$ of a cylinder with equal diameter D and height H.

Step 2: The approximate volume V of a 12″ sphere is equal to the base area of a cylinder with 12″ diameter D and a H $\frac{2}{3}$ that of its diameter D.

Step 3: The area of the base is $3\frac{1}{7} \times 6 \times 6$ and the volume V of the sphere is $3\frac{1}{7} \times 6 \times 6 \times 8 \simeq 905$ cu. in.
Your first thinking should be: $3 \times 6 \times 6 \times 8 = 18 \times 48 = 864$ cu. in. plus what?

4.23 TO FIND THE SURFACE AREA OF A SPHERE

The surface area of a sphere is approximately equal to the lateral surface of a cylinder of the same diameter D and height H as the sphere. The surface area of a 12″-diameter sphere is $12 \times 3\frac{1}{7} \times 12 \simeq 453$ sq. in.
Your first thinking should be: $12 \times 3 \times 12 = 432$ plus what?

4.24 TO ADD COLUMNS OF DIMENSIONED FIGURES

Neatly list the dimensions, then add them in pairs and pairs again to resolutions as follows:

$$
\begin{array}{l}
\text{ft.} \quad \text{in.} \\
\left.\begin{array}{r} 7 - 9\frac{1}{4} \\ 3 - 6\frac{1}{2} \end{array}\right\} 11 - 3\frac{3}{4} \\
\left.\begin{array}{r} 5 - 11\frac{5}{8} \\ 9 - 10\frac{1}{4} \end{array}\right\} 15 - 9\frac{7}{8} \\
\left.\begin{array}{r} 5 - 5\frac{1}{2} \\ 7 - 7\frac{1}{8} \end{array}\right\} 13 - 0\frac{5}{8} \\
\left.\begin{array}{r} 8 - 9\frac{5}{8} \\ 9 - 11\frac{1}{4} \end{array}\right\} 18 - 8\frac{7}{8}
\end{array}
$$

$\left.\right\} 27 - 1\frac{5}{8}$

$\left.\right\} 31 - 9\frac{1}{2}$

$\left.\right\} 58' - 11\frac{1}{8}''$

4.25 THE TWENTY-SEVEN TIMES TABLE

As an estimator you will frequently be converting cu. ft. to cu. yds. You should learn the 27 times table, which follows:

$1 \times 27 = 27$	$7 \times 27 = 189$
$2 \times 27 = 54$	$8 \times 27 = 216$
$3 \times 27 = 81$	$9 \times 27 = 243$
$4 \times 27 = 108$	$10 \times 27 = 270$
$5 \times 27 = 135$	$11 \times 27 = 297$
$6 \times 27 = 162$	$12 \times 27 = 324$

4.26 TABLE OF NAIL SIZES WITH ALLOWANCES FOR DIMENSION LUMBER AND FINISHING CARPENTRY

Size and Kind of Material	Board Measure	Type of Nail	Length of Nail	Lb of Nails Required 12" o.c.	16" o.c.	24" o.c.
1 x 4 boards & shiplap	1000	common	$2\frac{1}{2}''$	60	48	30
1 x 6 boards & shiplap	1000	common	$2\frac{1}{2}''$	40	32	20
1 x 8 boards & shiplap	1000	common	$2\frac{1}{2}''$	31	27	16
1 x 10 boards & shiplap	1000	common	$2\frac{1}{2}''$	25	20	13
1 x 12 boards & shiplap	1000	common	$2\frac{1}{2}''$	31	24	16
1 x 4 T & G blind nailed	1000	common	$2\frac{1}{2}''$	30	24	15
1 x 6 T & G blind nailed	1000	common	$2\frac{1}{2}''$	20	16	10
1 x 8 T & G blind nailed and 1 face nail	1000	common	$2\frac{1}{2}''$	31	27	16
1 x 10 T & G blind nailed and 1 face nail	1000	common	$2\frac{1}{2}''$	25	20	13
1 x 12 T & G blind nailed and 1 face nail	1000	common	$2\frac{1}{2}''$	21	16	11
2 x 4 to 2 x 16 framing	1000	common	$4''$	20	16	10
			$3\frac{1}{2}''$	10	10	6
			$3''$	8	6	4
Built-up beams	100 lin	common	$3\frac{1}{2}''$		3	
2 x 6 T & G flooring	1000	common	$4''$	35	27	18
2 x 8 T & G flooring	1000	common	$4''$	27	20	14
6" bevel siding	1000	siding	$2''$	15	13	
8" bevel siding	1000	siding	$2''$	12	10	
10" bevel siding	1000	siding	$2\frac{1}{4}''$	45	35	
12" bevel siding	1000	siding	$2\frac{1}{4}''$	60	50	
1 x 3 softwood flooring	1000	Floor brads	$2\frac{1}{2}''$	42	32	21
1 x 4 softwood flooring	1000	Floor brads	$2\frac{1}{2}''$	32	26	16
1 x 6 softwood flooring	1000	Floor brads	$2\frac{1}{2}''$	22	18	11
$1\frac{1}{2}''$ hardwood flooring	1000	flooring	$2''$	13	10	
2" hardwood flooring	1000	flooring	$2''$	11	8	
$2\frac{1}{4}''$ hardwood flooring	1000	flooring	$2\frac{1}{4}''$	20	14	
Base	100 lin	Finish	$2\frac{1}{2}''$		1	
Interior trim	100 lin	Finish	$2\frac{1}{2}''$		1	
Casing		Finish	$2\frac{1}{2}''$	1 lb per opening		
Outside trim	100 lin	Finish	$2\frac{1}{2}''$		$2\frac{1}{2}$	
Outside mouldings	100 lin	Finish				
Carpet strip	100 lin	Finish	$1\frac{1}{2}''$		$\frac{1}{2}$	
Metal lath	100 sq yd	Roofing	$2''$		$17\frac{1}{2}$	
Metal lath	100 sq yd	Staples	$1''$		12	
Gypsum lath	100 sq yd	Gyproc	$1\frac{1}{2}''$		10	
Gyproc wall board	100 sq yd	Gyproc	$1\frac{1}{2}''$		10	
Plywood $\frac{1}{4}''$	1000 sq ft	Finish or common	$1\frac{1}{2}''$		9	
Plywood $\frac{3}{8}''$	1000 sq ft	Finish or common	$2''$		12	
Plywood $\frac{5}{8}''$	1000 sq ft	Finish or common	$2\frac{1}{2}''$		20	
Shingles	per M	Shingle	$1\frac{1}{4}''$		$4\frac{1}{2}$	

4.27 NUMBER OF NAILS TO THE POUND

Size	Length	Common	Finishing	Casing	Size	Length	Common	Finishing	Casing
2d	1"	850			10d	3"	65	107	96
3d	1¼"	550	640		12d	3¼"	50		60
4d	1½"	350	456		16d	3½"	40		50
5d	1¾"	230	328		20d	4"	31		
6d	2"	180	273	228	30d	4½"	22		
7d	2¼"	140	170	178	40d	5"	18		
8d	2½"	100	151	133	50d	5½"	15		
9d	2¾"	80	125	100	60d	6"	12		

4.28 TABLE OF WOOD SCREWS

Length	Gage — Steel screws	Gage — Brass screws
¼	0 to 4	0 to 4
⅜	0 to 8	0 to 6
½	1 to 10	1 to 8
⅝	2 to 12	2 to 10
¾	2 to 14	2 to 12
⅞	3 to 14	4 to 12
1	3 to 16	4 to 14
1¼	4 to 18	6 to 14
1½	4 to 20	6 to 14
1¾	6 to 20	8 to 14
2	6 to 20	8 to 18
2¼	6 to 20	10 to 18
2½	6 to 20	10 to 18
2¾	8 to 20	8 to 20
3	8 to 24	12 to 18
3½	10 to 24	12 to 18
4	12 to 24	12 to 24
4½	14 to 24	14 to 24
5	14 to 24	14 to 24

Gage number	Actual size	Decimal	Approx. Fraction	Drill or Auger Bit size A	B	C
0	O	0.060	1/16	1/16		
1	O	0.073	5/64 −	3/32		
2	O	0.086	5/64 +	3/32	1/16	3/16
3	O	0.099	3/32	1/8	1/16	4/16
4	O	0.112	7/64	1/8	1/16	4/16
5	O	0.125	1/8	1/8	3/32	4/16
6	O	0.138	9/64	5/32	3/32	5/16
7	O	0.151	5/32 −	5/32	1/8	5/16
8	O	0.164	5/32 +	3/16	1/8	6/16
9	O	0.177	11/64	3/16	1/8	6/16
10	O	0.190	3/16	3/16	1/8	6/16
11	O	0.203	13/64	7/32	5/32	7/16
12	O	0.216	7/32	7/32	5/32	7/16
14	O	0.242	15/64	1/4	3/16	8/16
16	O	0.268	17/64	9/32	7/32	9/16
18	O	0.294	19/64	5/16	1/4	10/16
20	O	0.320	21/64	11/32	9/32	11/16
24	O	0.372	3/8	3/8	5/16	12/16

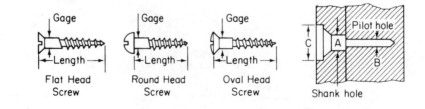

Flat Head Screw   Round Head Screw   Oval Head Screw   Pilot hole / Shank hole

4.29 TABLE OF ALLOWANCE FOR WASTE IN WOOD

A 1 x 8 common board is understood to be 1" x 8" board (end section) as it leaves the saw. The finished product, after having been through the planing machine, will finish $\frac{3}{4}$" x $7\frac{1}{2}$". As a consequence, an allowance must always be made for waste, both in production, and in application, on the job by the carpenters. The following table gives the allowance for some of the wood products after application.

	Add
Common 1 x 8 boards laid square	10%
Common 1 x 8 boards laid diagonally	17%
Shiplap 1 x 6 laid square	20%
Shiplap 1 x 6 laid diagonally	25%
Shiplap 1 x 8 laid square	15%
Shiplap 1 x 8 laid diagonally	20%
Shiplap 1 x 10 laid square	13%
Shiplap 1 x 10 laid diagonally	17%

Drop siding 6"	20%
Lap siding 6" with 4" to the weather	50%
Lap siding 6" with $4\frac{1}{2}$" to the weather	25%
Flooring T & G (Tongue and Groove) 6"	20%

4.30 ALLOWANCE FOR WASTE IN HARDWOOD FLOORING

	Add
$1\frac{1}{2}$" face	50%
$1\frac{3}{4}$" face	40%
2" face	38%
$2\frac{1}{4}$" face	33%
$2\frac{1}{2}$" face	30%
$1\frac{1}{2}$" face $\frac{3}{8}$" thickness	33%
$1\frac{3}{4}$" face $\frac{3}{8}$" thickness	30%
2" face $\frac{1}{4}$" thickness	25%
Building paper (400 sq. ft. roll) allow	8%
(Note that building paper is purchased by weight per roll.)	

4.31 TABLE OF ROOM WALLS PLUS THE CEILING AREAS

Dimension of Room	8 ft Ceiling Requires Square Feet	9 ft Ceiling Requires Square Feet	10 ft Ceiling Requires Square Feet	Dimension of Room	8 ft Ceiling Requires Square Feet	9 ft Ceiling Requires Square Feet	10 ft Ceiling Requires Square Feet
8' x 8'	320	352	384	11' x 16'	608	662	716
8' x 10'	368	404	440	11' x 18'	662	720	778
8' x 12'	416	456	496	12' x 12'	528	576	624
8' x 14'	464	508	552	12' x 14'	584	636	688
8' x 16'	512	560	608	12' x 16'	640	696	752
8' x 18'	560	612	664	12' x 18'	696	756	816
9' x 10'	394	432	470	13' x 14'	614	668	722
9' x 12'	444	486	528	13' x 16'	672	730	788
9' x 14'	494	540	586	13' x 18'	730	792	854
9' x 16'	544	594	644	14' x 14'	644	700	756
9' x 18'	594	648	702	14' x 16'	704	764	824
10' x 10'	420	460	500	14' x 18'	764	828	892
10' x 14'	472	516	560	15' x 16'	736	798	860
10' x 14'	524	572	620	15' x 18'	798	864	930
10' x 16'	576	628	680	16' x 16'	768	832	896
10' x 18'	628	684	710	16' x 18'	832	900	968
11' x 12'	500	546	592	17' x 17'	833	901	969
11' x 14'	554	604	654	18' x 18'	900	972	1044

4.32 MISCELLANEOUS CAPACITIES, MEASURES, AND WEIGHTS

Water

(a) An American standard gal. (gallon weighs 8.337 lbs. and contains 231 cu. in., and there are 7.48 gals. in 1 cu. ft.—say $7\frac{1}{2}$ gals. per cu. ft.

(b) An Imperial gal. weighs 10 lbs. and contains 277.418 cu. in., and there are 6.2321 gals. in 1 cu. ft.—say $6\frac{1}{4}$ gals. per cu. ft.

(c) *Cement* weighs 94 lbs. per sack, which contains 1 cu. ft.

(d) *Concrete* (cement, sand, and stone) weighs 144± lbs. per cu. ft.

(e) *Gravel* or screened crushed stone weighs 2500 lbs. per cu. yd.

(f) *Earth* after excavation swells 5 percent to 50 percent according to character. Study soil-analysis reports in specifications.

CHAPTER 4 REVIEW QUESTIONS

1. Make a freehand sketch of a typical wall section for a single story brick veneer-wood-framed house with a concrete basement and arrow-name all the parts. See page 67 for an example.

2. Multiply 2345 by 6789 (freehand) and check your answer by the casting out of nines. See page 72 for examples.

3. By mental process multiply 31 × 31; 43 × 43; 26 × 26. See page 72 for examples.

4. By mental process, what is the approximate area of each of the three following circles with 14"; 18"; and 3'-4" diameters respectively? See page 72 for examples.

5. An ellipse has a major and minor axis of 40" and 20" respectively. What is (a) the approximate area, and (b) the approximate circumference? See page 000 for examples.

6. Write from memory the 27 times table from 1 × 27 = 27 to 27 × 12 = 324.

7. When building contracting, anything forgotten on the original successful bid is a four way loss; name them!

8. List eight temporary but necessary city amenities when building.

9. What is meant by attending on other trades, and whose responsibility is it?

10. What are the decimal equivalents of $\frac{7}{16}''$, $\frac{11}{16}''$, and $\frac{15}{16}''$?

11. Name two advantages and two disadvantages of balloon framing.

12. Name two advantages and two disadvantages of Western (or Platform) framing.

13. Name two advantages of brick veneer finish to wall framing against that of a wood product. Name one disadvantage of brick veneer finish.

14. What is the difference in architectural drawings between: (a) a plan; (b) an elevation; (c) a typical wall section?

15. In architectural drawings, which takes precedence in interpretation—the drawings or the specifications? Give reasons for your answer.

5

Preliminary Building Operations

In this chapter we discuss some factors that may influence a contractor either to build or not build a residence or residences on a particular property. Some of the things that he must take into account are as follows: the location of the site; its proximity to schools, churches, hospitals, shopping centers, and its accessibility to existing roads; topographical features; available utilities; subsurface exploration and construction; his ability to finance the project (financing in the construction industry, because of high risk, is more difficult than in most others); availability of men, materials, also equipment, either contractor owned or hired; volume of work already on hand; weather conditions such as rain, snow, ice, permafrost, flash flooding, relative humidity and their effect upon men and machines; and the storage of materials.

5.1 UNDERGROUND CONDITIONS AND DEWATERING

Sometimes, due to unpredictable underground conditions, either natural ones, or, in older cities, pre-existing man made structures, such as sewer mains, city utilities, or underground cables, it is not uncommon for the successful bidder for work to be performed above ground to be awarded a contract on a cost plus basis for work to be performed below ground. Cost plus is the actual cost of the work plus overhead expenses, plus a percentage for the contractor's fee, say, ten percent.

There are more hazards and unpredictables in excavating and in the underpinning of buildings than in any other feature of the construction industry. While the contractor can be reasonably sure of all operations above ground, there is never any certainty about conditions below ground.

When it is necessary to excavate below the natural water level (unless tremie concrete is to be placed), the basic methods of water control are reducing the flow by diversion — thus keeping the water out of the excavated area — or designing a system of well points and pumps. Apart from being an accident hazard to workmen, seepage of water into excavated areas may cause sloughing of the walls of the excavation, the heaving of the bottoms, and unstable underpinning.

Where foundation work must be carried out in the "dry" (on firm, unyielding foundations), it may be necessary to engage the services of a professional engineer for the dewatering of the area of operations. This will involve an evaluation of field conditions. Some of the considerations to be taken into account are:

(a) The size and depth of the excavation.

(b) The length of time that the area must be kept de-watered.

(c) The conditions necessary for the completion of the structure.

(d) A careful survey of the subsurface geological conditions.

(e) The proximity and sources of water.

(f) The variation in the water table; artesian pressures caused by seasonal conditions and variations in rivers, lakes, tides, and freezing conditions; the influence of these conditions on the sides and bottoms of excavated areas.

5.2 SUBSURFACE EXPLORATION

There are a number of methods that may be used for making an inspection of subsurface conditions. The method adopted will depend upon the depth to which the test is to be made, the probable nature of the soil and the weight of the structure to be imposed upon the earth. Methods include sounding rods, augers, wash borings, rock drillings, and test pits.

Test Pits: These are expensive since they are often dug by hand; they usually measure about 4'0" square and 9'0" deep. In the case of wet soil, sheet piling is driven in advance of pit excavation. This, in itself, would be indicative to the contractor that sheet piling may have to be provided over some or all of the area for mass excavations. Some of the advantages of test pits are that visual inspection may be made of the soil layers; compactness and water content of the layers may be checked; undisturbed samples may be taken for laboratory investigation; load tests may be made on any of the soil at any depth desired. For deep test pits, machine excavating would have to be resorted to.

5.3 SANITARY FILL

Many authorities bring low lying land to a desired grade by sanitary filling the depression. First the top soil is stockpiled on the site; then household refuse is compacted in the depression to a depth of several feet; this is followed with several feet of clean gravel or other approved clean inert compacted fill. This sandwiching is continued until the replaced top soil brings the whole area to the desired grade.

Within a few years the area is perfectly sanitary and ready for building or recreational uses. Records of reclaimed land, under their jurisdiction, are main-tained by local authorities and are available for study to the public.

Estimating volumes of earth is not difficult. Nevertheless, even though the architect/engineer supplies soil data in the contract documents, there is always a risk in ground conditions—from rock to water. See Sect. 3.5.

The importance of locating the correct corner stakes and the equal importance of the contractor making his own subsurface exploration cannot be over emphasized. There is more money lost through these hazards than on any other facet of building construction, with the exception of incorrectly locating the building, according to local ordinances.

5.4 INSPECTING THE BUILDING SITE

Before purchasing a lot for building residences, or before submitting a bid to erect a residential building, it is important to identify, and inspect thoroughly, the proposed site.

Incorrectly Located Buildings are, daily, being either torn down or removed from an incorrect lot to a correct one, or are being relocated properly on the correct lot. *Be careful and employ the services of a registered surveyor if you have any doubts on these matters.*

The following example (among others) of an incorrectly placed building operation came to my attention. A contractor was well advanced on the building of a city home. His operations were on an open staked new subdivision, and he believed that he was working on 15th Street and 18th Avenue N.W. of a certain city. There were no other houses in the area at that time. The building inspector arrived at the correct lot to make his first inspection and discovered a vacant lot. Immediately due north he saw a house in the course of construction and ready for its first inspection; this was situated due north on 15th Street and 19th Avenue N.W. The building inspector informed the builder of his mistaken site operations.

Imagine the critical blow that this could have been to the contractor. In this case, the rightful owner very graciously consented to exchange lots—with the provision that the contractor bear the total cost of the legal expenses for conveyance of titles and without any further penalties. (The conveyance of title to property means the official re-registration of the title of a parcel of land from one registered owner to another.)

A surveyor's fee would have been much cheaper; it would not have involved any risk of building on the incorrect lot. It would not have delayed building operations for several weeks. Furthermore, the professional charges of the surveyor would have been reflected in the contractor's fee for the building of the home.

Demolition or Removal of Existing Buildings must be authorized by the local authority. In addition, all service lines below and above ground from the city mains to such buildings must be completely removed from the property by the contractor. That is to say that water, sewer, power, telephone and cable vision lines must be detached from the building, unearthed and removed back to the city mains. This is understandable; otherwise disused underground lines over a long period of time would become confusing and could lead to accidents.

5.5 CITY OFFICES

The estimator or contractor should become familiar with all the city departments which may be helpful to him in his business.

Trouble Calls

Some city departments provide twenty-four-hour service. These include: electric light and power, gas, sewer, garbage, streets, fire, police, and ambulance.

You as a builder may want at any time one or more of these services. Their telephone numbers should be kept in a ready file. Assume that your street has caved in because of a flash flood. What are you going to do about it?

City Planning Department

Before purchasing building land, the contractor should visit the planning department. Here may be seen the proposed plans for the future development of the city, showing future freeways, expressways, airport developments, bridges, recreation grounds, suburban development for residential construction, schools, shopping centers, churches, libraries, hotels, and motels. Also shown will be the very important zoning regulations such as districts for single dwelling units, two units, three units, and multiple dwelling units such as apartment blocks. In some areas the heights and floor areas of buildings will be restricted. In the better-class districts, residential areas may have all the powerlines buried and have very decorative standard street lighting. These services will be reflected in the city taxes.

Future industrial developments will also be shown

with road and railway trackage and available services.

Some housing project promoters purchase large areas of land which they develop with roads and city utilities. They divide the land into individual building lots; erect modern houses, and sell them to the public. All this development must conform and be approved by the city engineers department and the city planning department. An advantage to both the vendor and the purchaser of such units is that the utilities—that is, roads, sidewalks, boulevards, gas, power, water, telephone, and cable television—are included in the original cost of the property. The purchaser is only responsible for making payments to one person each month.

When a smaller housebuilder offers a new home for sale, it is quite usual that the purchaser pays to the vendor each month a certain sum to cover the mortgage interest and repayment plus a separate annual tax bill to the city. In many instances, however, the purchaser may have to wait a long period of time before all the utility amenities are installed.

It is from the city planning department that a contractor might want to seek to have a parcel of land rezoned from a single-dwelling-unit area to a multiple-dwelling area for the erection of an apartment block. *You must visit your city planning department. The officers there will be very happy to show you through and answer your questions.* From the information given and a further talk with the taxation department, you may get a fair assessment of future property taxes in the area in which you wish to build.

City Engineers Department

The city electrical engineers department is usually separate from the remainder of the engineering services. The city engineers are responsible for streets, sewers, garbage, building inspections, waterworks, and testing of building materials such as concrete. They will have their engineers yard and also an industrial development coordinator, license and building permit bureau, construction and maintenance division, and a traffic engineers department.

Building Permits

A proposal for a new building permit must first be submitted to the city planning department, which examines the plans and specifications to determine whether they will conform to the planned pattern of the area in which the building is proposed to be erected.

When approved by the planning department, the

application is passed to the city engineers for structural examinations. If they approve, a building permit is issued upon payment of a scaled fee, depending on the type of building. This building permit must be prominently displayed at the new building site.

Other permits may include water, street closing, permission to use city water hydrants, and connections to city sewer line, and so on.

5.6 SURVEY

Some cities may, for a charge, officially survey your land. *Where this is not done and there is the slightest doubt about the location of the original survey stakes, employ the services of a registered surveyor.* While you are reading this chapter, someone is having a partly (or completely finished) house removed from one building lot to a correctly located one. This is avoidable. However, discuss this with building inspectors, and they will give you many instances where people have built on the incorrect lot.

Occasionally, even the surveyor can make a mistake, but he is bonded against such an eventuality, and thus you are protected. The advice, therefore, is to employ a registered surveyor who will correctly locate your lot, put up the batter boards, and establish your levels. When this is done professionally, you can commence your building operations with every confidence as to location and levels. (See Figs. 6.5 and 6.12.)

5.7 DRAWINGS AND SPECIFICATIONS

The best residential designs are prepared by architects who specialize in this field. Their duties fall under four main divisions: preliminary sketches; working drawings and specifications; obtaining and the letting of contracts; and the supervision of the structure. Such residences will be designed to suit: soil conditions; lot; view; compass points; prevailing wind; and all to the satisfaction of the local building code and the needs and purse of the owner. When interpreting drawings and specifications, specifications take precedence.

Government Sponsored Drawings and Specifications are available through the state and also manufacturers' associations. The U.S. Department of Agriculture Forest Service, Forest Products Laboratory, Madison, Wisconsin, is an excellent source for obtaining draw-

ings and specifications for inexpensive homes. They will supply drawings and specifications at little cost. The Canadian Government sponsors a nonprofit making Central Mortgage and Housing Corporation from whom may be obtained many free publications about housing, together with a publication depicting hundreds of architect designed homes. Four sets of drawings for any of these homes may be purchased for less than $20.00. These drawings are acceptable across the country by all local authorities. You may want to build speculatively one or two of these homes, or you could recommend this publication to a prospective customer. Many of these drawings lend themselves to enlargement, but any alterations must have the approval of the issuing body and of the local building engineering department.

Real Estate Companies usually have a great number of house drawings from which may be selected the type of home either you or your prospect would like to build. A great advantage here is that you may go to see an occupied home built to the design and specifications which interest you. The owners may let you look through their home. This could be a very profitable visit. You may find out not only what the owners like about it, but how they would improve on it if they were building another similar home. *It is most important that you build the type of home that people want at the price they can afford to pay.*

Distribution of Copies of Drawings and Specifications are as follows:

(a) The owner.
(b) The authority having jurisdiction. That one will be the Master Copy on which the authority will enface any corrections or special requirements to be complied with. (Such amendments should be neatly copied onto all other copies.)
(c) The lending organization. (This organization may also make periodic inspections.)
(d) The builder.

General Conditions of Contract always embrace the following:

(a) The contractor and all subcontractors shall comply with all local, state, provincial, or government rules, regulations and ordinances. They will prepare and file all necessary documents and information, pay for and obtain all licenses, permits and certificates of inspection as may be specified or required.

Wherever municipal bylaws or state or provincial legislation require higher standards than those set forth, they shall govern. Note: Assume that there is some peculiar subsoil in a certain area. It is fair to assume that the local authority, knowing the conditions, may require some special precautions to be taken that are more stringent than those set forth in the general specifications. In all cases, the highest building standard shall prevail.

(b) By executing the contract documents for the contract for construction, the contractor represents that he has personally visited the site, familiarized himself with the local conditions under which the work is to be performed, and correlated his observations with the requirements of the contract documents and the local building code.

Building Site Drawings show the corners of the site; the existing contours of the land; the new grade to which it is to be leveled; and objects such as existing buildings, fences, service lines, trees that are to be removed. The *architectural drawings* will show the elevations of the service lines, foundation levels, the first floor and other floor levels, and the roof level.

The corners of the property will be identified usually by a wooden stake with three faces painted white, which may bear some letters such as I.P. meaning iron peg. Close to the stake may be found an iron peg which had been sunk into the ground by a previous official survey team. (See Fig. 5.1.)

It is an offense, under law, to remove such an officially placed and registered iron peg. It is very important that the position of such markers be related to the drawings. The exactness of their positions should be confirmed by both the owner of the land (or his agent) and the purchaser or contractor who intends to build on the property. The same parcel of land may have been surveyed many times, and it is possible that there may be several stakes located near the one placed by or for the owner. (*A parcel of land is an area of land held as one unit under single ownership.*)

The specific corner stakes pointed out by the owner or his agent should be clearly marked in his presence, and this operation should be witnessed by a third party. It is better to employ the services of a registered surveyor to confirm that the stakes purporting to be the exact corner locations of the property are correctly placed. It is from these stakes that the contractor or his surveyor will set the batter boards (see Fig. 6.5) for the building lines to conform with the local building code.

A building line is the prescribed minimum distance set back from the street or avenue fronting the lot; it is illegal to build between that line and the street. It is also the prescribed minimum distance that the structure must be kept from other buildings or boundaries of the property. A house that is built infringing on these regulations may have to be moved to its correct place on the lot. *Be careful, this is very important!* A builder may recover from all sorts of mistakes, but an avoidable error of this nature may force a builder out of business immediately if it occurs.

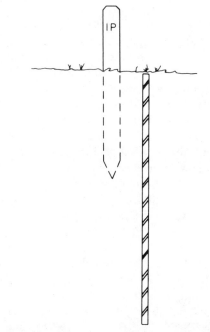

Fig. 5.1 Wooden stake indicating official survey metal pin.

5.8 DATUM AND BENCH MARKS

A very important post held by one of the officials at the city engineers department is the "official grade striker." There is probably more grief over incorrect grades in building construction than in any other phase of the operation. After the survey of the land, but before the contractor builds, the bench mark or datum must be established.

The official grade striker will come to the land and set up a peg showing the depth of the sewer line and another peg showing the height of the city sidewalk curb. This is assuming that you are building on virgin ground. If you set your house too low in the ground, you may have difficulty in getting a fall to the sewer line of the city. If your house is built too high, you may find later that the city will cut a road and leave your house far too high, making it difficult to drive onto your lot.

A *Bench Mark* is a surveyors symbol marked thus ⊼ . The center of the horizontal line indicates the actual bench mark, and when such symbols are enfaced on public monuments or stone steps, they indicate the elevation, at that point, above or below sea level. Internationally, the sea level is accepted as a datum. The location of official city bench marks are listed in the city engineers department. They enable architects/engineers to establish elevations for cut and fill; road grades; sewer lines; footings for foundations; and floor and roof levels for building projects.

Building lines are determined from site boundary pegs; elevations of buildings are determined from bench marks. Often it is necessary for a surveyor to establish a temporary bench mark on a building site. This must be durable enough so that its elevation shall not change during the total building operation. Such a bench mark may be established from a stout wooden or concrete post, set into the ground, enfaced with its symbol and surrounded with a fence to keep unauthorized persons or animals away. A manhole cover in the middle of the road may be taken as an initial reference. It is of vital importance that the elevation of a building be correct; the engineering servicing for sewers and storm flooding depends upon this contingency. A speculative builder may get information from the city hall, which for a nominal charge will supply a profile drawing of the road, sewer, and water lines adjacent to the property.

5.9 BUILDING INSPECTORS

It is in the public interest that progressive official inspections be made of new structures; for example, faulty plumbing could be a severe public hazard. It is obligatory for contractors to request that inspections be made at prescribed times. All work—such as on the sewer line or electric circuits—must be left open until an official certificate of approval has been given by the building inspector.

Building inspectors specialize in their own fields. You must be aware that inspections have to be arranged ahead of time so that the work schedule will not be delayed.

5.10 BUILDING IN FRIGID CONDITIONS

In some areas, especially in Canada, special consideration will have to be given to maintaining continuous work under frigid working conditions by using winter enclosures. The National Research Council, Division of Building Research, Ottawa, Canada, has done considerable research in this field. You may get from them at nominal cost authoritative information on: enclosed scaffolding; covered-in suspended scaffold; plastic covered arch or dome; air supported tent, and so on. The manufacturers of polyethylene market their product in 6 mil, 10 mil and 20 mil thicknesses to meet the needs of builders for use as vapor barriers, crawl space dampproofing, concrete curing, weather protection for men and materials, swimming pool covers, air supported structures, and so on.

5.11 SINGLE AND MULTIPLE RESIDENTIAL UNITS

The decision to build certain types of homes will be influenced by local appeal—single dwellings with basements or single dwellings without. The former are particularly recommended where severe weather conditions prevail, the basement being useful for children's game rooms, for storage, and especially for the installation of heating units for the home.

Consideration may be given to the building of side by side, two story duplexes with or without basements or to the building of apartments. The advantage of this type of construction (where it is acceptable according to the local building code) is that only one foundation or basement has to be dug; only one set of service lines installed; the workmen are under one roof for a longer period of time; there is less moving of equipment and temporary buildings.

It is important to remember that actual building costs are the same for similar residences on expensive or cheap poor lots. It is a good policy to produce homes to complement building lots and patterns of existing homes.

5.12 THE COMMITMENT TO BUILD

Once a determination has been made either to sign a contract to build a residence, or to build on speculation of selling, immediate consideration must be given to the labor force and from where it will be recruited; who will run the job; what kind of equipment will be needed.

Some stationary equipment will have to be assembled such as: cut-off saws; hand electric saws; electric drills; ladders; scaffold; picks; shovels; wheelbarrows; wrecking bars. (See Guide List art 4.9.)

Temporary Buildings will have to be provided and they may include some mobile units for offices and first-aid rooms. Remember that if a mobile unit costs $4,000 and is used forty times, the actual cost per usage is only $100 plus its maintenance and haulage from job to job. Information is available on depreciation of equipment and leasing.[1]

5.13 LEGAL CONSIDERATIONS

It is important that you consult with your attorney before, not after, signing any legal documents; his fee should be considered as an overhead expense like that of the surveyor; such fees will of course be reflected in the end price of the product. The builder must be aware of many points of law and a definition of a foreclosure is offered as a guide.

Foreclosure is a legal proceeding by which the equity of redemption of a mortgagor in and to a mortgaged property is extinguished. Foreclosure is in order when a mortgagor fails to pay principal and interest monies as stipulated, or to comply otherwise with the conditions of mortgage such as paying taxes on mortgaged property, etc. In the event of foreclosure sale, the mortgagor is entitled to receive any excess funds produced, by the sale, over the amount due to the mortgagee and over the legal cost and expenses. In many states and provinces of Canada, the method is to enter a foreclosure suit in court. The sheriff advertises and sells the mortgaged property. Be aware that methods may differ in different jurisdictions. It is important to remember that the courts require a complete statement of all monies paid by the mortgagor to the mortgagee. *Keep good records!*

CHAPTER 5 REVIEW QUESTIONS

1. List ten considerations that a speculative builder should take into account before erecting residential units in any particular area.

2. List three advantages to a contractor by hand digging test pits for subsurface exploration.

3. Briefly describe sanitary fill and give three advantages and three disadvantages of this method of land fill.

4. List the legal steps you would take before commencing to demolish or remove an existing building in a city, and state what earthwork would be required of you.

5. The four copies of drawings and specifications for a residence that is to be speculatively built but partly financed through a lending institution are distributed as follows:

    (a)   (b)
    (c)   (d)

6. What is the minimum distance for a new residence to be set back from either the street or avenue in any new subdivision of your area at the present time?

7. What is (a) a datum line, and (b) a bench mark?

8. What is the function of an official building inspector, who pays him, and what is expected of him?

9. Define a mortgage, a lien, and a foreclosure on real estate.

10. State four distinct advantages in building multiple residential units against single ones.

11. The local authority has competent persons making periodic inspections of buildings under construction. How many different types of inspections for such buildings are made in your area?

12. What city departments are on a 24-hour service in your area?

13. List ten features that may be seen on the drawings for proposed future developments at your city planning department.

14. When interpreting drawings and specifications for a proposed new building, which takes precedence—the drawings or the specifications? Give reasons for your answer.

---

[1] Alonzo Wass, *Construction Management and Contracting* (Englewood Cliffs, N.J.: Prentice-Hall, Inc., 1972), Ch. 9.

6

Foundations, Excavations, and Depreciation

In this chapter we present: how to use the right angle triangle in building construction; building lines; batter boards; the builders square; builders equipment and depreciations.

6.1 THE RIGHT ANGLE TRIANGLE

Let us refresh our memories of school days when we were taught that: *The square on the base plus the square on the height equals the square on the hypotenuse.*

It will be seen that the squares on the base (3 × 3 = 9) added to the squares on the height (4 × 4 = 16) equals the square on the hypotenuse (5 × 5 = 25). See Fig. 6.1.

Any convenient multiple of these basic unit measures are equally valid as the following examples show:

Base	Height	Hypotenuse	
3	4	5	
6	8	10	
9	12	15	
15	20	25	
30	40	50	*and so on*

In the construction industry we talk about the 30:40:50 principal of triangulation, and these measures in feet are used extensively for the laying out of 90° angles for foundation work and so on.

Assume that a building lot has a (survey pegged) straight line 80'-0" frontage from which we are to lay out a 90° angle. See Fig. 6.2.

Step 1: Set a nail at the exact spot on each peg so that the distance between them is exactly 80'-0", as at A-B.

Step 2: Using A-B as a base line, from -A- lay off and stake a distance of 30'-0" as shown at -a-. At the exact measure of 30'-0" set a nail on the top of the stake.

Step 3: Using two steel tapes, hook each one to one of the protruding nails at stakes -A- and -a-, pay out the tapes to read 50'-0" on the one for the hypotenuse, and 40'-0" for the one forming the height of the right angle triangle. Traverse the two tapes until they are at 40'-0" and 50'-0" respectively.

Step 4: Drive a stake at the intersection of the two tapes; then spot a nail at the **exact** intersection of the two tapes. This is the completion of the layout for a 90° angle as used in building layouts.

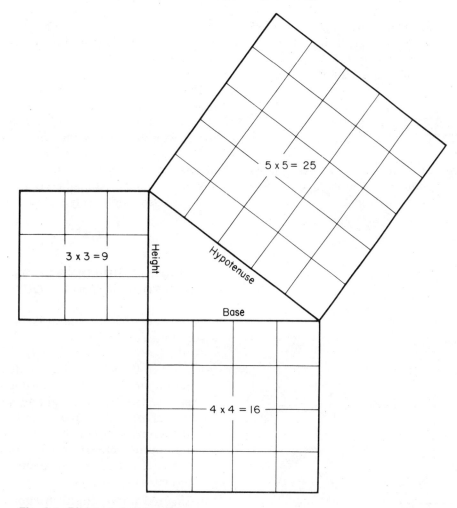

Fig. 6.1. Right angle triangle.

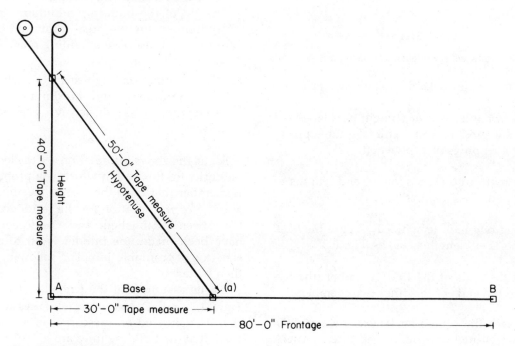

Fig. 6.2. Foundation layout.

The larger the units of measure used, the more exact the layout. As an example (if it is not windy) you could use 60'-0", 80'-0" and 100'-0", or in restricted areas you could use 15'-0", 20'-0" and 25'-0" measures or less.

6.2 THE BUILDER'S FOUNDATION LAYOUT SQUARE

For convenience you may make and use a large wooden square for some foundation layouts. See Fig. 6.3.

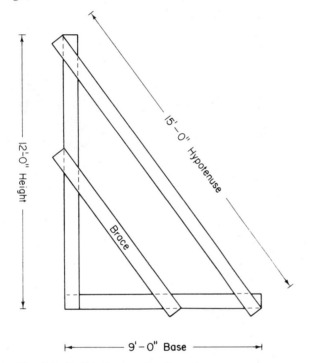

Fig. 6.3. Builder's wooden layout square.

To make the square take the following steps:

Step 1: Take three lengths of straight parallel edged pieces of 1 x 4 pine, or better still take three pieces of trued 1 x 4 rippings of $\frac{3}{4}$" plywood.

Step 2: Cut them into **exact** squared-end lengths of 9'-0", 12'-0" and 15'-0" respectively.

Step 3: Using one nail, temporarily secure the 9'-0" length to the 12'-0" length as at -A-, Fig. 6.3.

Step 4: Nail the ends of the 15'-0" member (the hypotenuse) to the ends of the other two members.

Step 5: Recheck the square for its exactness as a 90° angle; then well nail and clench the three corners. Nail in place an intermediate brace to stiffen the

frame. Trim off the corners from the hypotenuse member and fix the stiffening brace—the foundation square is ready for use.

The number of times that such a square may be used would depend on the finish and care of the instrument. It may be painted red so that anyone seeing it would respect it as being an important tool. Such a square has many uses for small insets and offsets in the layout of housing units. A similar square of smaller dimensions (say 3'-0", 4'-0" and 5'-0") is used extensively in the building trade.

6.3 GRADING: BUILDING LINES AND FOUNDATION LAYOUT

Before any layout is undertaken the builder must assure himself that he has:

(a) Checked the soil conditions, depth of frost penetration which will affect the depth of foundations and service lines, height of present or future centers of public roads and sidewalks, street furniture requirements (decorative street lighting, buried electric power, cable vision lines, and so on) and special treatment for sidewalks and curbs.

(b) Checked for zoning restrictions such as—single or multiple housing area, maximum permitted height of the building, minimum floor area requirement for the area, minimum setbacks for building lines from adjoining roads and boundaries of the lot.

(c) Checked the grading restrictions which may impose minimum and maximum gradients for property walks, driveways and swales (low places).

For all the above information see the local planning authority for the area in which you propose to build. Remember that once the correct location of the site and the correct location for the residence on the site have been established, and the foregoing checks have been made, the builder is off to an excellent start for a profitable job. The following steps may then be taken:

Let us assume that the irregular shaped lot as in Fig. 6.4, A-B-C-D, **has surveyed stakes at each corner,** that it is situated on the corner of a main and side road, that the building lines are to be 20'-0" from the main road and 12'-0" from the minor road. See Fig.

6.4. To establish the building lines from the known information, take the following steps:

Step 1: Clear the top soil and stockpile it ready for dispersing on the site for landscaping.

Step 2: Reduce the lot to the correct grade and remember that the land should fall away from the residence on all sides for drainage.

Step 3: String a taut mason's line between (nail spotted) points A-B-C. See Fig. 6.4, south and east sides of the lot.

Step 4: On line B-C, from -B- measure 20'-0" (for the frontage setback to the building line) and drive a peg and spot a nail on the top of the peg to mark the **exact** measurement as at -b-.

Step 5: Using A-B as a base line, place the base of the builder's wooden square (as described in art. 6.2) as shown at Fig. 6.4. Draw a steel tape tight and parallel to the blade (long side) of the square and measure 20'-0" setback. Drive a stake, and spot a nail to mark the **exact** measurement as at -a-.

Step 6: String a taut mason's line from stake -b- over the spotted nail of stake -a-, and continue the line to a new peg to be placed at -c-.

Step 7: Using A-D as a base line, place the base of the wooden square at a convenient place near the stake -D- (northwesterly) as shown at Fig. 6.4, and drive stake -d- and spot the **exact** point with a nail.

Step 8: Using A-D as a base line, place the base of the wooden square at a convenient place from stake -c-, and set stake -e-.

Step 9: String a taut line from -d- over stake -e- to a new stake -f- on the frontage line.

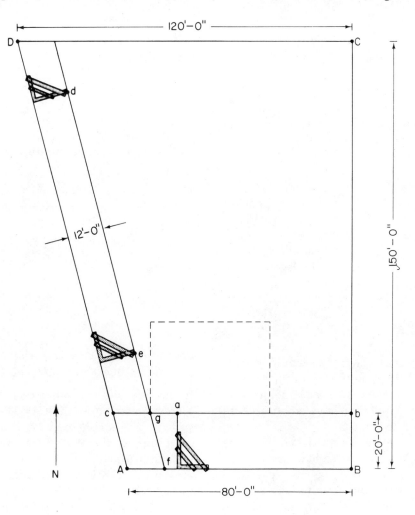

Fig. 6.4. Layout of building lines.

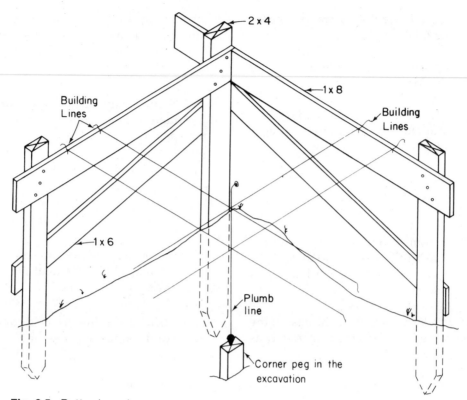

Fig. 6.5. Batter boards.

Step 10: At the **exact** intersection of the two building lines, drive the last stake -g- and spot the exact position with a nail. This last stake -g- is the critical focal point for the layout of the rectangular residence shown in broken lines.

To get the most out of this exercise, get a piece of paper and draw freehand the irregular shaped lot; then follow the text (imagine that you are out in the field) and do the layout step by step.

Once a proposed building has been correctly located for its place and elevation on its legal lot, the building is off to a good start.

6.4 BATTER BOARDS

Batter boards are frames placed adjacent to (but on the outside corners of) proposed excavations over which a taut mason's line (or wire for larger jobs) is strung for delineating building lines when excavations are completed. See Fig. 6.5.

Let us assume that we have established the front building line stakes for a residence having a rectangular plan of 30'-0" x 48'-0". See Fig. 6.6. Note carefully that for small offsets from an otherwise rectangular plan, the wooden square may be used as described in Fig. 6.2.

To lay out the batter boards, proceed as follows:

Step 1: Using the methods of triangulation described in the foregoing articles, lay out the four corner stakes for the above mentioned residence as in Fig. 6.6.

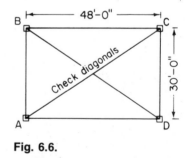

Fig. 6.6.

Step 2: Remember the importance of checking by steel tape all diagonals in field layout. They should be equal.

It once came to my notice that the building lines for a rectangular house 30'-0" x 90'-0" were 2'-0" out of square. The layout had been done using a transit level, but the diagonals were not checked for equality of length (diagonals must always be checked). The error was not discovered until the carpenters tried to lay the subfloor with 4'-0" x 8'-0" sheets of plywood. *Be careful! Be very careful!* You cannot rush this kind of work.

Step 3: With two men working together, one may sight over stakes A and B to align stake -b- which the second man will place about 3'-0" from stake B. See Fig. 6.7.

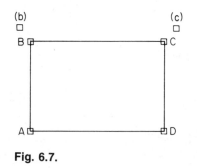

Fig. 6.7.

Step 4: This operation may be repeated with one man now sighting, over stakes B and C, while the other places stake -c- (about 3'-0" from C), and so on, until two stakes have been placed adjacent to each original corner stake. See Fig. 6.8.

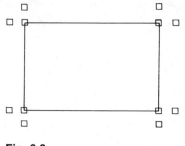

Fig. 6.8.

Step 5: The original stakes A-B-C and D may now be removed leaving the area free for the excavator. Some small jobs may be accomplished using stakes only. See Fig. 6.9 which shows the original corner stakes removed and the mason's lines drawn over the outer stakes. Remove the lines for excavating, then replace them, dropping a plumb line from each of their intersections to reestablish the stakes in the bottom of the excavation. See Fig. 6.5.

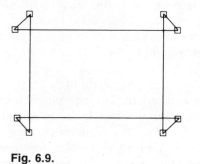

Fig. 6.9.

Step 6: It will be seen from Fig. 6.10 that a more durable job of establishing the building lines may be made by using batter boards on the outside corners of proposed excavations. Remember that the further away that batter boards are placed from an excavation, the longer the arms must be. If the batter boards are placed 3'-0" clear from the proposed excavation, the arms must be at least 4'-0" in length.

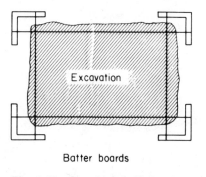

Fig. 6.10. Checking building layout.

Step 7: Set the arms of the first batter board with its top edges level to each other, and placed at a determined height, say, the top of the basement wall. The three other batter boards must be placed and leveled to the first.

Step 8: To level from one batter board to another, use a long straightedge of wood (say, 14'-0" in length) and a spirit level, the longer the better. See Fig. 6.11. Place one end of the straightedge on the first batter board; level the straightedge with the spirit level as shown in Fig. 6.11; and set a stake, at the other end of the straightedge, exactly level with the batter board. Turn the straightedge and spirit level, end for end, and recheck the work. The bubble of the spirit level should read the same for both ways. Then level from one stake to another until the next batter board is leveled from the first.

It is recommended that the first level be made to the center of the area to be excavated. Then from the accuracy of height of this stake, it may be used as a reference to radiate all the remaining batter boards. The less times the straightedge has to be used, the less the risk of error.

Step 9: A saw kerf may be cut into the tops of the arms of each batter board indicating the building lines. On large jobs, the footing lines and wall widths may also be cut on the arms of the batter boards.

Step 10: It is important that the excavation be made well clear of the building lines to give room for men to work between the basement underpinning and the walls of the excavations. Some jurisdictions have stipulated minimum clearance for such working

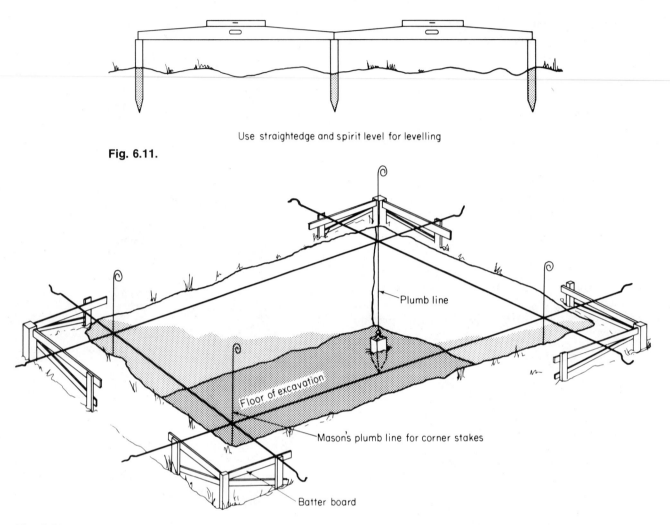

Use straightedge and spirit level for levelling

Fig. 6.11.

Plumb line

Floor of excavation

Mason's plumb line for corner stakes

Batter board

Fig. 6.12.

areas. Get a copy of the local building code from your local authority.

Step 11: When the area has been excavated, the building lines may be established in the excavation by dropping a plumb line from each corner intersection of the building lines above. See Fig. 6.12, also Fig. 6.5.

6.5 EQUIPMENT: SCHEDULES AND MAINTENANCE

Many general contractors have large investments in construction equipment which must always be kept in first class working order; the equipment, also, must be rigidly accounted for on the jobs on which it is used. As with all schedules, there are various ways

of making and keeping them up-to-date. The following is offered as a guide:

The head of the maintenance department would be responsible for the filing of all owners maintenance manuals issued by the manufacturer of each piece of equipment. He would keep on file a case history of each machine, the date it was bought, whether new or used, the latest book value, what major repairs have been done to it and when. He would determine the regular oiling, greasing, and maintenance done in the field and the shop.

Over the life expectancy of equipment, maintenance costs will run from one to two times the initial cost of the equipment. The equipment must be kept in good condition whether in use or not, and it must always be ready for the owner's use, leasing, or for sale. The timing of major repairs, for years of high income, may lessen the tax charges against the company for such years. Maintenance costs tend to increase as a company ages and needs more maintenance crews and technicians. Large companies may set up a separate company to handle all major equipment; they may also operate a rental service to others.

The head of the maintenance department would maintain a filing system showing the dates that each piece of equipment is "charged out" to each job. This means that each individual job "overhead expense account" would be charged a daily rate for each piece of equipment detained on the job, whether in use or not. The job superintendent would show in his daily job report, each piece of equipment on his temporary charge; it would be in his interest to return the equipment as soon as possible and thus keep his job overhead expenses at a minimum.

Some large contractors engage the services of one of the equipment dealers to service and maintain their products. The dealers provide continual supervision of their products. They maintain a 24-hour service, and it is claimed by some that they can keep equipment in working order for more hours per year than could the construction contractor himself. Remember that depreciation goes on whether or not the machines are in use. Management must decide whether to concentrate its undivided attention on construction contracting only, or on construction, plus servicing and maintenance of its own equipment and the possibility of leasing some of its own equipment to others. In short, in what business or businesses does the contractor wish to operate.

When large (international) construction contractors bid on large jobs, the unsuccessful bidders often lease their equipment and crews to the successful bidder. No company can afford to have millions of dollars-worth of equipment kept idle for any length of time waiting for a successful bid.

Leasing Contractor's Equipment

There are many advantages to a contractor leasing major items of equipment. Some of them are as follows:

(a) Leasing, instead of purchasing, frees more funds for working capital.

(b) It virtually eliminates maintenance crews and is known in the industry as "saving another headache."

(c) It reduces accountancy entries, and the contractor knows exactly what his charges will be in one statement.

(d) It reduces the time in keeping track of insurances and depreciation.

(e) It may be easier to lease than to find financing for new equipment.

(f) There is little idle time with leased equipment.

(g) The time saved may be used for the purchase of new equipment or the disposing of the old.

(h) Estimators can accurately price the cost for the use of heavy equipment on any specific job.

(i) Working the machines overtime may reduce overall costs of hiring instead of owning.

(j) The owner of leased equipment must keep up-to-date with all the latest models to remain competitive.

(k) Renewal of leasing the same equipment may be cheaper to the contractor.

(l) The leasing of automobiles can only be assessed against the anticipated annual mileage; the greater the mileage, the more advantageous the leasing.

6.6 INDUSTRIAL EQUIPMENT

To remain competitive a contractor must keep his industrial equipment up-to-date: by wise and discriminate purchases; by assurance of available parts servicing agencies; by efficient maintenance; and by a knowledge of methods of depreciating equipment (see the local federal taxation department for this information).

Before making a purchase, a contractor should see a demonstration of the piece of equipment in action, doing the type of job for which he would use it; further, he should see a demonstration of a similar piece of equipment from a competitor doing the same work on similar terrain to the first. It pays to shop around!

Remember that depreciation commences from the moment of transfer of ownership.

A further important point is to determine at what geographical place the transfer of ownership will occur. It once came to my notice that an expensive piece of equipment had been lost (through horseplay) in the bottom of a deep lake. The equipment was insured and replacement was made at the stated contracted place according to the terms of the insurance contract, but the contracted place was more than 1,000 miles from the scene of the loss. Be careful!

As an alternative to the purchase of heavy equipment a contractor may consider leasing or renting equipment, with the advantage of little or no responsibility for the maintenance of the equipment, nor for accounting purposes other than the initial fee.

6.7 DEPRECIATION

In accounting for contractors' equipment, the depreciation of an asset represents a loss in its worth from its last *assessed book value*. The book value of an asset is not necessarily the actual worth of said asset on the market. Equipment such as woodworking machines, tubular scaffolding, trucks, tractors, and so on, are all subject to allowable depreciation for income tax purposes. It should be noted that there is a difference between "depreciation" and "depletion"; the latter refers to a wasting asset through dissipation, such as a mine or an oil well which is worked to complete exhaustion.

There are several acceptable methods used to evaluate depreciation, and each year governments make tax guides available outlining such methods. However, it must be remembered that *tax laws may be changed at any time by an Act of Congress in America, or an Act of Parliament in Canada*. You may obtain a copy of "Depreciation Guidelines and Rules" (Revenue Procedure 62-21) from the Superintendent of Documents, U.S. Government Printing Office, Washington, D.C. 20402. Price 35 cts., U.S. Currency. Since section 521 of the Tax Reform Act of 1969 made changes in U.S. law relating to depreciation with respect to taxable years ending after 24 July 1969, a copy of Public Law 91–172 may be obtained from the same source for one dollar U.S. currency.

The Canadian Government issues a General Tax Guide (free) that may be obtained from any federal taxation office. For more details see one of the revenue department officials of the country concerned.

There are different taxation schedules showing the lifetime expectancy for different kinds of assets. Certain equipment may have a lifetime expectancy of five years *or less.* On the other hand, a building may have an allowable lifetime of fifty years. Using the "straight line method of depreciation" the former allows for a 20 per cent per annum depreciation (over a period of five years) of the original cost less the salvage value. The latter allows for a 2 per cent depreciation (over a period of fifty years) from the original cost less the salvage value.

In all cases the salvage value of an asset is not allowed "by law" to fall below a reasonable value. In case of doubt as to what is meant by a "reasonable value", contact an Internal Revenue Department official.

In cases where major repairs and/or modifications have been made to an asset which has increased its value and utility, the remaining book value at the end of the year may be more than its value at the previous year end. The improvement then becomes a further capital cost and the asset *appreciates* in the year end value.

In unusual cases, depreciation can result from an asset becoming obsolete. Assume that a new piece of machinery could outperform older, existing equipment in quality, quantity of work performed, relative prime cost, and maintenance expenses. The older machine may be declared obsolete because the owner could not possibly compete on equal terms with the owners of new equipment. However, a very convincing argument would have to be presented before the officials of the Internal Revenue Department would accept it.

The basic premise for depreciation allowances arises from the fact that in using a piece of equipment to achieve a desired end, such equipment is subject to wear, tear, and cost of maintenance, and that such costs are recoverable by the owner from the revenue of the work performed. Thus the owner recovers the depreciated value of the equipment for the period of time during which it is operated.

Depreciation may be stipulated in terms of a year, a month, a day, or even an hour. Assume that the installed cost of a piece of equipment is $9,250.00, with a salvage value of $750.00 over a five year life. Using the "straight line method of depreciation" (which follows), the asset would depreciate at the rate of 20 per cent per annum from its original cost less salvage value. Thus in five years $8,750.00 ÷ 5 = $1,700.00 per annum depreciation. Further, suppose the working time for the equipment to be 2,000 working hours per annum. Then the actual depreciation of the equipment would be $1,700.00 ÷ 2,000 = $0.85 per hour. Estimators use this method when pricing certain jobs.

Straight Line Method of Depreciation

Given the following charges for the purchase and installation of a new piece of equipment: original cost $8,450.00; taxes $173.00; transportation charges from point of ownership between vendor and purchaser $272.00; unloading $150.00; and cost of assembly and installation $205.00. The total cost would be $9,250.00. Note that all the above costs would be incurred legitimately by the purchaser before he could start to earn money from his investment. Assuming the life expectancy of the asset to be five years, and allowing for a final salvage value of say $750.00, the straight line method of depreciation would be as follows.

Step 1: List and find the total cost of the equipment to be depreciated.

Purchase price	$8,450.00
Taxes	173.00
Transportation	272.00
Unloading	150.00
Installation	205.00
	$9,250.00

Step 2: Deduct from the total cost the final salvage value of the asset, say $750.00; then $9,250 - 750 = $8,500.00.

Step 3: Divide the cost of the asset less its salvage value by the number of years over which the equipment is to be depreciated. Thus $8,500.00 over a life expectancy of five years would be $8,500 ÷ 5 = $1,700.00 depreciation per annum. This is the straight line method of depreciation.

Step 4: Lay out a table as follows:

Original total gross cost of the asset $9,250.00, salvage value $750.00, life expectancy, five years.

Year	Depreciation per Annum	Cumulative Depreciation	Year End Value
0	$1,700	$ 0	$8,500
1	1,700	1,700	6,800
2	1,700	3,400	5,100
3	1,700	5,100	3,400
4	1,700	6,800	1,700
5	1,700	8,500	0

Cumulative depreciation	$8,500.00
Salvage value	750.00
Original cost	$9,250.00

Summary: The straight line method of depreciation may be adopted where a regular flow of business is expected and where recovery of the cost of the asset is made in annual equal amounts during the life of the equipment.

Declining Balance Method of Depreciation

This method differs from the straight line method of depreciation in that, instead of depreciating the value of the asset by a fixed-dollar value each year, the depreciation is made by a fixed annual percentage reduction of the year end value of the asset. Using the same cost and condition of installing a piece of equipment as on page 94, the declining balance method of depreciation is as follows.

Step 1: Find the rate percentage that you would use on the same asset by the straight line method of depreciation. This was 20 percent per annum of the original installed cost (less salvage value) spread over a period of five years.

Step 2: Double that rate making 20 percent 40 percent.

Step 3: Depreciate each *year end book value* of the asset by 40 percent.

Step 4: Using this method, *the salvage value of the asset is not immediately taken into account,* but the balance remaining at the end of the fifth year will represent the salvage value with this qualification: The Internal Revenue Department says that you are never allowed to depreciate the value of an asset below a reasonable salvage value. The term "reasonable value" may have to be determined through an interview with an official from the taxation department.

Step 5: Using the total cost and life expectancy of the asset as being the same as that shown in the straight line method, the following table would result.

Step 6: Lay out the table as follows:

Original total gross cost of the asset is $9,250.00. Salvage value equals the remaining balance at the end of five years of depreciation.

Year	Depreciation per Annum from Year End Book Value (40%)	Cumulative Depreciation to Salvage Value	Year End Value of Asset
0			$9,250
1	$3,700	$3,700	5,550
2	2,220	5,920	3,330
3	1,332	7,252	1,998
4	799	8,052	1,198
5	479	8,531	719

Cumulative depreciation from column 3	$8,531.00
Salvage value remaining column 4	719.00
Original Cost	$9,250.00

Analysis: Depreciation for the first year, 40 percent of $9,250.00 taken without any deduction at this time for salvage value of the asset, is $3,700.00. Deduct this amount from the original cost of the asset as shown at column 4, leaving $5,550.00 as the book value of the asset at the end of the first year.

The second year depreciation, 40 per cent of $5,550.00, is $2,220. Deduct this amount from the year end book value of the asset as shown at the end of the second year column 4, leaving $3,330.00.

The cumulative depreciation for the first two years is $3,700.00 plus $2,220.00 which equals $5,920.00 as shown for the second year in column 3, and so on until there is a final year end value of the asset (Salvage) remaining of $719.00 as shown at the fifth year column 4.

Summary: The declining balance method of depreciation may be adopted when a contractor wishes to take the greatest amount of depreciation during his early years of ownership of the asset, to compensate for high annual profits during that projected period.

Sum of the Years-Digits Method of Depreciation

This method yields high rates of depreciation during the early years of the asset's life. Use the same equipment and conditions as follows.

Step 1: Find the sum of the digits of the years of life expectancy of the asset.

Step 2: The sum of the years digits $= 1 + 2 + 3 + 4 + 5 = 15$ digits.

Step 3: After deducting the salvage value of the asset, the depreciation is calculated by deducting the following fractional amounts each year.

First year:	$\frac{5}{15}$ of the original cost of the asset
Second year:	$\frac{4}{15}$ of the original cost of the asset
Third year:	$\frac{3}{15}$ of the original cost of the asset
Fourth year:	$\frac{2}{15}$ of the original cost of the asset
Fifth year:	$\frac{1}{15}$ of the original cost of the asset
Total digits	$\frac{15}{15}$

Step 4: Lay out the table as follows.
Original total gross cost of the asset $9,250.00; salvage value $750.00; life expectancy of five years.

Year	Depreciation per Annum	Cumulative Depreciation	Year End Book Value
0			$8,500
$1\frac{5}{15}$	$2,833	$2,833	5,667
$2\frac{4}{15}$	2,267	5,100	3,400
$3\frac{3}{15}$	1,700	6,800	1,700
$4\frac{2}{15}$	1,133	7,933	565
$5\frac{1}{15}$	567	8,500	0

Cumulative depreciation	$8,500.00
Salvage value	750.00
Original cost of asset	$9,250.00

Summary: Using the sum of the years-digits method of depreciation, the greatest recovery is made during the early years of the life of the asset. This method may be adopted when the contractor wishes to take advantage of prospective high earnings during the first few years of the life of the asset (and thus save in income tax charges).

Comparative Depreciation Table

The following comparative depreciation table shows the relative amounts that may be depreciated for the same equipment for the same period of time as outlined in all the previous articles in this chapter.

Comparative Depreciation Table

Year	Straight Line Method 20%	Declining Balance Method 40%	Sum of Years-Digits Method
1	$1,700	$3,700	$2,833
2	1,700	2,220	2,267
3	1,700	1,332	1,700
4	1,700	799	1,133
5	1,700	479	567
Salvage	750	719	750
	$9,250	$9,250	$9,250

Analysis: The depreciation for the first and last two years of the asset are as follows.

Straight line (20%)	$3,400	$3,400
Declining balance (40%)	$5,920	$1,278
Sum of the years-digits	$5,100	$1,700

Summary: The highest depreciation occurs during the first two years (of a five year period) on the declining balance method, and the same method also shows the lowest depreciation for the last two years. Where earnings are expected to be high during the early life of an asset either the declining balance method or the sum of the years-digits method would be advantageous from a tax point of view. Where a steady turnover is expected, a regular straight line method may be adopted.

CHAPTER 6 REVIEW QUESTIONS

1. Make a freehand sketch to define the 30:40:50 principle of triangulation as used in land and building layout.

2. Assume that you have set the four corner pegs for a rectangular basement 30'-0" wide and 65'-0" long, how would you check that the four corners are true right angles?

3. Quote four different zoning restrictions (for buildings) in your area.

4. List six major considerations that a builder should take into account before purchasing a piece of heavy equipment.

5. State four advantages and four disadvantages for a builder owning heavy mobile equipment.

6. State four advantages and four disadvantages for a builder renting heavy mobile equipment.

7. List eight checks that should be made before any site layout, such as building lines and foundation layout, is made on a building site.

8. Sketch and define a batter board.

9. Define the general meaning of *depreciation.*

10. From what source may you obtain authoritative *"Depreciation Guide Lines and Rules"*?

11. Assume that you purchase a piece of equipment to be installed in a building at a certain price; list five items that may affect the installed cost and which should be taken into account for calculating its depreciation.

12. Name three distinct methods of depreciation that may be used in industry.

7

Foundations, Piles, Chimneys, and Slab on Grade

In this chapter we shall discuss footings for foundation walls, pier pads, chimney foundations and the number and weight of bricks in chimney stacks; why step footings or piles and ground beams are used. The chapter concludes with a discussion of slab on grade panel heating.

7.1 UNDERPINNING

The stability of a building depends upon the adequacy of its foundation. In all cases excavations shall extend down to undisturbed soil unless the foundation is specifically designed for the existing soil conditions. *Study the local building code!*

Special care must be taken in designing pier pads for columns; foundations for chimneys; and foundations for pilasters (a pilaster is a column or pier forming an integral part of a wall, and partly projecting from the face of it). In each of the foregoing cases, the superimposed weight of that part of the structure to be imposed on the earth is proportionately greater than that of the main foundation footings.

7.2 PIER PADS

Pier pads at the foot of columns must be designed for surface area, depth, types of concrete reinforcing

each way (EW) in the footing. Unless an engineering design shows otherwise, footings (pier pads) supporting piers or columns should be at least four square feet in one story buildings, six square feet in two story buildings, and eight square feet in three story buildings.

Consider the weight of the superimposed weight on the pier pad, as in Fig. 7.1, where 5,000 lbs. are imposed centrally on a beam between each wall and the column. How much weight is imposed on the pier pad? Now consider the weight imposed on each pier pad, as in the plan at Fig. 7.2, where 5,000 lbs. are imposed, centrally, between the walls and the columns. How much weight is imposed on the pier pads shown on the plan?

7.3 CHIMNEY FOUNDATIONS

Foundations for chimneys must be designed for surface area; depth; types of concrete; and reinforcing EW in the footing so that it will support the weight of the masonry in the chimney without any settlement causing fractures in the stack itself, or in the adjoining walls, or in the flashing at roof level causing leaks. Remember also that a fractured chimney is a serious fire hazard. For a better appreciation of the methods of constructing, estimating quantities

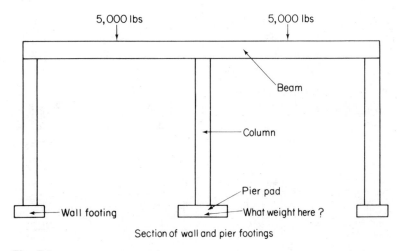

5,000 lbs   5,000 lbs

Beam

Column

Pier pad

Wall footing   What weight here ?

Section of wall and pier footings

Fig. 7.1.

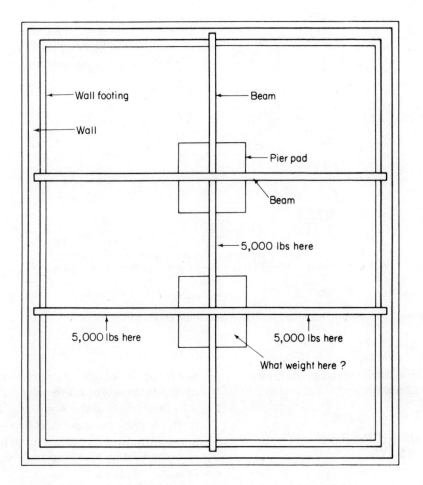

Wall footing   Beam

Wall

Pier pad

Beam

5,000 lbs here

5,000 lbs here   5,000 lbs here

What weight here ?

Fig. 7.2. Plan of wall footings, beams, and pier pads.

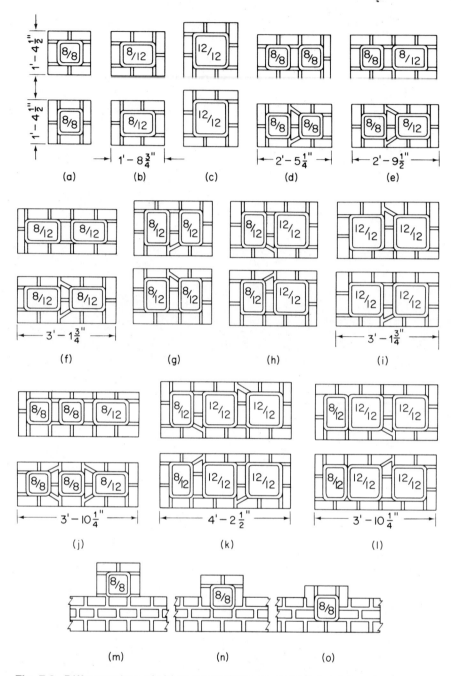

Fig. 7.3. Different sizes of chimneys and flues with their brick bonds.

of materials, and estimating the weights of chimneys, study the following two articles.

7.4 ESTIMATING THE NUMBER OF BRICKS IN CHIMNEYS

The usual method for finding the number of bricks required in a chimney is to find the number of bricks per ft. in the height of the chimney and to multiply by the total height taken in ft.

In order to find the number of bricks required per ft. in height, we must first know how many bricks are required for *one course* in height of the chimney, and then how many courses there are in 1 ft. in height.

In Fig. 7.3 is shown various sizes of chimneys and flues and the way the bricks are bonded in each case. For each different size of flue and chimney, two layers of courses of brick are shown. This is done in order to show how the courses will bond. In some cases, one course will require more bricks than the other, as, for example, chimney (f). This will require 12 bricks for one course and 13 bricks for the other course. This makes an average of $12\frac{1}{2}$ bricks per course.

From Fig. 7.4 we can see how many courses make up 1 ft. in height. We may also figure out how many courses make up 1 ft. in height. The standard brick is $2\frac{1}{4}''$ thick and the average joint is $\frac{1}{2}''$ thick. This makes $2\frac{3}{4}''$ in height for every course. In 1 ft. or 12″ in height, there are $12 \div 2.75 = 4.36$ courses.

Chimney (a) in Fig. 7.3 is shown to have 6 bricks to each course, which would require $6 \times 4.36 = 26.16$, say, 27 bricks per ft. in height. Chimney (f), which we found above to have an average of $12\frac{1}{2}$ bricks per course, would require $12\frac{1}{2} \times 4.36 = 54.5$ bricks per ft., say, 55 bricks per ft. in height. To find the total bricks in a chimney, we must multiply the bricks per ft. in height by the total height. Thus a chimney of the type (a) requiring 27 bricks per ft. in height, if built 30-ft. high, would require $30 \times 27 = 810$ bricks.

Chimneys of types (m), (n), and (o), which are built as part of a wall, do not require as many extra bricks. Type (m) requires 4 extra bricks for each course. Type (n) requires only 2 extra bricks, that is, 2 more bricks for each course than would be required if the wall were built straight without the chimney.

How to Estimate Fireplaces

Fireplaces for chimneys of irregular shape that cannot be figured by lin. ft. in height are often figured by the cu.-ft. method. By this method the total

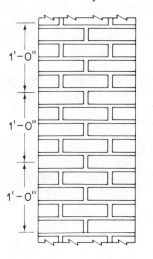

Fig. 7.4. Showing number of courses in one foot of height.

cu. ft. of brickwork required is figured and the openings for the flues or ash pits are deducted. *(See also Sect. 16.5.)*

7.5 AVERAGE WEIGHT OF SOLID BRICK WALLS

Brick assumed to weigh $4\frac{1}{2}$ lbs. each — $\frac{1}{2}''$ mortar joints

Area in Sq. Ft.	4-Inch Wall	8-Inch Wall	12-Inch Wall
1	36.782 lb.	78.808 lb.	115.414 lb.
10	368	788	1,154
20	736	1,576	2,308
30	1,103	2,364	3,462
40	1,471	3,152	4,617

Estimating Firebrick

For estimating on fire-brickwork, use the following figures: From 400 to 600 pounds of high temperature cement or fire clay are enough to lay one thousand nine-inch straight brick.

1 square foot $4\frac{1}{2}$-inch wall requires 6 nine-inch straight brick.

1 square foot 9-inch wall requires 12 brick.

1 square foot $13\frac{1}{2}$-inch wall requires 18 brick.

1 cubic foot of fire-brickwork requires 17 brick.

1 cubic foot of fire-brickwork weighs 125 to 140 pounds.

1000 brick (closely stacked) occupy 56 cubic feet.

1000 brick (loosely stacked) occupy 72 cubic feet.

7.6 WALL FOOTINGS

It is important to study closely with this section the footings and typical walls as shown in Chapter 4.

In some areas where the soil (ground) is firm enough, it may be permitted to dig trenches for wall footings and directly fill them with concrete without any formwork as in Fig. 7.5. When this method is used, it is recommended that stakes be driven, centrally, every 10'-0″ apart in the footings, and leveled throughout. A screed may then be activated, from stake to stake, to level the whole surface area of wet concrete.

A better and cleaner job may be done by placing two inch planks of suitable depth to edge the footings. These members are secured in place and kept level with stout stakes nailed every six feet apart; braced across the top, and supported at the sides, as shown in Fig. 7.6.

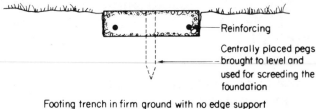

—Reinforcing

Centrally placed pegs brought to level and used for screeding the foundation

Footing trench in firm ground with no edge support
no scale

Fig. 7.5.

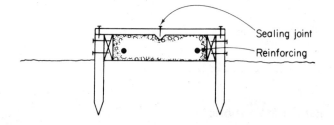

Sealing joint
Reinforcing

Fig. 7.6.

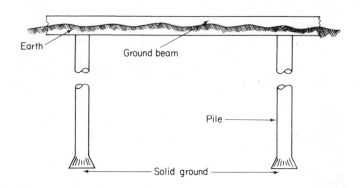

Fig. 7.7. Step footings.

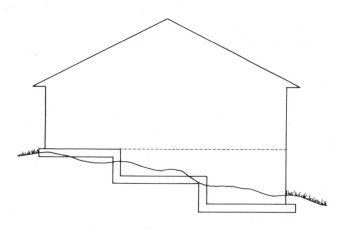

Earth
Ground beam
Pile
Solid ground

Fig. 7.8.

7.7 STEP FOOTINGS

To prevent the slippage of buildings erected on sloping ground, it is mandatory that step footings be provided, as shown in Fig. 7.7. In solid rock, such horizontal step footings may be cut where convenient with indeterminate heights of risers between adjoining horizontal members. In other types of soils, the rise of any step shall not exceed the length of either of its adjoining horizontal steps. This is an important phase of construction and it is important to know your local building code.

7.8 PILES AND GROUND BEAMS

In all cases foundations shall extend to solid ground below frost line. In some cases a builder may wish to develop land for housing units on ground that has previously been filled (such land is also known as "made" ground). City authorities, usually, have a history of such filled land, and profiles of it may be seen at the city engineer's office. In such cases it may be economical to have (engineer designed) piles and ground beams for underpinning. Expensive houses erected in coastal areas affording spectacular views may also use this system. See Fig. 7.8.

7.9 SLAB ON GRADE FLOOR PANEL HEATING

This article is included at the suggestion of Mr. Harry A. Panton, P.E., who reviewed the outline for this book, and is presently living in a ranch style basementless house with floor panel heating which he recommended.

The following notes on the preparation of a concrete slab on grade are offered as a guide, *but builders must familiarize themselves with the local building code.*

(a) The ground should gently slope away from the concrete slab on all sides, and the top of the slab should not be less than eight inches above the surrounding ground.

(b) The perimeter excavation for wall footings should extend to solid ground below the frost line.

(c) The footings should be reinforced with three No. 6 rods, with lappings of at least twenty-four bar diameters. It is recommended that footings and wall be monolithic (poured with concrete at one time).

(d) To minimize heat loss all outside walls should be protected between the slab and the wall, and down to the footings with waterproof insulation such as: cellular-glass insulation board available in 2, 3, 4, and 5 inch thicknesses; glass fibers with plastic binder available in thicknesses of $\frac{3}{4}''$, 1", and $1\frac{1}{4}''$; foamed plastic (polystyrene) and others in varying thicknesses; insulating concrete such as expanded mica aggregate in proportion of 1 part cement to 6 parts aggregate; also concrete made with lightweight aggregate of expanded slag.

(e) Soil beneath the slab should be compacted, and provided with five inches of coarse clean granular material with not more than 15% by weight of material passing a #10 sieve. The fill should be compacted and brought to a level of not less than six inches from the top of the foundation wall.

(f) Metal pipes passing under or through cinder or other corrosive material should be protected by a heavy coating of bitumen, or encased in concrete, or otherwise protected against corrosion. Install sewer, water, cable vision, and power lines *before placing the concrete slab.*

(g) A vapor barrier should separate the inert fill from the *initial* three inches of concrete slab, and this slab should be reinforced with wire fabric 6 x 6 — 6 x 6. A spaded finish is acceptable for the initial slab. The types of vapor barriers include: 55 lb. roll roofing or heavy asphalt laminated duplex barrier; heavy plastic film such as 6-mil polyethylene; three layers of roofing felt mopped with hot asphalt; heavy impregnated and vapor-resisting rigid sheet material with sealed joints; or other acceptable material to meet the local building code.

Mesh: Steel-Welded Reinforcing

Mesh: Steel-Welded Reinforcing is used largely in concrete floors, driveways, and roads. It is also used in other areas of building construction as temperature reinforcing. Some of the main things that an estimator should know about this type of reinforcing are as follows:

(i) It is fabricated in both square and rectangular mesh.

(ii) The longitudinal gage of the wire may be of a heavier gage than the transverse wire.

(iii) When there is any difference in the gages of the wire for any one type of mesh, the longitudinal wire is the heavier gage.

(iv) The style of the mesh may be described as 4" x 4" — 9 x 12 (or 44 — 912) which means that the area of each mesh is 4" x 4", and that the longitudinal gage of the wire is No. 9, and that the transverse wire is No. 12 (for a 4" x 8" — 9 x 12 or 48 — 912 welded steel mesh, each mesh is 4" x 8" and the gages of the wire are No. 9 and No. 12 respectively).

(v) The smaller the number of the gage, the thicker the wire.

(h) The slab should be laid out into its planned areas, and the plates secured to the initial concrete slab with concrete nails.

(i) On the initial slab the heating man may now arrange the soft copper tube in a serpentine manner. See the following *Installation Details* for floor panels and Fig. 7.9.

(j) After all the copper tubing is placed, it must be thoroughly pressure tested, then the final $1\frac{5}{8}''$ concrete slab may be poured and brought to a level trowelled finish.

INSTALLATION DETAILS

Floor Panels

The following material has been excerpted from the second edition of a twenty-eight-page booklet titled, *A Simplified Design Procedure For Residential Panel Heating,* prepared by the Research Department Staff and Consultants of *Revere Copper and Brass Incorporated,* 605 Third Avenue, New York, N.Y. 10016 and is reproduced here with the permission of the copyright holder.

Recommended tube spacing for floor panels is 9" or 12" on centers.

Coils should not be installed in cinder base concrete or on cinder fill. The tube can be fastened to reinforcing wire mesh or supported on blocking to establish the desired depth of bury.

Note that instead of the suggested location of the balancing cock and cut-off valve, they could be located near the heater, provided individual return lines from each panel are run to that point.

Shown is one method of insulating a floor panel against heat loss through the foundation walls and into the ground.

These illustrations show a preferred method of insulating floor slabs in a Radiant Heating installation. Note that in this case the insulation extends under the entire panel as well as around the edge of the slab in contact with exterior foundation walls.

INSTALLATION DETAILS

Venting, Controls, Testing, Starting, Balancing

Venting: The system must be properly vented so that positive circulation through all coils will be possible.

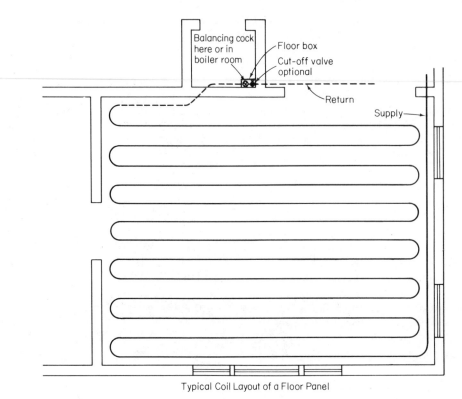

Typical Coil Layout of a Floor Panel

Fig. 7.9.

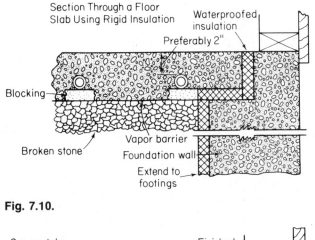

Fig. 7.10.

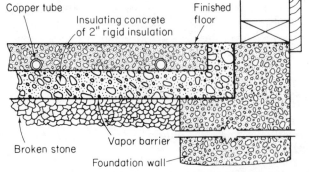

Fig. 7.11. Section through a floor panel using insulating concrete.

Filling the system with water should be done slowly with all vents open so that the air in the system will be completely expelled. The vents should be closed as soon as water issues from them. Occasional checking of air vents and expansion tank is advisable.

Controls: In selecting the controls for a radiant heating system, the designer has a choice of several methods of control. The control of a radiant heating system is substantially the same as that of a conventional two-pipe forced circulation hot water system. Controls can be simple or relatively complex, depending in most cases on the requirements of the particular installation. Continuous circulation and modulation of the water temperature with changes in outdoor temperature conditions will produce more uniform indoor conditions. Excellent literature has been published by control manufacturers who are willing to assist designers in the proper application of their products.

Testing the System: The piping system and panel coils should be thoroughly pressure tested before any tube is embedded in the plaster ceiling or concrete floor, to make sure there are no leaks in any of the joints. This should be done before any connections are made to the boiler, circulator, or other devices so that excess pressure will not affect these parts.

Fig. 7.12.

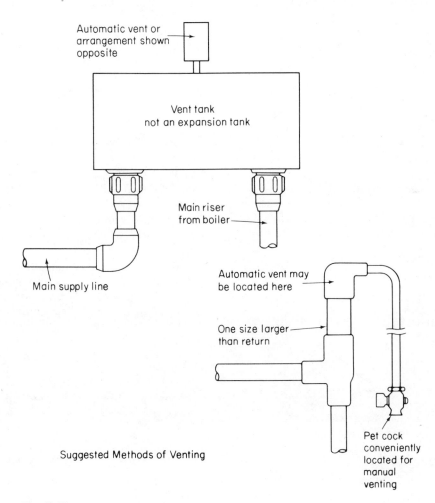

Automatic vent or arrangement shown opposite

Vent tank
not an expansion tank

Main riser
from boiler

Main supply line

Automatic vent may
be located here

One size larger
than return

Pet cock
conveniently
located for
manual
venting

Suggested Methods of Venting

Fig. 7.13.

It is suggested that the system first be tested with air at fifty pounds pressure or more. This permits any leaks that may occur to be repaired in a dry condition which facilitates the operation and eliminates the necessity of draining the system.

A second test with water at a pressure of 150 pounds per square inch can also be applied to the system for a period of 4–8 hours and all joints carefully checked for leaks.

Starting the System: The procedure in starting a radiant heating system for the first time, and in making subsequent adjustments, is similar in most respects to a two-pipe forced circulation hot water system.

Before starting the system, the plaster panels or concrete floor panels must be thoroughly dry. The drying period may extend over three or four weeks, but in no case should the radiant heating system be used to hasten the drying of the panels. It is recommended that when water is first circulated through the coils, its temperature does not exceed 90° for the first day or so, and then, increased gradually until design water temperature is reached.

All controls, circulators and other devices that are a a part of the system should be installed and adjusted according to manufacturers' recommendations.

Balancing: In a radiant heating system, in an average age residence, little, if any, trouble should be experienced in balancing the flow of water through the various parts of the system. After the system has been in operation, however, it may be necessary to make some adjustments in balancing due to heat regain in the structure, exposure, or wind conditions. Balancing the system is best accomplished on a cold day.

At the start, all balancing cocks should be wide open. If the temperature conditions in any room seems to be higher than in others, the balancing cocks for those panels concerned should be adjusted so as to reduce the flow of hot water through them. Adjustments should be continued until temperature conditions in all rooms seem to be satisfactory.

7.10 REINFORCING-STEEL CONTRACTORS' SPECIFICATIONS

The following clauses are typical extracts from the specifications for the reinforcing-steel contractor:

(a) Reinforcing steel shall be stored on racks or skids to protect it from dirt and to keep its fabricated form.

(b) All reinforcing steel shall be placed by experienced steel men, and shall be wired in position, and shall be approved by the architect or his representative before concrete is placed.

(c) Reinforcing bars shall be of medium-grade deformed steel suitable for working stress of 20,000 psi and shall conform to meet ASTM and CESA standards. *Note that* ASTM *is American Society for Testing Materials and* CESA *is Canadian Engineering Standards Association.*

(d) Reinforcing steel shall be bent cold and shaped as shown or required, accurately spaced and located in forms, and wired and secured against displacement before concrete is placed.

(e) Place reinforcing so that the distance from the face of steel to the nearest face of the concrete is not less than 1 diameter nor in any case less than the following:

Footings	3″	Walls	2″
Columns	1½″	Slabs	¾
Beams	1½″		

(f) Surface of bars shall be absolutely clean and free from mill scale, loose rust, oil, paint, and so on. Wire for tying shall be 18 U.S.S.G. annealed. *(Note: U.S.S.G. is United States Standard Gage.)*

(g) Bend horizontal wall steel around corners and continue 40 diameters of the bar. Support reinforcing on steel chairs and space with bar spacers. Support footing steel on brick or stone.

(h) Necessary splices not shown on the drawings shall be made by lapping and wiring adjacent bars. Splices and adjacent bars shall be lapped at least 24 bar diameters.

7.11 COMBINED SLAB AND FOUNDATION

The following material has been excerpted from a book titled *"Canadian Wood-frame Construction."* It is recommended reading for house builders and may be obtained from the Central Mortgage and Housing Corporation, Ottawa, Canada.

Combined Slab and Foundation. The combined slab and foundation, (Fig. 7.14) sometimes referred to as the thickened-edge floating slab, consists of a shallow perimeter reinforced footing or beam placed integrally with the slab. The bottom of the footing should be at least 1 foot below the natural gradeline and be supported on solid, unfilled, and well-drained ground. The slab is usually designed to support interior bearing partitions and a masonry chimney or fireplace. Anchor bolts are provided for securing the sill plates to the slab.

CHAPTER 7 REVIEW QUESTIONS

1. Fill in the blanks in the following question. Unless an engineering design shows otherwise,

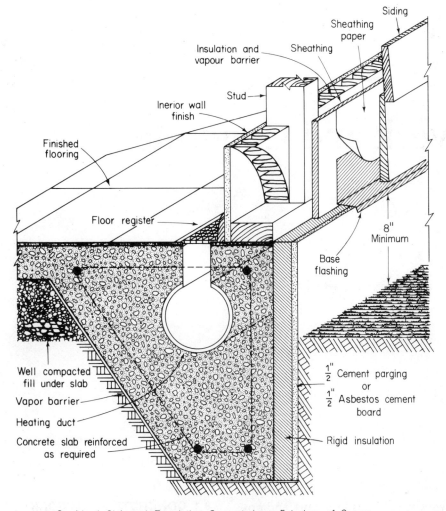

Combined Slab and Foundation Supported on 5 Inches of Coarse
Granular Fill, Well Compacted.

Fig. 7.14.

pier pads supporting columns should be at least
_____ ft. sq. in one story buildings, _____
ft. sq. in two story buildings, and _____ ft. sq.
in three story buildings.

2. Sketch a step footing and quote the minimum
   construction requirements at each step as stated
   in your local building code.

3. Define a ground beam and state what economies
   may be made by using them in certain ground
   conditions.

4. Give four advantages and four disadvantages of
   slab on ground, floor panel heating.

5. Make a sketch plan of a slab on ground floor
   panel heating system and support your like or
   dislike of the system in not more than 150 words.

6. Why should masonry chimney stacks be provided with special foundation pads?

7. Describe how to estimate the number of bricks
   required for a brick chimney.

8. Define a pilaster.

9. Define: underpinning, and pier pads.

10. State briefly how you would estimate the num-
    ber of bricks required to build a chimney 32'-0"
    in height, with a 12/12 flue, and coursing as
    shown in Fig. 7.3 (c).

Formwork and Concrete

8

In this chapter we shall discuss formwork and on-the-job quality-controlled concrete by volume mix, and how to determine the correct amount of cement, sand, gravel, and water, that should be hand fed into any concrete mixer to produce a predetermined strength of concrete. It has been my experience that many technicians know a great deal about concrete, but they are often at a loss to determine the exact volumes of materials to be fed into a specific concrete mixer to produce a specified strength of concrete to conform with the local building code for residential construction. For this reason, several handwritten examples showing how to determine volume mixes are given.

There is ample literature about ready-mixed concrete which may be ordered to meet any specification. The contractor is urged to take an extension course on the design and control of quality concrete. The Portland Cement Association is constantly doing research and making authoritative publications available to engineers, contractors, and farmers in language suitable to occupation. Your personal library should be well-stocked with reference books.

8.1 CONCRETE FORMWORK: BASEMENT WALLS

There are many types of semipermanent patented forms used in concrete work. The following example is for a contractor making his own forms for reuse by his own men. The number of times that such forms may be used depends upon the organizing ability of the job superintendent. *Some men would only get five uses out of them while others would use them up to fifty times.*

Design of Forms

(a) Forms should be substantial and sufficiently tight to prevent leakage of mortar. They should be properly braced and tied together so as to maintain their position and shape. If adequate foundations for shores cannot be secured, trussed supports shall be provided.

(b) Snap ties should be used for internal ties so arranged that when the forms are removed no metal shall be within 1″ of any surface. Wire ties will be permitted only on light and unimportant work; they must not be used where discoloration would be objectionable.

(c) Shores supporting successive lifts should be placed directly over those below, or so designed that the load will be transmitted directly to them.

(d) Forms should be set to line and grade, and so constructed and fastened as to produce true lines. Special care should be taken to prevent bulging.

(e) Forms should be lined with plywood for main-entrance surroundings.

Specifications

The specifications for a set of semipermanent $\frac{3}{4}$″ plywood forms frames with 2 x 4 dimension lumber are as follows:

(a) 80 forms 4′-0″ x 8′-0″ S1S $\frac{3}{4}$″ plywood. (S1S means solid one side.)

(b) 16 forms 2′-0″ x 8′-0″ S1S $\frac{3}{4}$″ plywood.

(c) 8 forms 1′-0″ x 8′-0″ S1S $\frac{3}{4}$″ plywood.

(d) Provide one top plate and one bottom plate for all forms, and space the studs 16″ OC for the 4′-0″ x 8′-0″ forms, all others to have stud spacings of 12″ OC.

(e) Construction-grade dimension lumber shall be used.

(f) Frames shall be secured with $3\frac{1}{2}$″ common nails.

(g) Plywood shall be secured to frames with $2\frac{1}{2}$″ common nails.

(h) Sheet-metal strapping shall be secured to all outside corners of forms (see Figs. 8.1 and 8.2).

8.2 DESIGN MIX BY WEIGHT

Modern concrete mixes are designed by weighing the component parts of water, cement, sand, gravel, and any other additives necessary to achieve the designed results. This is done so that a predetermined strength of concrete may be obtained. Sand and gravel are called aggregates; sand is fine aggregate (FA), and crushed rock is called coarse aggregate (CA). The final strength of concrete is known as the psi, indicating the number of pounds per sq. in. that a test cylinder of concrete 12″ long and 6″ in dia. will withstand (after a stated period of time in moist curing) under hydraulic pressure without fracturing.

Test specimens are taken while the concrete is being placed. The test specimens will be required by the local authority, the architect, and by the contractor for his own satisfaction. The cylinders are

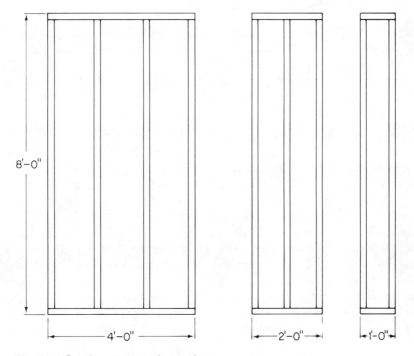

Fig. 8.1. Semipermanent forms for concrete.

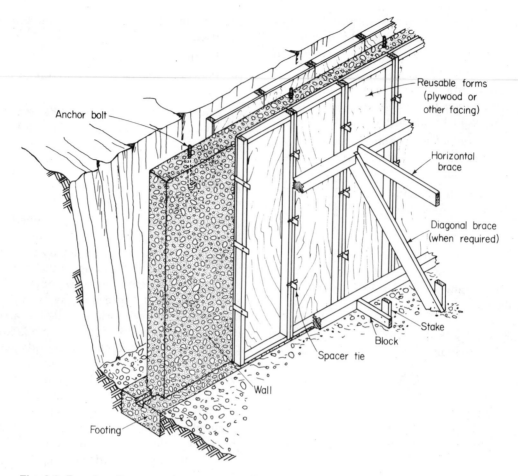

Fig. 8.2 Forming for poured concrete walls *(courtesy of Forest Products Laboratory, U.S. Department of Agriculture).*

kept under a fine spray of 70°F water for a stated time — usually up to a maximum of 28 days — and then subjected to testing for the psi. Concrete will attain a much higher psi strength when kept moist; this accounts for the curing measures to be taken by the contractor as specified by the architect.

The contractor must remember that the curing process requires equipment both for summer and winter curing, and that these items must be allowed for on the estimate. The cost of the tests must be allowed for on the estimate. The specifications always state that such tests shall be made at the expense of the contractor.

The basic components of concrete are clean, drinkable water, cement, sand, and gravel or crushed rock. Other additives may also be included, such as an agent to accelerate the setting of concrete or an air-entraining agent which will give a more porous result; coloring matter may also be used. There are many other additives too. See any reputable publication for further study of the different designs for specific functions of concrete. *Remember: water may have to be purchased.*

Few building materials lend themselves to so many abuses in the hands of untaught and inexperienced hands as concrete. This happens in spite of the fact that concrete has been in use for thousands of years.

8.3 FIELD BATCHING PLANT

To control the design mix of concrete in the field, as for road-making, large buildings, and so on, field batching plants are used. The cement, sand, and gravel are loaded into large metal hoppers, and the proportions required for each mix are weighed and delivered to the batch mixer, where measured water is added. This is to control the quality so that every batch is of similar composition.

It is a cardinal rule, always, to have the batching operations as close as possible to the area where the concrete is to be placed, and it is on this supposition that estimates are made. On large jobs, it would require a survey to be made for the most advantageous placing of the plant and the runways for buggies and barrows. *The cost of ramps and runways must be estimated.*

8.4 READY-MIXED CONCRETE

Great progress has been made, especially during this century, in the design and mix of concrete. In all our cities, concrete manufacturers will deliver (ready-mixed) designed concrete to meet specified requirements. There is a maximum free time allowed for unloading ready-mixed concrete from the delivery vehicle. After the allowed free time has elapsed, a charge is made for each fraction of an hour delay. For small jobs and for delivery of ready-mixed concrete to ordinary householders, the merchants will deliver in units of 1 cu. yd. and to the next higher $\frac{1}{3}$ cu. yd. They will also place and finish the concrete for an inclusive price. *Mass concrete is charged by the cu. yd., and finished concrete—as for concrete floors and sidewalks—is charged by the sq. ft. placed and finished.*

8.5 DESIGN MIX BY VOLUME

In this chapter, it is accepted that the most accurate method of designing concrete mixes is by weight; but very great quantities of concrete are placed every day in all countries by volume mix. This is especially true on small jobs which are too far from a ready-mix plant.

Satisfactory concrete can be made with a volume mix, and there are many examples of concrete structures made this way that are still standing after many hundreds of years. As a notable example, the great dome of the Pantheon in Rome was built of concrete over 1800 years ago and is still standing.

Some building codes state, "If a proportion by volume of 1:2:4 (cement: sand: gravel) is mixed with just sufficient water to make a plastic mix, a 2000-psi concrete will result providing that the aggregates are good." The words "just sufficient water" imply that the sand may be wet, and therefore just sufficient water would be less water than that required for very dry sand. Such a mix would be suitable for foundations for small light buildings, such as housing.

We must be aware that any of us could find ourselves in an undeveloped country, having to do a job of building, using such local labor, equipment, and materials as we find on hand. In any case, all estimators and builders must be very resourceful men.

8.6 WATER

Water suitable for concrete should be fit to drink. When it is drawn from sloughs in the country, it should be tested for silt, bacteria, and so on—all of which would be injurious to the concrete.

The estimator must constantly keep in mind that water is an integral part of the concrete-making, so therefore, it may have to be purchased by piping from the city mains, or obtained by digging a well, by pumping from a river or slough, or by hauling it by water tanker. Any one of these methods will affect the cost and must be allowed for on the estimate. For many jobs, it is necessary to make a survey of the area. Even with the simplest country job, it may require considerable lengths of $\frac{3}{4}''$ hose to run a water supply from the nearest source of supply.

8.7 CEMENT: STANDARD AND HIGH EARLY STRENGTH

In the manufacturing process, cement is so finely ground that 94 to 97 percent will pass through a sieve having 40,000 openings per sq. in. Each particle is manufactured so small that it will readily attract moisture. The volume of one sack of both American and Canadian cement is approximately 1 cu. ft.

High early-strength cement is a special type which is designed to reach as great a strength in 3 days as ordinary cement will reach in 8 to 10 days. It is more expensive than ordinary cement, but from an estimating point of view it might be advantageous to use under special circumstances. Assuming that a contractor was erecting a 10-story building with concrete floors, it would be a distinct advantage for the contractor to use high early-strength cement so that the concrete forms could be removed in a shorter time. Some contracts stipulate a time for the completion of a building, and where time is an all-important consideration, although high early-strength cement may not be called for on the specifications, it may be very profitable to use.

Great use is made of high early-strength cement where maintenance gangs may move into a factory area on a Friday and have a complete new concrete floor in operation on the following Monday.

8.8 STORAGE OF CEMENT

Since good cement is so dry and finely ground, it follows that it is very susceptible to dampness. If it goes lumpy while in storage and will not revert to its powdered form when rolled in the sack, it should be discarded. The addition of moisture sets up a chemical reaction, and once this reaction has started, the cement will not revert to its powdered form, and so should be discarded entirely. On many jobs, the inspector will insist that lumpy cement be hauled away from the building site.

Cement must be stored under very dry conditions. No amount of careful estimating can prevent a loss unless every precaution is taken to keep the cement (and indeed all building materials) secure against injurious hazards.

When it is necessary to haul cement onto the job some time before it will be used, it should be enveloped in polyethylene. First a wood platform should be provided about 0'-6" above the ground level. This may consist of a few pieces of 4 x 4 common lumber supporting any rough loose boarding such as subfloor material. Then the polyethylene sheet is placed over the platform and the stack of cement is enveloped by the sheet. *Remember that if you are using tarpaulins or polyethylene sheets, the cost will have to be allowed for on your estimate.*

8.9 WATER-CEMENT RATIO

Mixed water and cement is known as cement paste. The whole concept of concrete making is based on the relationship of gallons of water to 1 sack of cement. The ratio of 5 gals. of water to 1 sack of cement is known as a 5-gal. paste. Similarly, the ratio of 6 or 7 gals. of water to 1 sack of cement is known as 6- or 7-gal. pastes, respectively. It would follow that the thicker (stronger) the paste, the stronger and more durable the concrete; and the thicker the paste, the more expensive the concrete. With this in mind, the estimator would expect to find concrete mixed with a very strong paste to be used in the most psi-demanding places on a building. The concrete specifications always state the mix of concrete to be used in each section of the building. The estimator must examine the drawings and specifications very carefully to see where different psi mixes are required, since the cost of the concrete will vary with the different psi requirements.

When the cement paste is intimately mixed with the aggregates, every particle of sand and crushed rock must be completely coated with cement paste. Some architects specify the minimum allowable time for the mixing of each batch of concrete. This time must be given careful thought—especially since the number of men in the crew and the total time required to place the concrete will be governed by the mixture time allowed per batch. Assume a specification called for a 2-min. mix for 1-cu.-yd. delivery batch mixer as against another specification calling for a 1½-min. mix for the same machine. The saving in crew time is 25 per cent with the latter

time factor. This part of the specifications must be given very careful study before determining the labor cost of placing concrete.

8.10 MANUFACTURED AGGREGATES: CRUSHED ROCK

Clean, well-graded, hard, and rounded manufactured crushed rock, ranging in size from ¼" and up to a maximum grading limit of 1½", is known as coarse aggregate (CA). In the crushing process, too many fine particles are produced and some of these have to be eliminated to produce a well-graded aggregate.

You will have noticed that with both fine and coarse aggregates, rounded material is recommended; this is because such shaped pieces of rock have the least surface area.

Example

Examine the three solids shown in Fig. 8.6, on page 118—each has the same cubic capacity, but the shape and surface area of each is different.

The cu. capacity of (a) is 64 cu. in., and the surface area is 96 sq. in.
The cu. capacity of (b) is 64 cu. in., but the surface area is 112 sq. in.
The cu. capacity of (c) is 64 cu. in., but the surface area is 168 cu. in.

In each instance, the cu. capacity is the same but the surface area is different. Imagine the greatly increased surface area of aggregates containing many long slivers. This accounts for the specification calling for coarse aggregate from ¼" and up to 1½" that will grade to 1½" with a maximum of 15 percent slivers.

8.11 BANK OR PIT-RUN AGGREGATES

Bank or pit-dug aggregate is the natural virgin product of sand and gravel—discovered in most places by digging a pit.

The common characteristics of this class of aggregate are:

(a) It usually contains far too high a ratio of fine-to-coarse aggregate (i.e., fine rock (sand), to rock over ¼ in. in size).

(b) It often is dirty and unfit for use in concrete without washing.

(c) It may contain organic or acidic matter.

8.12 SURFACE AREA OF AGGREGATE

It will be recalled that the determining factor in the making of concrete is the water-cement ratio (cement paste). The difference in quality between, say, a $4\frac{1}{2}$-gal. and a 7-gal. paste is, simply, that in the latter case the yield of cement paste is 55 percent greater than in the former, and the latter is consequently a weaker paste and only suitable for restricted uses.

The finer the aggregate, the more surface area of aggregate to be covered.

Example

Step 1: A 12" cube has a surface area of 12 x 12 x 6 sides = 864 sq. in.

Step 2: A 6" cube has a surface area of 6 x 6 x 6 sides = 216 sq. in.

Step 3: The cu. capacity of the 12" cube is equal to the sum total cu. capacity of eight 6" cubes.

Step 4: The surface area of eight 6" cubes is 6 x 6 x 6 sides, which is 216; this times 8 cubes is a sum total surface area of 1728 sq. in. which is double the surface area of a 12" cube.

The estimator, when making a survey of an area for sand and gravel, should keep firmly in mind that the materials must be of a rounded nature. Unless materials are rounded, the mix will require more cement per batch to cover the increased surface area; or, worse yet, the specimen tests may not stand up to the psi specifications and the work may be condemned.

Coarse aggregates contain very little moisture compared with sand, and so their effect may be disregarded.

Conclusion

Any cube cut into 8 equal parts doubles the surface area of the original cube. This situation is equally evident with any sphere remolded into 8 equal-size spheres. This will approximately double the surface area of the original sphere.

The estimator must keep in mind the enormous increase in surface area of aggregate where the amount of fines (sand and very fine gravel) is out of all proportion to the requirements of evenly graduated material. The more evenly graduated the sizes of aggregate, the greater the economy of cement paste.

In recapitulation, the estimator should try to insure:

(a) That rounded aggregates are used with not more than 15 percent slivers.

(b) That aggregates with not too much fine grade are used.

In both cases if the surface area of the aggregates is increased, the result will be that either more cement is required for a given mix, or else the design of the mix may fall below the psi requirements of the specification. The permissive amount of slivers in aggregates is about 15 percent.

8.13 ESTIMATING QUANTITIES OF DRY MATERIALS FOR ONE CUBIC YARD OF WET CONCRETE

The estimator must discipline himself to remember that on all estimates for volume-mix concrete he should add one-half as much more material in the dry state by volume than the capacity of the finished volume of concrete.

Note that the additions of cement and sand for two mortar-design mixes shown in the right-hand columns of Table 8.1 add up to 36 and 37 cu. ft, respectively. This is $1\frac{1}{3}$ (plus) bulk volume of dry materials to wet placed mortar. Allow $1\frac{4}{9}$ dry for 1 wet.

Table 8.1 Estimating Materials: Quantities of Cement, Fine Aggregate, and Coarse Aggregate Required for 1 cu. yd. of Compact Mortar or Concrete

	Mixtures			*Quantities of Materials*	
Cement	FA (Sand)	CA (Gravel or Stone)	Cement in sacks	FA, cu ft	CA, cu ft
1	2	—	12	24	—
1	3	—	9	27	—
1	1	$1\frac{3}{4}$	10	10	17
1	$1\frac{3}{4}$	2	8	14	16
1	$2\frac{1}{4}$	3	$6\frac{1}{4}$	14	19
1	$2\frac{3}{4}$	4	5	14	20

8.14 ESTIMATING QUANTITIES OF DRY MATERIALS FOR A CONCRETE FLOOR AND FOR EACH BATCH OF THE MIXER TO BE USED

Example

You are referred to the completed work-up sheets, after studying the following problem, which should be carefully followed step-by-step. See the problem on page 117.

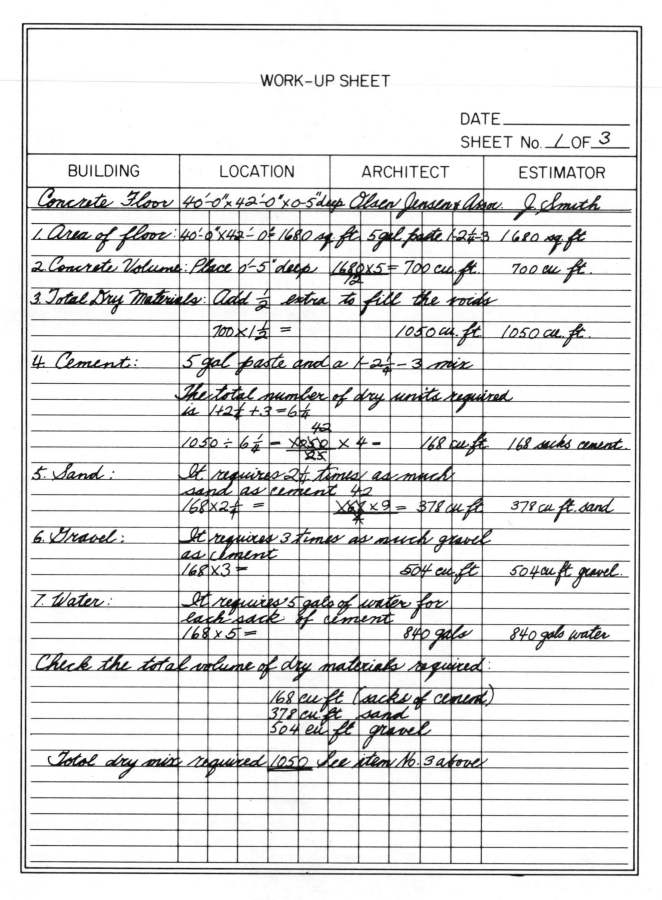

WORK-UP SHEET

DATE_____

SHEET No. _1_ OF _3_

BUILDING	LOCATION	ARCHITECT	ESTIMATOR
Concrete Floor	40'-0"x42'-0"x0-5"deep	Olsen Jensen & Asso.	J. Smith

1. Area of floor: 40'-0"x42'-0"= 1680 sq. ft. 5 gal paste 1-2¼-3   1680 sq. ft

2. Concrete Volume: Place 0'-5" deep   1680×5 = 700 cu. ft.   700 cu ft.
                                          12

3. Total Dry Materials: Add ½ extra to fill the voids

              700×1½ =   1050 cu. ft.   1050 cu. ft.

4. Cement: 5 gal paste and a 1-2¼-3 mix

             The total number of dry units required
             is 1+2¼+3=6¼
                                    42
             1050 ÷ 6¼ = 1050 × 4 =   168 cu ft.   168 sacks cement.
                          25

5. Sand: It requires 2¼ times as much
           sand as cement 42
           168×2¼ =   168 × 9 = 378 cu ft   378 cu ft. sand
                            4

6. Gravel: It requires 3 times as much gravel
            as cement
            168×3 =   504 cu. ft   504 cu ft gravel.

7. Water: It requires 5 gals of water for
           each sack of cement
           168 x 5 =   840 gals   840 gals water

Check the total volume of dry materials required:

                    168 cu ft (sacks of cement)
                    378 cu. ft. sand
                    504 cu ft gravel

   Total dry mix required 1050 See item No. 3 above

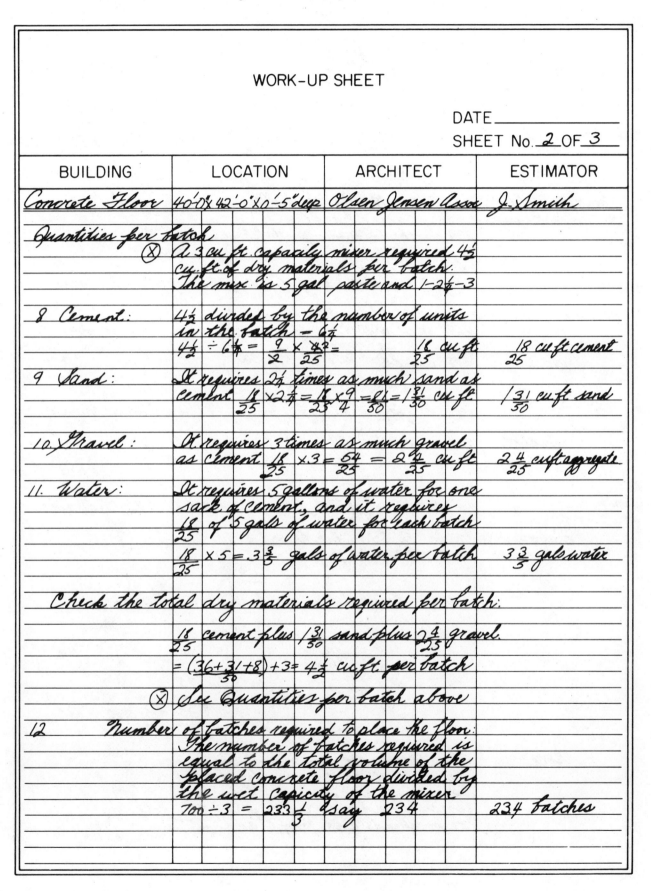

WORK-UP SHEET

DATE _____

SHEET No. 2 OF 3

BUILDING	LOCATION	ARCHITECT	ESTIMATOR
Concrete Floor	40'-0" x 42'-0" x 0'-5" deep	Olsen Jensen Assoc	J. Smith

Quantities per batch

⊗ A 3 cu. ft. capacity mixer required 4½ cu. ft. of dry materials per batch. The mix is 5 gal paste and 1-2½-3

8 Cement: 4½ divided by the number of units in the batch = 6¾

$4\frac{1}{2} \div 6\frac{3}{4} = \frac{9}{2} \times \frac{4}{25} = \frac{16}{25}$ cu ft   **18/25 cu ft cement**

9 Sand: It requires 2½ times as much sand as cement $\frac{18}{25} \times 2\frac{1}{4} = \frac{18}{25} \times \frac{9}{4} = \frac{81}{50} = 1\frac{31}{50}$ cu ft   **1 31/50 cu ft sand**

10 Gravel: It requires 3 times as much gravel as cement $\frac{18}{25} \times 3 = \frac{54}{25} = 2\frac{4}{25}$ cu ft   **2 4/25 cu ft aggregate**

11. Water: It requires 5 gallons of water for one sack of cement, and it requires $\frac{18}{25}$ of 5 gals of water for each batch

$\frac{18}{25} \times 5 = 3\frac{3}{5}$ gals of water per batch   **3 3/5 gals water**

Check the total dry materials required per batch:

$\frac{18}{25}$ cement plus $1\frac{31}{50}$ sand plus $2\frac{4}{25}$ gravel.

$= \frac{(36+31+8)}{50} + 3 = 4\frac{1}{2}$ cu ft per batch

⊗ See Quantities per batch above

12   Number of batches required to place the floor. The number of batches required is equal to the total volume of the placed concrete floor divided by the wet capacity of the mixer.

$700 \div 3 = 233\frac{1}{3}$ say 234   **234 batches**

WORK-UP SHEET

DATE _____

SHEET No. 3 OF 3

BUILDING	LOCATION	ARCHITECT	ESTIMATOR
Concrete Floor	40'-0" X 42'-0" X 5" deep	Olsen Jensen Assoc	J Smith

13. Time required to place the floor:

234 batches @ 2 min per batch = 468 min

468 ÷ 60 = 7 hrs 48 min
Say 9 hrs including start and clean up   **9 hours**

14 The estimated cost to place the floor:

Cement: 168 sacks @ $1.50 per sack   **$252.00 cement**

Sand: 378 cu ft   27)378 (14 cu yds
    27
    108

14 cu yds @ $3.40 per cu yd $47.60   **$47.60 sand**

Gravel: 504 cu ft   27)504 (18⅔ Say 19 cu yd
    27
    234
    216
    18

19 cu yds of gravel @ $3.85 per cu yd   **$73.15 Gravel**

Labor: 1 man @ $5.80 per hr = $5.80
4 men @ $4.40 per hr = 17.60
    23.40

9 crew hrs @ $23.40 per hr $210.60   **$210.60 Labor**

Summary: Cement   252 00
    Sand   47 60
    Gravel   73 15
    372 75
    Labor   210 60
    $583 35   **$583.35 Cost**

Note: This estimate does not cover labor and materials for forms, runways, equipment, traveling time, overheads and profit. For overheads and profit add 25%

Problem

A concrete floor 40'-0" x 40'-0" is to be placed with 0'-5" of finished concrete. The mix is to be by volume. Using a 3-cu.-ft.-capacity concrete mixer, a 5-gal. paste, and a 1:2¼:3 mix, estimate the following:

1. The area of the floor.
2. The quantity of placed concrete required for the floor.
3. The total volume of materials in the dry state required to place the floor (allow 1½ times the volume of dry materials to place the volume of concrete required).
4. The number of sacks of cement required to place the floor.
5. The number of cu. ft. of sand required to place the floor.
6. The number of cu. ft. of gravel required to place the floor.
7. The quantity of water required to place the floor.
8. The quantity of dry cement required per batch.
9. The quantity of sand required per batch.
10. The quantity of gravel required per batch.
11. The quantity of water required per batch.
12. The number of batches required to place the floor.
13. The time it would take to place the floor, allowing 2 mins. per batch and allowing at least 1 half-hour extra for starting and cleaning up afterwards.
14. The estimated cost to place the floor, using the following price list:

Materials: Cement $1.50 per sack
Sand $3.40 per cu. yd. ⎱ take up to the
Gravel $3.85 per cu. yd. ⎰ next higher cu. yd.
Labor: 1 man $5.80 per hour
4 men $4.40 per hour
Note: **Add for inflation.**

In addition to these costs, an allowance must be made for lumber, screeds, and framework, and possibly for constructing a wheelbarrow ramp. Make an allowance also for depreciation of plant, maintenance, and gas. If the project is on the outskirts of a town, you may have transportation costs to meet for the crew.

8.15 MEASURING THE CORRECT VOLUMES PER BATCH

The resolved problem of the volumes of cement, sand, gravel, and water for a 5-gal. paste and a 1:2¼:3 mix per batch for a 3-cu.-ft.-capacity mixer is $\frac{18}{25}$ sack of cement, $1\frac{31}{50}$ cu. ft. of sand, $2\frac{4}{25}$ cu. ft. of gravel, and $3\frac{3}{5}$ gal. of water per batch. The standard American gal. weighs 8.377 lbs. and the Canadian Imperial gal. weighs 10 lbs. *See the work-up sheets.*

Cement

To measure the correct amount of cement per batch, take an empty 5-gal. drum and pour into it 68 lbs. of cement. This poundage in each case is $\frac{18}{25}$ of the sacks respectively. Calibrate a drum at the height to which it is to be filled with dry cement by indenting it with a nail set or marking it by a paint line. Keep this drum only for measuring the dry cement.

Sand

The volume of sand required will be $1\frac{31}{50}$ cu. ft. per batch. This may be measured by making a 1-cu.-ft. measuring box; use 1 full box measure plus $\frac{31}{50}$ (say, $\frac{3}{5}$) of another, or take $1\frac{31}{50}$ of the weight of 1 cu. ft. of sand ($1\frac{31}{50} \times 94$ lbs). Calibrate a drum to measure the sand. One 5-gal. drum may be used if calibrated to hold just one-half the quantity of required sand. Use two measures for this since it would be too heavy for one man to be lifting $1\frac{31}{50}$ cu. ft. of sand all day. Sand weighs approximately 94 lbs. per cu. ft. Be certain that the men do not guess the quantities.

A Wood Buck. A temporary wood frame which is used in concrete work to form an opening in a wall for a door frame, window frame, or small door such as a crawl space; the latter should not be less than 2'-0" x 2'-0". A crawl space is the space between the lowest floor member of a house and the ground beneath, usually not more than two or three feet. The bucks have bevelled cleats nailed to their sides which remain in the concrete and afford good nailing for the final frame. Nail the cleats from the inside walls of the bucks to the cleats, with double headed nails. See Figs. 8.3 and 8.4.

When pouring concrete into walls that have bucks, it is important to keep pouring in one place until the concrete flows under the frame and emerges at the other end after completely filling the wall with concrete to the underside of the buck. If wet concrete is poured on both sides (at or about the same time), it will form an air-lock and create a void under the sill of the frame. Should this happen, bore a few holes through the formwork at a point under the sill, release the air and work the concrete until it fills the void. Remember that all openings in concrete walls weaken the structure unless reinforcing is introduced. Check!

Fig. 8.3. Section of wood window buck.

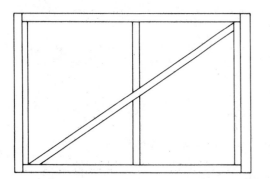

Fig. 8.4. Elevation of wood window buck with diagonal brace and center support against deflection under load of wet concrete.

Stripping and Reconditioning of Forms. As soon as the forms are taken from the concrete, they should be cleaned of all loose and clinging concrete, repaired, oiled, and level-stacked ready for shipping or carrying to the next job. Note carefully that no allowance is made in this exercise for shipping costs from job to job. This must be allowed for according to distances and conditions.

Form Hardware. The following material has been excerpted from "Fir Plywood Concrete Form Manual," by permission of the publisher, Plywood Manufacturers of British Columbia, 14477 West Pender Street, Vancouver 5 B.C.

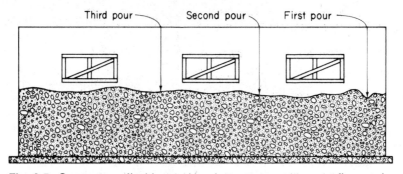

Fig. 8.5. Concrete wall with wood bucks; wet concrete must flow under each window and fill up to the underside.

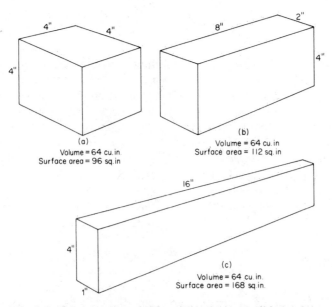

Fig. 8.6. Comparative surface area of three solids having the same cubic capacity.

WHAT GRADE OF PLYWOOD TO USE

Fir plywood is made in a variety of grades, virtually all of which are suitable for concrete form use. Contractors should study grades available in the table below and select one which will best meet the requirements of the job at hand.

TABLE 1	Job Requirements	Recommended PMBC Plywood Grade & Description	Comments
	The best concrete surface and/or largest number of re-uses: i.e. an architectural finish.	HIGH DENSITY OVERLAY GRADE (commonly called plastic overlay). A resin-impregnated cellulose fibre sheet is bonded to plywood face. Overlay is translucent, hard and smooth. Bond between overlay and plywood is equal to waterproof glue line between veneers.	Faces do not require sealing and provide excellent surface for parting agents. *Edges should be sealed either at factory or by contractor. Care desirable to avoid chipping of overlay by careless handling or vibrating.
		Medium Density Overlaid Plywood is not recommended for use as form material.	
	A good smooth concrete surface when both plywood faces will be used.	GOOD TWO SIDES (G 2 S) Each face smooth and sound, no knots or open defects, may contain neatly made patches. Waterproof glue.	Recommended for repetitive formwork. *Edge sealing recommended. Re-oil following each use.
	A good smooth concrete surface when only one plywood face will be used.	GOOD ONE SIDE (G 1 S) Face smooth and sound, no knots or open defects, may contain neatly made patches. Back may have limited size knot holes or other defects which have no material effect on strength or serviceability. Waterproof glue.	Recommended for liners and general formwork. *Edge sealing recommended. Re-oil following each use.
	A good smooth concrete surface when both plywood faces will be used.	CONCRETE FORM GRADE 2 SIDES Each face solid, contains neatly made patches, tight splits, small sound knots, with reasonable amounts of rough or torn grain. Both faces sanded, edges sealed with green coloured compound.	Recommended for majority of repetitive formwork. Re-oil following each use. Panels may be factory-oiled.
	A good, smooth concrete surface when only one plywood face will be used.	CONCRETE FORM GRADE 1 SIDE Face solid, contains neatly made patches, tight splits, small sound knots, reasonable amount of rough or torn grain. Back may contain limited size knot holes and other defects which have no material effect on strength or serviceability. Both faces sanded, edges sealed with green coloured compound.	Recommended for liners and general formwork. Re-oil following each use. Panels may be factory-oiled.
		SOLID 2 SIDES GRADE Each face solid, contains neatly made patches and small sound knots. Both faces sanded, slight roughness permitted.	*Recommend edges be sealed either by factory or contractor. Re-oil following each use.
		SOLID 1 SIDE GRADE Face solid, contains neatly made patches, and small sound knots. Back may contain limited size knot holes and other defects which have no material effect on strength or serviceability. Both faces sanded, slight roughness permitted.	*Recommend edges be sealed either by factory or contractor. Re-oil following each use.
	A good concrete surface permitting some surface irregularities.	SELECT SHEATHING GRADE One face has no open defects except for limited number of splits. Back may have limited size open defects which have no material effect on strength or serviceability. Both faces unsanded.	*Edge sealing not required unless plywood will be subjected to many re-uses. Oil forms for best results.
	For hidden or buried concrete surfaces where finish not important with few re-uses required or where strong backing for curved forms is required.	SHEATHING GRADE Each face may have limited size open defects which have no material effect on strength or serviceability. Both faces unsanded.	*Edge sealing not required unless plywood exposed to moisture for extended periods or many re-uses desired. Oil forms for best results.

*Note: Edge Sealing, where required, is recommended to reduce moisture picked up by wood, not because glue lines require protection.

COATED CONCRETE FORM PANELS

These are plywood panels with resin-fibre overlays or surface coatings of epoxy resins, modified poly-urethanes, or other proprietary compositions developed in the research laboratories of the plywood industry. They simplify form stripping and cleaning, improve the quality of concrete finish, and extend the service life of the form. The overlays and coatings seal the plywood surface and make it impervious to oil and water. Many of the compositions are self-lubricating and the frequency of application of a release agent is decreased.

Coated concrete form panels are variations from the standard grades shown in Table 1 and are marketed under the manufacturers own brand name. For complete list of coated concrete form panels write Plywood Manufacturers of B.C.

FORM HARDWARE

Contractors should obtain their hardware only from recognized suppliers who can furnish proof of the strength of their accessories by laboratory or field tests. While different types of hardware have been developed, all standard types are suitable for use with plywood. Of particular interest as far as plywood is concerned are various kinds of high tensile steel ties or bolts, some of which have been specifically designed for plywood.

Snap-ties made from high tensile wire are also popular as an inexpensive method of form support. These ties are designed to prevent the forms from separating and achieve their ultimate strength on a gradual increase in pressure. Stripping the forms should be done only after concrete has set. With forms removed, a tie breaker is applied to exposed end of snap-tie and bent at right angles as close to the wall surface as possible. The unit is then turned with a clockwise circular motion until the snap-tie breaks at break-off point.

This type of tie, designed to be used where breakback is not required, consists of a straight unthreaded pencil rod with buttons. Clamps are slipped over the rod and bear against the walers.

A separate spreader must be used.

A bar tie may be used with wedges through the slots in the end of the tie.

For most commercial concrete finishes a rod or bar tie with integral spreaders is customary. Such ties generally have a notch or similar reduction in section which allows the tie to be broken back a set distance from the wall surface.

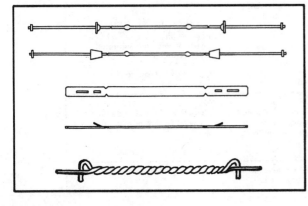

Rod ties are shaped to prevent rotation in the concrete when the rod is snapped off by twisting.

The bar tie shown on the left has a breakback notch and spurs which act as a spreader.

Twisted galvanized wire tie with integral spreader and breakback features.

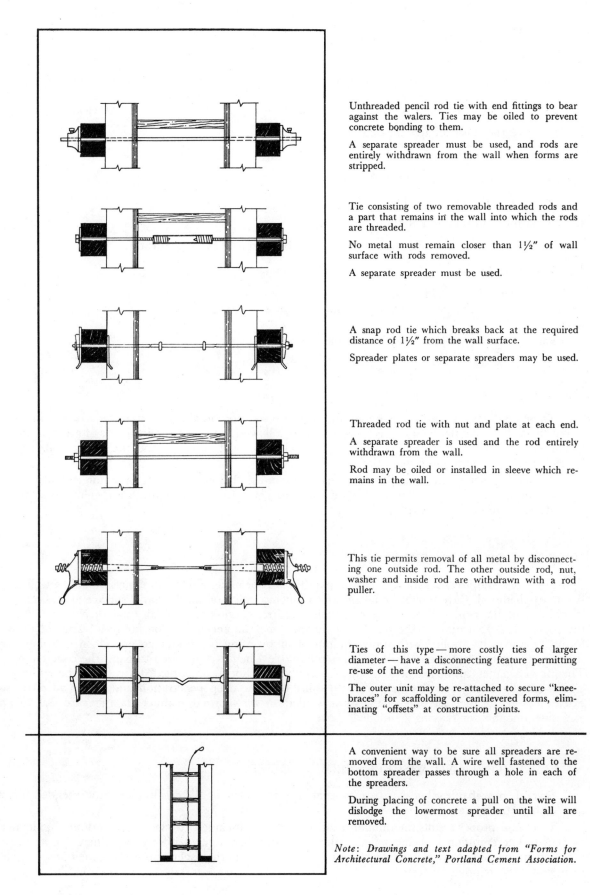

Unthreaded pencil rod tie with end fittings to bear against the walers. Ties may be oiled to prevent concrete bonding to them.

A separate spreader must be used, and rods are entirely withdrawn from the wall when forms are stripped.

Tie consisting of two removable threaded rods and a part that remains in the wall into which the rods are threaded.

No metal must remain closer than 1½″ of wall surface with rods removed.

A separate spreader must be used.

A snap rod tie which breaks back at the required distance of 1½″ from the wall surface.

Spreader plates or separate spreaders may be used.

Threaded rod tie with nut and plate at each end.

A separate spreader is used and the rod entirely withdrawn from the wall.

Rod may be oiled or installed in sleeve which remains in the wall.

This tie permits removal of all metal by disconnecting one outside rod. The other outside rod, nut, washer and inside rod are withdrawn with a rod puller.

Ties of this type — more costly ties of larger diameter — have a disconnecting feature permitting re-use of the end portions.

The outer unit may be re-attached to secure "knee-braces" for scaffolding or cantilevered forms, eliminating "offsets" at construction joints.

A convenient way to be sure all spreaders are removed from the wall. A wire well fastened to the bottom spreader passes through a hole in each of the spreaders.

During placing of concrete a pull on the wire will dislodge the lowermost spreader until all are removed.

Note: Drawings and text adapted from "Forms for Architectural Concrete," Portland Cement Association.

PARTING AGENTS

A GOOD PARTING AGENT SHOULD PERFORM THE FOLLOWING FUNCTIONS:

A   Reduce bond between form face and concrete surface, permitting form to break clean from concrete.

B   Improve the finish of the concrete surface.

C   Extend the life of the form faces.

D   Increase resistance to alkali water penetration.

E   Be non-corrosive to metal fasteners.

F   Be harmless to skin and clothing.

G   Resist wear by foot traffic without creating a dangerously slippery surface.

H   Reduce grain rise.

A parting agent should always be applied to the plywood face if forms are to be used more than once or twice. Good maintenance and proper application of a good parting agent will be amply repaid by longer form life.

Parting agents available may be classified into three types:

(a) LACQUERS AND PAINT

These coatings are applied as liquid and dry into waterproof, alkali-resistant films of varying degrees of hardness. The manufacturers' directions should be followed as proper application is of great importance. Grain rise and face checking will be reduced, form cleaning will be easier and the plywood itself will be protected from large moisture variations.

A good lacquer coating should last three or four re-uses before re-coating is required, but it may last as many as thirty re-uses if form oil is used each time. Generally the coatings are coloured so that untreated areas are easily noticed.

(b) FORM OILS AND GREASES

Form oils and greases generally have a paraffin wax or mineral oil base. Panels should be well cleaned before applying oil. Care should be taken not to put too much oil on the panel surface as this may result in staining of the concrete. Most oils eventually oxidize, causing them to dry up, or else they soak into the plywood. Application of the oil should therefore not be performed too far in advance of using the forms. Oils may be applied with spray equipment or by using brushes or rollers. If oil is to be used as the parting agent, then a paraffin base oil is preferable when it is intended to paint the concrete surface. The life of a panel can be extended by initially applying a liberal amount of form oil to the entire panel. This is usually done just a few days before first use to allow deep penetration and drying of oil to some extent. For re-use the face only of the form is oiled in the normal manner, taking care not to apply excessive amounts for reasons mentioned above.

(c) EMULSIONS AND MISCELLANEOUS AGENTS

Several emulsifiable oil or wax base concentrates are available. Generally these are intended for dilution with water or kerosene.

To obtain proper results the dilution proportions recommended by the manufacturer should not be exceeded. The emulsion is then applied with each re-use like a regular form oil. There are also several proprietary compounds available which react with chemicals in the concrete to form a membrane intended to permit easy stripping of forms.

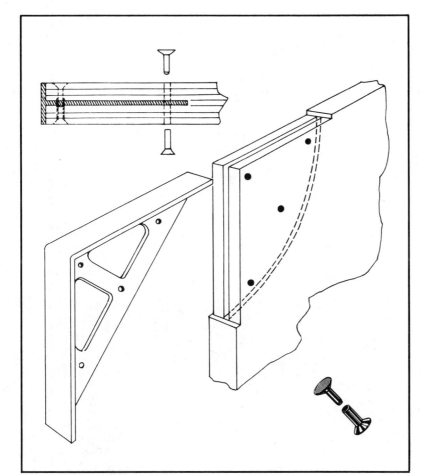

As panel corners are subjected to the hardest use, protective corner brackets should more than repay the extra initial cost in greater re-use of material and lower maintenance.

Another way to obtain longer panel life is to trim off damaged panel ends.

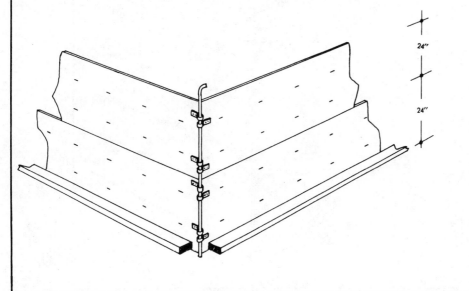

For walls 4' high: panels should be stacked horizontally two high, using top alignment waler and braces every 8'.

Use three rows of ties for a 3' wall and four rows for more than 3'.

Corners should be plumbed and braced. Top waler is placed after ties and bars have been placed.

This is another arrangement for 4′ high walls using full 4′ x 8′ panels.

Other details are identical with those in the previous drawing.

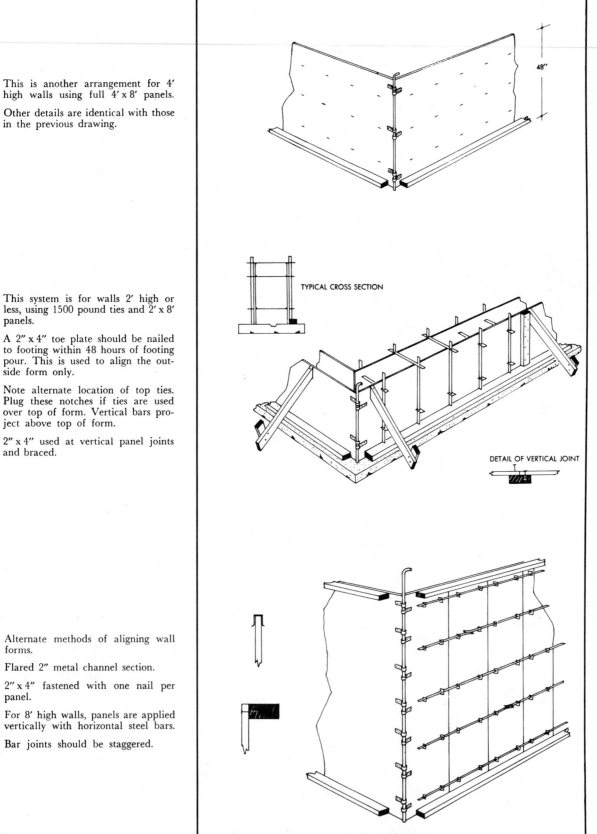

This system is for walls 2′ high or less, using 1500 pound ties and 2′ x 8′ panels.

A 2″ x 4″ toe plate should be nailed to footing within 48 hours of footing pour. This is used to align the outside form only.

Note alternate location of top ties. Plug these notches if ties are used over top of form. Vertical bars project above top of form.

2″ x 4″ used at vertical panel joints and braced.

Alternate methods of aligning wall forms.

Flared 2″ metal channel section.

2″ x 4″ fastened with one nail per panel.

For 8′ high walls, panels are applied vertically with horizontal steel bars.

Bar joints should be staggered.

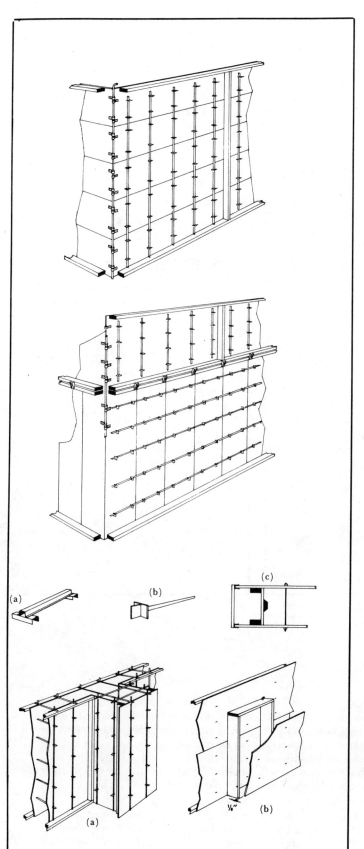

For wall heights up to 16': this system can have all horizontal panels or a combination of horizontal and vertical panels.

When panels are used horizontally, nail 2" x 4" studs at all vertical panel joints and braces.

When panels are used in combination horizontally and vertically use a double waler with standard snap ties at the break between the horizontal and the vertical.

Rod hole notched or drilled in lower form only.

(a) Channel top tie permits accurate spacing and alignment at top panel joints. Re-usable.

(b) Cleanout tool for cleaning tie slots of excess concrete after form removal.

(c) Suggested detail for wall bulkheads.

FORMING PILASTERS:

(a) For pilasters deeper than 10", ties must be used at right angles to the projections as illustrated.

(b) Leave-outs (or bucks) should be at least 1/8" narrower than finished wall dimensions to facilitate threading bars through ties.

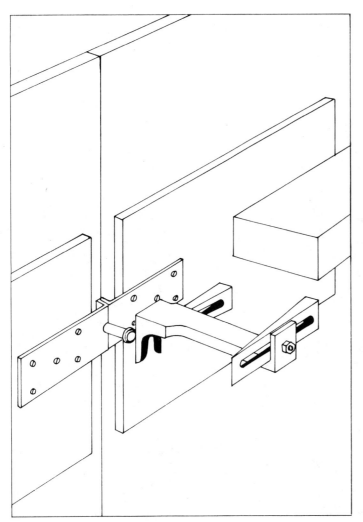

OTHER TYPES OF UNFRAMED PLYWOOD FORM PANELS

This type of panel has a combination clamp which draws panels together into a tight lock, secures the snap tie and lines up panel faces. It also provides shelf and wedge for walers.

The primary material is 1⅛" Concrete Form Grade fir plywood to which are glued and nailed ½" fir plywood plates.

Another patented form panel system using fir plywood is shown below.

Steel clamps are bolted to the plywood panels.

Panels contain a special slot which holds the snap tie.

Manufacturers say assembly process is especially applicable to stepped-up footings, pilasters, columns and beams.

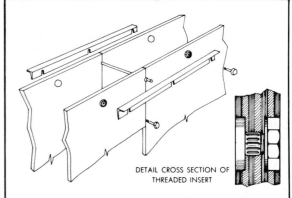

DETAIL CROSS SECTION OF THREADED INSERT

FRAMED FORMS

These forms consist of a plywood face from ½" to ¾" thick with a wood or metal frame. Plywood dimensions may be 4' x 8' or 2' x 8' with miscellaneous filler pieces. A great variety of hardware is available. Ties pass between or through the panels depending on panel size. Framed forms are widely used for all straight wall foundation work, residential and commercial; all above ground concrete work, including walls, columns, slabs, spandrel beams and joists; heavy industrial concrete work; and architectural concrete. They assist in the production of the highest quality wall obtainable, and by the nature of their construction withstand rougher use than unframed forms. The framed form system lends itself to gang forming. Fewer but stronger ties are required.

Framed forms 32" x 8' are sometimes used to effect economies under certain combinations of pour, tie and form conditions and/or module considerations. Such panels placed horizontally result in vertical tie spacing of 32". These 32" panels can be made from 2 sheets of plywood (the 2 fall-off strips are combined to make up a third form).

FABRICATING FRAMED FORMWORK

Framed form panels, for reasons of economy and quality, should be fabricated in the contractor's yard under supervision rather than under field conditions.

METAL FRAMES

Metal frames are usually attached with split rivets. Since these frames are proprietary articles no attempt has been made to discuss them in detail. A wide variety is available from various manufacturers.

WOOD FRAMES

A well equipped shop will require a cut-off saw and a rip saw—or one that will perform both these functions. A drill press and dado equipment may also be needed.

If a 2' x 8' panel is to be used, the panels should be checked for squareness after they have been cut from the full 4' x 8' sheet.

Good quality, straight lumber framing is cut to length and carefully nailed together into a complete frame. The plywood is then fastened to the frame by gluing, screws or nails.

The following drawings and photographs illustrate typical fabricated framed form panels which are in use throughout North America.

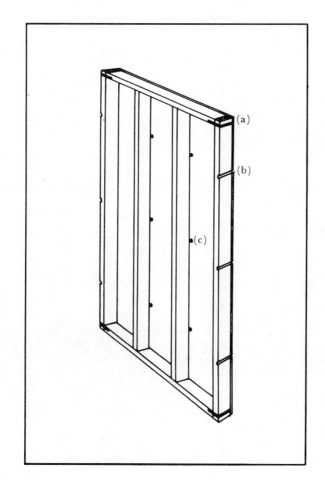

A typical prefabricated plywood form panel with ¾" fir plywood and 2" x 4" framing used by many contractors for residential and light frame construction.

(a) reinforcement straps for corners.

(b) slots for ties.

(c) tie holes.

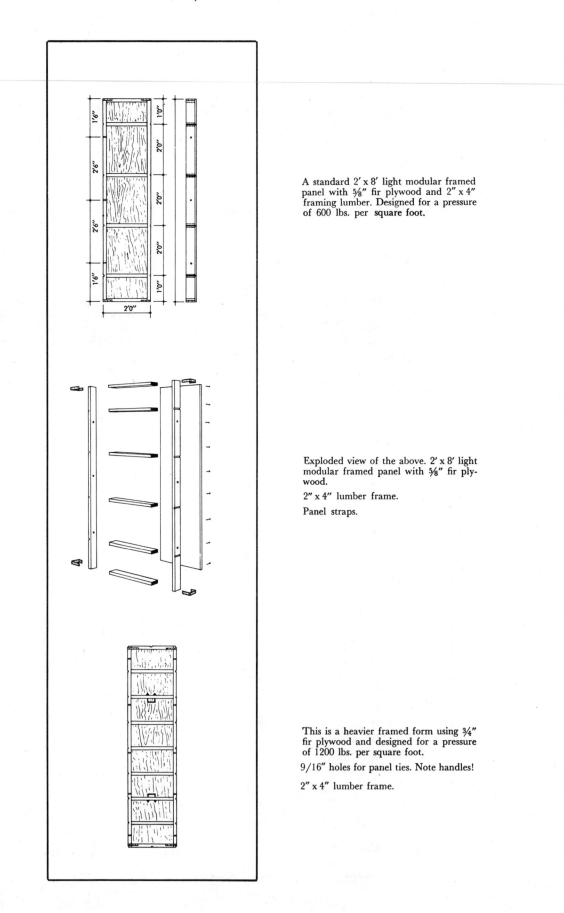

A standard 2' x 8' light modular framed panel with ⅝" fir plywood and 2" x 4" framing lumber. Designed for a pressure of 600 lbs. per square foot.

Exploded view of the above. 2' x 8' light modular framed panel with ⅝" fir plywood.

2" x 4" lumber frame.

Panel straps.

This is a heavier framed form using ¾" fir plywood and designed for a pressure of 1200 lbs. per square foot.

9/16" holes for panel ties. Note handles!

2" x 4" lumber frame.

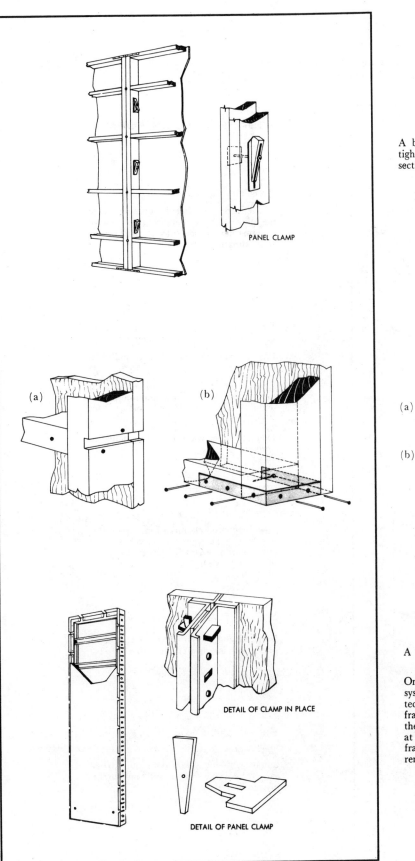

PANEL CLAMP

A bolt and wedge device to ensure a tight fit between pre-framed panel sections.

(a)

(b)

(a) Suggested nailing detail. Toe nail with one nail. Two nails through side frame into 2" x 4" member.

(b) Panel corner detail showing application of steel strap.

DETAIL OF CLAMP IN PLACE

DETAIL OF PANEL CLAMP

A typical 2' x 8' metal framed panel.

One of the advantages of these patented systems is that they offer extra protection to the edges of the plywood. The frames themselves have a long life and the plywood can be replaced or reversed at minimum cost. Initial cost of metal framed forms is high and they are rented in many instances.

COLUMNS

Several methods of forming square and rectangular columns from plywood are illustrated below.

Column sizes should, wherever possible, be selected which will permit use of standard panels or minimize cutting waste.

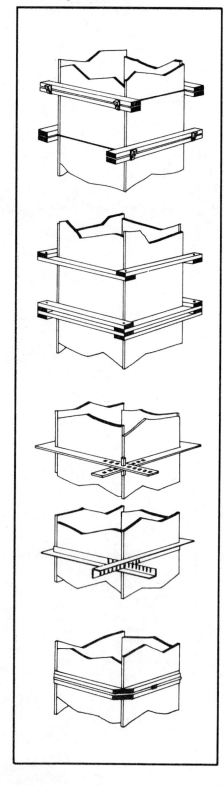

SQUARE COLUMN FORMS

Waler and tie rod.

As tie rod does not adequately brace the plywood this method is limited to columns having one side not wider than approximately 12".

For square or nearly square columns this is an alternate arrangement.

Single timber walers bolted or nailed at corners.

Strength of single waler and corner joint limits this method to a maximum face width of 16" for 2" x 4" and 20" for 2" x 6".

Double timber waler bolted or nailed at corners.

Steel column clamps. Flat bar type which may be rented or purchased for any column size and serves to brace plywood as well.

Notched angle type. May be rented or purchased for any column size.

Steel strapping.

Lumber bracing is always required because the strap alone will not brace plywood form panels.

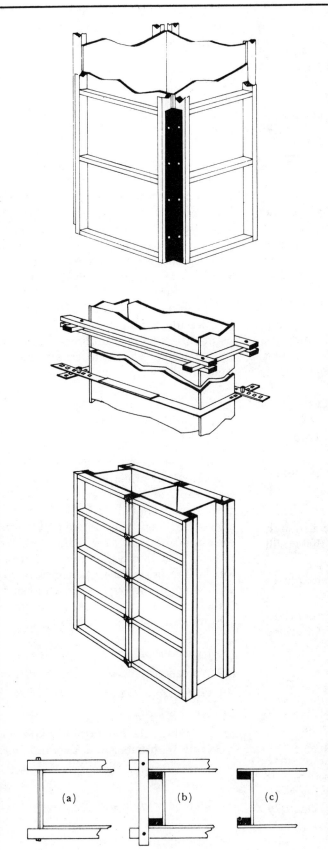

Standard wall panels. Corners are tied by angle irons.

RECTANGULAR COLUMN FORMS

Double walers can be of 2″ x 4″, 2″ x 6″, 4″ x 4″ or 2″ x 8″ lumber.

Special steel column clamps are readily available for column widths up to 6 feet.

Center section reinforced.

These standard wall panels use snap-back ties, plywood end plates and nailed vertical studs.

END PLATE DETAILS

(a) Steel bar tie.

(b) Bolted lumber cross bar.

(c) Nailed vertical studs.

Tie rod can be substituted for timber if face width is less than 12″.

8.16 AMERICAN AND CANADIAN GALLON

Note: This problem has been worked out using the standard gallon of the United States, which is the old English wine gallon containing 231 cu. in. and weighing 8.377 lbs. The Canadian Imperial gallon contains 277.418 cu. in. and weighs 10 lbs. For Canadian use, in Step 8 express $12\frac{5}{6}$ American gals. in pounds and ounces and use the same weight of water per batch.

Problem

How much water, cement, sand, and gravel will be required per batch of concrete when using a 5-gal. paste and a $1:2\frac{1}{4}:3$ mix in a 16-S concrete mixer? See 8.21.

8.17 RUNWAYS AND RAMPS

Nearly all concrete jobs require runways, ramps, lumber for screeds, and so on. Every job has its own estimating problems which may require a survey to be made of the area.

The following points should be remembered:

(a) Estimate for the lumber, nails, ties, and labor to construct runways and ramps.

(b) Allow material for screeds; this lumber cannot be reused.

(c) Allow for scaffolds.

(d) Remember that formwork must be strong enough to withstand the concrete pressures so that it will not deflect under load.

(e) Floors will have to support wet concrete, power buggies, men, and tools.

(f) Estimate whether or not any of the lumber used (a), (b), or (c) may be reused. Most specifications clearly state that all lumber on the job may be used for one purpose only.

8.18 MACHINE MIXING TIME

The machine mixing time is reckoned from about 1 min. per batch for a small-batch mixer up to 2 mins. per batch for a large unit. Opinions vary, but architects may specify the minimum mixing time for each batch. Study this mixing time very carefully indeed, since the whole operation of placing concrete is entirely dependent upon the delivery time

per cu. yd. from the batch mixer. The maximum allowable time for placing concrete is usually about 45 mins. after mixing.

8.19 WEIGHT OF CONCRETE

One cu. ft. of concrete weighs about 140 lbs. If the first floor of a building is 40'-0" x 60'-0" and is placed to a depth of 0'-6" with concrete, the weight of supported concrete would be 168,000 lbs. or 84 tons. This does not include the weight of the forms, nor of the men, nor of the tools, nor of the loaded buggies during placing. The estimator must get the "feel" of a job when estimating, in the same way that the job superintendent must be aware of all the hazards inherent in the administration of the physical work to be performed.

No amount of careful estimating can offset the inefficient running of a job. The chief ingredient of estimating and job-running is the morale of all persons engaged. The estimator and the job superintendent should have comparable job knowledge.

8.20 CURING

The object of curing is to keep the newly placed concrete from either drying out too fast or, even more important, to keep it from freezing. In the former case, it may have to be covered with polyethylene sheets; or it may be specified that it must be cured by the application of a very fine water spray on the surface for 7 to 28 days, according to the nature of the job. In the latter case, it may have to be protected with straw, burlap, tarpaulins, and so on. In all cases it requires that both labor and materials be estimated.

Some of the agents for curing are:

(a) Heat to repel freezing conditions.

(b) Water-spraying for warm conditions.

(c) Continuous water-saturated covering of sand.

(d) Wet burlap.

(e) Spraying the flat surfaces with water and covering with balsawood panels.

(f) Water-sprayed surfaces covered with cotton mats.

(g) Water-sprayed surfaces covered with polyethylene sheets.

(h) Sealing compounds.

Most specifications call for exposed surfaces of newly placed concrete to be kept moist for a minimum of 7 days; some require 14 days.

Water is applied on trowelled surfaces as soon as the concrete has set sufficiently so that the cement will not wash away. For untrowelled surfaces, water may be sprayed on as soon as the forms are removed.

The beginning estimator is strongly urged to take a course on design and control of concrete mixtures.

Gravel

After the correct proportions of cement and sand have been fed into a 3 cu. ft. mixer, it is not necessary to measure the gravel. Just fill the mixer to capacity. If the mixer is $3\frac{1}{2}$-capacity side delivery, check that it is on solid and level ground and fill to capacity, less $\frac{1}{2}$ cu. ft. measured out. Observe the quantity of the mixer filled to 3 cu. ft. and ensure that it is always filled to this amount every time. *Be certain that the mixer is always on level ground. See 8.15.*

Water

You should have one special drum for water, and this must be calibrated to gallons. *Water is the determining factor in the strength of all concrete.* If the sand is wet, it will require less water and more sand because of the free water contained in the sand, which causes bulking in volume.

Checking the Amount of Used Cement While Placing

When delivered, stack the sacks of cement on four separate piles (close together). While placing concrete, make checks of the amount used against the amount of concrete placed at intervals of quarters of work completed. This may be done by observation. Careless work can and will ruin your estimates.

Concrete Mixer

This must be in first-class shape at all times. Check the oil, gas, and belts and be sure that sufficient tools and/or spares are ready to use in the event of breakdown. Ensure that the mixer is thoroughly cleaned after use. You can make competitive estimates only with an efficient work force.

8.21 SIZES OF CONCRETE MIXERS

Some standard types of concrete mixers follow:

Type of Mixer	Cu. Ft. of Wet Concrete Capacity
3-S	3
6-S	6
11-S	11
16-S	16

Note: 3-S means side delivery, and so on.

8.22 DRY MATERIALS AND WATER PER BATCH

Problem

Assuming a concrete foundation is to be placed, using an 11-S mixer with a specified 7-gal. paste and a 1:3:5 mix, how much water, cement, sand, and gravel will be required per batch?

Example

Step 1: An 11-cu.-ft. side-delivery concrete mixer will require $1\frac{1}{2}$ times its wet delivery capacity of dry materials to be fed into it per batch.

Step 2: $11 \times 1\frac{1}{2} = 16\frac{1}{2}$ cu. ft. of dry materials to produce 11 cu. ft. of wet concrete.

Step 3: The mix is 1:3:5 with a 7-gal. paste. The total units of dry materials per batch are 9, of which cement is one.

Step 4: $16\frac{1}{2} \div 9 = 1\frac{5}{6}$ *sacks of cement per batch.*

Step 5: It will require 3 times as much sand as cement: $1\frac{5}{6} \times 3 = 5\frac{1}{2}$ *cu. ft. of dry sand per batch.*

Step 6: It will require 5 times as much gravel as cement: $1\frac{5}{6} \times 5 = 9\frac{1}{6}$ *cu. ft. of gravel per batch.*

Step 7: Check the total cu. ft. of dry volumes required per batch: $1\frac{5}{6}$ cement, $5\frac{1}{2}$ sand, and $9\frac{1}{6}$ gravel added together equal $16\frac{1}{2}$ cu. ft. of dry materials (see Step 2).

Step 8: It requires 7 gals. of water per sack of cement: $7 \times 1\frac{5}{6}$ sacks of cement equals $12\frac{5}{6}$ gals per batch.

8.23 MAKING QUALITY CONCRETE

The following material has been supplied by the Portland Cement Association and is here published by their permission.

Concrete—A Versatile Building Material

Concrete has several unusual properties that make it a versatile and widely used building material. For example, it sets or hardens in the presence of water. This property is of great importance in the construction of buildings or parts of structures, such as footings and foundations, built on the earth or in wet locations.

Freshly mixed concrete can be formed into practically any shape. Many materials can be molded only when heated and they require elaborate casting facilities, but concrete can be molded at normal temperatures, which gives it tremendous flexibility.

Since concrete is composed of mineral materials, it is impervious to bacterial attack and the resultant decay. It also resists vermin, termites and rodents. Concrete is made of materials that cannot burn. Portland cement is manufactured at temperatures up to 2,700 deg. F. Aggregates are also incombustible. Thus concrete has a high fire resistance so important in rural areas where fire protection is often inadequate.

The basic ingredients of concrete—portland cement and aggregates—are available almost everywhere at reasonable prices. Hence, concrete is economical in all parts of the country.

These many properties combine to give concrete structures and improvements low initial costs, minimum maintenance requirements and long life.

What is Concrete?

Concrete is a mixture of portland cement, water and inert materials called aggregates. Aggregates are commonly divided into two sizes, fine and coarse. Fine aggregate is sand, and coarse aggregate is usually gravel or crushed rock.

During mixing, the cement and water form a paste which coats the surface of every piece of aggregate. Usually within two to three hours the concrete starts to set due to hydration, a chemical reaction between the cement and water. As hydration continues, the cement-water paste hardens much like glue and binds the aggregate together to form the hard, durable mass that is concrete.

Thus the quality of concrete is directly related to the binding qualities of the cement paste. Concrete of various degrees of strength and durability can be produced by changing the proportion of water to cement. As the amount of water per sack of cement is reduced, the strength of the paste increases and the concrete is stronger and more durable.

Well-graded aggregate has particles of various sizes. This photo shows 1½-in. maximum size coarse aggregate. Pieces vary in size from ¼ to 1½ in.

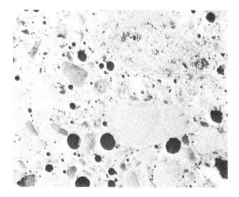

A greatly magnified section of air-entrained concrete. Black round areas are air bubbles formed by air-entraining agent. Pieces of aggregate visible are sand particles of various sizes.

The Ingredients of Quality Concrete

Portland Cement—The Magic Powder

Portland cement was patented by Joseph Aspdin, an English mason, in 1824. He called his product portland cement because it produced a grey concrete that resembled a fine building stone quarried on the Isle of Portland. In 1872 the first cement plant was built in America at Coplay, Pa., to give birth to the portland cement industry in the United States.

Cement is sold in bags and in bulk. Each bag of cement holds 1 cu.ft. and weighs 94 lbs. Many large contracting firms, block manufacturers and ready-mixed concrete producers buy cement in bulk. Large quantities are delivered by rail or truck and are quickly unloaded by automatic equipment into silos or hoppers for storage.

White Portland Cement

Most portland cement is grey in color. However, white portland cement is manufactured from raw materials that have a pure white color. It is used for paints, decorative work and other special jobs.

Mixing Water

Water to be used for concrete should be clean and free of acids, alkalis, oils, sulfates and other harmful materials. Many specifications for construction work state that the water used should be fit to drink. This is a good rule for use on all farm jobs, too.

Aggregates

Aggregates are used in concrete to provide volume at low cost. They comprise 66 to 78 per cent of the volume of concrete and can be called a filler material. Aggregates must be clean and free of materials such as dirt, clay, coal or organic matter. They must be durable, hard, and have few long, sliver-like particles.

Fine Aggregate

Natural or manufactured sands are the most common fine aggregates used in farm work. The fine aggregate should have particles ranging in size from ¼ in. down to some that will pass through a sieve having 100 openings to the inch.

Coarse Aggregate

Gravel or crushed rock are widely used coarse aggregates. The aggregate should have pieces in all sizes from ¼ in. up to the maximum size used for the job. The common maximum sizes are ½, ¾, 1, 1½ and 2 in.

Bank-Run and Commercially Recombined Aggregate

Bank-run material is aggregate that is used as taken from the quarry or gravel pit. The user should be certain that the material is clean and free from any ingredient that can harm the concrete. Many bank-run materials have a high percentage of sand, which makes it difficult to obtain a high yield of concrete per sack of cement and at the same time maintain the necessary strength. It is generally advisable to screen the bank-run material over a ¼-in. screen to separate it into fine and coarse aggregate. Some commercial firms also sell a mixed aggregate. Sand and gravel are separated and then recombined into the correct proportions for concrete.

Lightweight Aggregates

Lightweight aggregates are available in many parts of the country. They consist of cinders or expanded materials such as clay, shale or slag. They reduce the weight of concrete approximately one-third and improve the insulating qualities. Very lightweight aggregates, such as pumice or expanded mica, are also used for further weight reductions and greater insulation.

Size of Aggregate Is Important

Generally, the most economical mix is obtained by using the largest size coarse aggregate that is practical. This is usually considered to be about one-fifth of the thickness of vertically formed concrete and one-third of the thickness of flat concrete work such as floors or walks. When reinforcing steel is used, the largest size aggregate should not exceed three-fourths of the distance between bars. For example, if reinforcing steel is placed in a slab 1 in. from the bottom, the largest aggregate should be ¾ in.

Air—Another Important Ingredient

In the late 1930's it was discovered that air in the form of microscopic bubbles improves the durability of concrete. Concrete containing these air bubbles is called air-entrained concrete. Air entrainment virtually eliminates scaling due to freezing and thawing and salt action. Air-entrained concrete is more workable and cohesive, reduces segregation and bleeding and has improved sulfate resistance. Since it is more workable, it requires less mixing water—an added benefit.

Air-entrained concrete should be used where the hardened concrete is exposed to alternate freezing and thawing or to salt action. In northern areas it is recommended for practically all outside work. Because of its increased workability, it should be considered for many indoor jobs, too.

These tiny air bubbles in concrete are created by chemical compounds called air-entraining agents. These agents can be added at the time of mixing. Many cement manufacturers market portland cements that contain an air-entraining agent. Such cements are identified on the bag.

Preparing to Mix Concrete

The mixing site should be as close as practical to the point where the concrete is to be placed. The size of the mixer to use depends upon the size and type of the job and the manpower available. Concrete for large flat jobs is easy to place and a large mixer will help speed up the job. The mixer should be checked for capacity, which is usually given on an attached plate.

The aggregates should be placed so that they are convenient to the mixer operator. The cement must be stored in a dry location. If it is stored outside, it should be raised above the earth and covered with a tarpaulin, polyethylene or other suitable cover.

Ample water and containers to measure it accurately should be available.

How to Proportion Materials

The mix should be economical, and it must be workable so that it can be properly placed and finished. When the concrete has hardened, it must have strength to carry the loads, and it must be durable to resist elements to which it is exposed.

Select the Proper Water-Cement Ratio For Your Job

The strength and durability of concrete depend on the quality of the cement-water paste. To select the proper mix, refer to Table 1. First check the type of concrete work to be done to determine the gross amount of water to be used with each sack of cement. The first three columns of Table 1 give the net amount of water to use for sand of various degrees of wetness. Most sand as used on the job contains a surprising amount of water. Hence, the sand must be tested for water content and an allowance made.

For small jobs it is possible to estimate the amount of water in the sand by squeezing it in the hand. The photographs on page 6 show the appearance of damp, average wet and very wet sand.

After determining the relative moisture condition of the sand, refer to the proper column in Table 1 and find the net amount of water to use. For example, if a floor is being placed and the sand is wet (average), 5 gal. of water will be used with each sack of cement.

Table 1. Suggested mixes made with separated aggregates.

Kind of work	Gal. of water added to each 1-sack batch if sand is:			Suggested mixture for 1-sack trial batches††		
	Damp*	Wet** (average sand)	Very wet†	Cement, sacks (cu.ft.)	Aggregates	
					Fine, cu.ft.	Coarse, cu.ft.
5 gal. of water per sack of cement						
Concrete subjected to severe wear, weather, or weak acid and alkali solutions	With ¾-in. max. size aggregate					
	4½	4	3½	1	2	2¼
6 gal. of water per sack of cement						
Floors (such as home, basement, dairy barn), driveways, walks, septic tanks, storage tanks, structural beams, columns and slabs	With 1-in. max. size aggregate					
	5½	5	4¼	1	2¼	3
	With 1½-in. max. size aggregate					
	5½	5	4¼	1	2½	3½
7 gal. of water per sack of cement						
Foundation walls, footings, mass concrete, etc.	With 1½-in. max. size aggregate					
	6¼	5½	4¾	1	3	4

*Damp describes sand which will fall apart after being squeezed in the palm of the hand.
**Wet describes sand which will ball in the hand when squeezed but leaves no moisture on the palm.
†Very wet describes sand that has been subjected to a recent rain or recently pumped.
††Mix proportions will vary slightly depending on gradation of aggregates.

Mixing water must be carefully measured to obtain quality concrete. Here the mixer-operator marks correct waterline for a one-half sack batch.

Make a Trial Batch

After the net water-cement ratio has been determined, a trial batch of concrete should be made. Table 1 gives suggested proportions of cement, sand and coarse aggregate to try. These proportions are based on experience with typical aggregates. Note that the amount of water suggested is for a one-sack mix. Use proportionally smaller amounts if smaller mixers are used. Mix a batch using the correct amount of water and the proportions of other materials suggested.

Discharge a sample of the concrete from the mixer and examine it for stiffness and workability. If the mix is too wet, add a little more sand and coarse aggregate. If it is too stiff, the mix contains too much sand and coarse aggregate, and it will be necessary to reduce the amount of aggregate in subsequent batches. *Never add more water. Adjust the consistency of the mix by varying the sand and coarse aggregate.*

In some cases the mix may be too sandy or too stony. When this occurs, it is advisable to make a second trial batch and to vary the proportions of sand and coarse aggregate until the desired workability is obtained.

Concrete should be mixed at least 1 minute and preferably for 3 minutes after all materials have been placed in the mixer. There is little advantage in mixing over 3 minutes.

Compressive strength – lb. per sq. in.

Concrete* made with 6 gal. water per sack of cement

Same mix made with 7 gal. water per sack of cement

Same mix made with 8 gal. water per sack of cement

Water – gallons per sack of cement

*Type I portland cement, constant proportions of cement, sand and coarse aggregate. Moist cured at 70° for 28 days.

Fig. 1. The amount of water used per sack of cement is the key to quality concrete. Note the effect of additional mixing water on the strength of the concrete when the proportions of cement, sand and coarse aggregate are held constant.

"Damp sand" falls apart after being squeezed in the hand.

"Wet sand," which describes most sands, will form a ball when squeezed in the hand. It leaves no noticeable moisture on the palm.

"Very wet" describes sand that has been exposed to a recent rain or washing. When squeezed in the hand, it forms a ball and leaves moisture on the palm.

Placing Concrete

Concrete Requires Good Forms

The forms should be carefully set, accurately plumbed and leveled, and adequately braced. Fresh concrete exerts considerable pressure against the formwork. This pressure increases as the height of the form increases. The forms should be oiled with commercial form oil or clean motor oil before the concrete is placed so they may be easily removed.

Flat Concrete Work

The supporting soil for all flat concrete work should be adequately compacted to prevent unequal settling. Prior to placing the concrete, the earth or granular sub-base should be dampened to prevent it from drawing water from the freshly placed concrete. Concrete should never be placed on frozen earth or earth that is flooded with water.

Placing Concrete in Walls

Concrete for walls should be placed in 12- to 18-in. layers around the entire wall. The concrete should be spaded with a flat scraper or other thin-bladed tool or mechanically vibrated. This is done to eliminate a condition called "honeycombing," which occurs when coarse aggregate collects at the face of the wall. In inaccessible areas the forms can be lightly tapped with a hammer to achieve the same result.

Vibrators Are a Great Aid

Spud vibrators are excellent tools to consolidate fresh concrete in walls and other formed work. The spud vibrator is a metal tube-like device which vibrates at several thousand cycles per minute. When inserted in the concrete for 5 to 15 seconds, the spud vibrator consolidates it and improves the surfaces next to the forms.

Spud vibrators are often used to consolidate vertically formed concrete to obtain a smoother surface.

Finishing Concrete

How To Obtain the Texture That You Need

After the concrete has been placed, it is screeded or struck off with a straightedge by using a saw-like motion to level the concrete. The concrete is then sometimes floated with a long-handled float, called a bull float. This gives a more even finish than the screed. Further finishing should be delayed until the water sheen has disappeared from the surface of the concrete. Premature floating and steel troweling brings fine sand, cement and water to the surface and results in a surface with less wearability.

After the water sheen has disappeared, the concrete is floated with a wood, aluminum or magnesium float. Floating gives a gritty texture to the surface and is recommended for all areas where hogs and cattle walk. If a smoother surface is desired, the concrete should be steel troweled. Troweling is always done following the floating operation. Feed troughs and bunks should be steel troweled. Sprinkling cement on the surface of fresh concrete to absorb the water is not recommended.

A rougher texture can be obtained by drawing a broom across the surface of the concrete after floating. Pressure on the broom and the texture of the bristles determine the degree of roughness.

Newly placed concrete is first screeded or struck off with a straightedge to bring it to proper grade. One or two passes with the strike-off board is generally sufficient.

Screeding is followed with floating. Here a bull float, which is a wood or light metal float with a long handle, is being used to float the concrete.

Hand floats are often used for a final finish after screeding or bull-floating. A float finish has a gritty texture and is desirable for many farm jobs.

Where To Put Joints

Concrete expands and contracts slightly due to temperature differences. It may also shrink as it hardens. Joints should be put in the concrete to control expansion, contraction and shrinkage.

It is desirable to prevent slabs-on-ground, either inside or outside, from bonding to the building walls. Thus, the slab will be free to move with the earth. To prevent bonding, a continuous rigid waterproofed insulation strip, building paper, polyethylene or similar material, is placed next to the wall. These materials are also used next to other existing improvements, such as curbs, driveways and feeding floors. The continuous rigid, waterproof insulation strip acts as an isolation joint.

Wide areas, such as floor slabs and feeding floors, should be paved in 10- to 15-ft. wide alternate strips. A construction joint, also known as a key joint, is placed longitudinally along each side of the first strips paved. A construction joint is made by placing a beveled piece of wood on the side forms. This creates a groove in the slab edges. As the intermediate strips are paved, concrete fills this groove, and the two slabs are keyed together. This type joint keeps the slab surfaces even and transfers load from one slab to the other when equipment is driven on the slabs.

Contraction joints, often called dummy joints, are cut across each strip to control cracking. They do not extend completely through the slab, but are cut to a depth of one-fifth to one-fourth of the thickness of the slab, thus making the slab weaker at this point. If the concrete cracks due to shrinkage or thermal contraction, the crack usually occurs at this weakened section.

Contraction joints should be cut soon after the concrete has been placed in order to work the larger pieces

Brooming gives concrete a durable nonslip texture. On this job, watertight kraft paper was spread over the new concrete as soon as brooming was done.

of coarse aggregate away from the joint. A simple way to cut this type of joint is to lay a board across the fresh concrete and to cut the joint to the proper depth with a spade, axe or similar tool. A groover is then used to finish the joint. Contraction joints are generally placed 10 to 15 ft. apart on floor slabs, driveways and feeding floors. They are placed 4 to 5 ft. apart on sidewalks.

All open edges should be finished with an edger to round off the edge of the concrete slab to prevent spalling.

Since the development of special blades that can saw concrete, the practice of sawing contraction joints is becoming common. The proper time to saw joints is generally 18 to 24 hours after the concrete has been placed. A sawed joint is clean and attractive and works very well when cut to the proper depth. On larger jobs a special mobile concrete saw is used. On smaller jobs, such as sidewalks and driveways, a portable electric hand saw has been used with success. The operator must be careful to keep the saw blade straight so that it will not shatter.

Fig. 2. Joints are used to control contraction and expansion in concrete. They are easy to put in and should be used in all flat work such as floors, slabs, and walks.

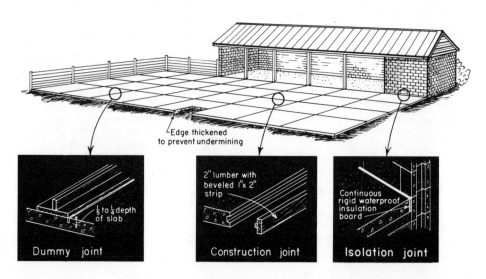

There are several ways to cure concrete. Polyethylene is being used here. It is moisture-tight and does an excellent job of preventing moisture from escaping from the concrete.

Cure Concrete for Best Results

Concrete should be cured for at least 5 days to prevent evaporation of the mixing water. If too much water escapes, there is not enough available to completely react with the cement, and the concrete does not gain the strength that it should.

Methods of Curing Concrete

Concrete can be cured by one of several methods. A common curing method is to cover the concrete with burlap, sand, straw or similar material. These materials are kept damp during the curing period. Another method is to cover the concrete with a vapor-sealing material, such as polyethylene or water-resistant kraft papers, to seal in the water and prevent evaporation. Commercial curing compounds can also be used. These are sprayed on the fresh concrete to seal the surface. In vertically formed concrete, a simple way to prevent the concrete from drying is to leave the forms in place.

Special Conditions

Cold Weather Concreting

Temperature has a considerable effect on the rate of hardening of concrete. The optimum temperature for placing concrete is assumed to be 70 deg. F. As the temperature drops, the rate of hardening slows. All new concrete must be protected from freezing. In buildings, heat is often supplied with an oil-fired stove called a salamander. The hot air should not be allowed to come into direct contact with the concrete as it will dry it out, and the area should be vented.

When concrete is mixed in cold weather, the water and aggregates are often heated. The water should not be heated to more than 150 deg. F. as overheating is likely to cause a flash set in the concrete.

High-early-strength portland cement is frequently used in winter concrete work because it sets more rapidly than normal portland cement. When normal portland cement is used, calcium chloride can be added to the mix in cold weather. It is not an antifreeze material but an accelerator that speeds up the chemical reaction between the cement and water. The quantity of calcium chloride should not exceed 2 lb. per sack of cement. It should be dissolved in the mixing water, not added in powder form to the mix. Use of calcium chloride is not a substitute for normal cold weather precautions.

Hot Weather Concreting

As the temperature rises above 70 deg., the initial rate of hardening of concrete increases. Evaporation of water from the concrete is also more rapid in hot weather. A combination of wind, high temperature, and low humidity dries water from concrete rapidly.

In extremely hot weather it may be necessary to reduce the temperature of the freshly mixed concrete. Aggregates should be stockpiled in the shade, if possible, and the coarse aggregate sprinkled with water to reduce its temperature. It is often advisable to delay placing concrete until late in the afternoon to take advantage of lower temperatures. Curing must be applied promptly to prevent the evaporation of water.

Estimating

How To Find the Amount of Concrete Needed

The unit of measure for concrete is the cubic yard, which contains 27 cu.ft. To determine the amount of concrete needed, find the volume in cubic feet of the area to be concreted and divide this figure by 27. The following formula can be used to determine the amount of concrete needed for any square or rectangular area:

$$\frac{\text{Width, ft.} \times \text{Length, ft.} \times \text{Thickness, ft.}^\circ}{27} = \text{Cubic yards}$$

For example, a 4-in. thick floor for a 30x90-ft. building would require:

$$\frac{30 \times 90 \times 0.33}{27} = 33.00 \text{ cu.yd. of concrete}$$

The amount of concrete determined by the above formula does not allow for waste or slight variations in concrete thickness. An additional 5 to 10 per cent will be needed to cover waste and other unforeseen factors.

How To Find the Amount of Materials To Order

Table 2 gives the number of sacks of cement and the amount of aggregate (in cubic feet and in pounds) needed to produce a cubic yard of concrete for water-cement ratios of 5, 6, and 7 gal. of water per sack of cement. Since it is generally impossible to recover all of the aggregate, a 10 per cent allowance has been included to cover normal wastage.

°The thickness dimension must be changed to feet or parts of a foot. The decimal part of a foot was used in the example. However, the fractional part may be used instead. Table 3 gives both the fractional and decimal parts of a foot for several common thickness dimensions.

Table 2. Materials needed per cubic yard of concrete made with separated aggregates.

Suggested mixture for 1-sack trial batches*			Materials per cu.yd. of concrete				
Cement, sacks (cu.ft.)	Aggregates		Cement, sacks	Aggregates			
	Fine, cu.ft.	Coarse, cu.ft.		Fine		Coarse	
				cu.ft.	lb.	cu.ft.	lb.
With ¾-in. maximum size aggregate							
1	2	2¼	7¾	17	1550	19½	1950
With 1-in. maximum size aggregate							
1	2¼	3	6¼	15.5	1400	21	2100
With 1½-in. maximum size aggregate							
1	2½	3½	6	16.5	1500	23	2300
With 1½-in. maximum size aggregate							
1	3	4	5	16.5	1500	22	2200

*Mix proportions will vary slightly depending on gradation of aggregate.

Table 3. How to change thickness in inches to fractions and decimal parts of a foot for use in calculating quantities of concrete.

Inches	Fractional part of foot	Decimal part of foot
4	4/12 or ⅓	0.33
5	5/12	0.42
6	6/12 or ½	0.50
7	7/12	0.58
8	8/12 or ⅔	0.67
10	10/12 or ⅚	0.83
12	1	1.00

CHAPTER 8 REVIEW QUESTIONS

1. Define bank or pit-run aggregates and give three common characteristics of this material.

2. How many cubic yards (approximately) of wet concrete could be produced from the following quantities of dry materials: 16 sacks of cement; 28 cubic feet of dry sand; and thirty-two cubic feet of gravel? Allow that the volume of one sack of cement is equal to one cubic foot.

3. Using a 6 gallon paste and a $1:2\frac{1}{4}:3$ dry mix, how much water, cement, sand, and gravel should be fed into a six cubic feet capacity concrete mixer to produce six cubic feet of wet concrete? Note: (a) A six gallon paste means using six gallons of water per sack of cement. (b) A sack of cement equals one cubic foot by volume. (c) Note the difference between the volume of one American gallon against one Canadian gallon. (d) See the pattern handwritten examples on pages 14–16.

4. A concrete floor 40'-0" x 60'-0" is to be placed with a 0'-4" depth of concrete. Using the same mix and mixer as in question three above, estimate the number of gallons of water, the number of sacks of cement, and the number of cubic yards of sand and gravel that would be required to completely place and finish the floor with concrete?

5. Sketch a wood buck for a window opening in a concrete wall.

6. Make a sketch of three different types of concrete form hardware as used in your area.

7. What should be done with lumpy cement?

8. Name five prerequisites for making quality concrete.

9. Define cement paste.

10. Which would make the most concrete (with the same slump test)—a five gallon paste or a seven gallon paste? Give reasons for your answer to this question.

9

Basement Walls, Columns, and First Floor Assembly

In this chapter we shall discuss below ground insulation, vapor barriers, walls, and floor assemblies. The drawings shown in this chapter are not exhaustive, and continuing research by builders is necessary to keep abreast of building techniques. Two books containing rich sources of information for builders are as follows:

---

1. *Wood-Frame House Construction*, U.S. Department of Agriculture, Forest Service, Agriculture Handbook No. 73. This book is prepared by the Forest Products Laboratory, U.S. Department of Agriculture, Madison, Wisconsin. It is constantly being updated, and is for sale by the Superintendent of Documents, U.S. Government Printing Office, Washington, D.C. 20402.

2. *Canadian Wood-Frame House Construction*, obtainable from any Canadian Office of the Central Mortgage and Housing Corporation. This book too is continually under revision. See the telephone directory for the address of the local office.

Both these books are authoritative and are recommended reading for house builders and contractors. Excerpts from these books are reproduced in this book with the permission of both the above-mentioned authorities.

9.1 A GLOSSARY OF UNDERPINNING MEMBER TERMS

Anchor bolts — A steel bolt deformed at one end and embedded in concrete to secure a sole plate. Steel bolts are placed at stipulated OC's (on centers) to one another.

Basement — Has one half or more of its height above ground.

Beams — Structural members supported at two or more points, but spanning unsupported distances.

I-beam — A steel beam with a cross section resembling the letter 'I'.

wood, solid — A beam of solid wood.

wood, built-up — Wood beam built up from smaller members spiked or glued together.

wood, laminated — Selected wood members glued together under controlled conditions.

Brick veneer — A facing of bricks or ceramic tiles attached to a wood or masonry wall.

Bridging — Wood or metal members used to stiffen wood floors.

Building paper	—Used in many areas of building construction, it may form a dust trap between a hardwood floor and its subfloor.
Cant strip	—A piece of wood used laterally at the base of siding to cant the first board outwards, in line with, or beyond, the rest of the siding. See Fig. 4.1 on page 63.
Caulking	—Sealing material between basement floor and wall forming a moisture barrier and termite protection.
Cavity wall	—Two separate leaves of masonry units forming a wall with a space of 2″ minimum between them and secured together with metal ties or bonding units. The leaves are called wythes or withes and may be of different sized units.
Cellar	—Has one half or more of its height below ground.
Column	—A vertical member transmitting its load to its base.
Crawl space	—Space between the lowest member of a floor and the ground beneath affording two or three feet of clearance.
Dampproofing	—Material used to render a surface impervious to the passage of damp by an emulsion or with special mortar for masonry units near ground level.
Drain tile	—Clay or concrete units 4″ in diameter and 12″ long, laid with open joints and covered with asphalt paper and placed outside footings to drain water away. See also noncorrode perforated piping in eight-foot lengths.
Fire cut	—An angular cut to a wooden member that fits into masonry walls, enabling the member to fall free in case of fire without fracturing the wall. See Fig. 9.7 on page 151.
Flashing	—Metal or other material used to shed water or exclude termites in walls; also used in roofs.
Floor sheathing	—Floor boarding, such as T & G (tongue and groove), or S.E. (Square-edged), or shiplap boards, or plywood.
Footings or Footers	—The concrete base upon which a building stands.
Frost line	—The depth of frost penetration below which footings and service lines should be placed.
Girths or Girts	—Wood members tightly fitted between wood walls as stiffeners and as fire traps. See Fig. 9.11, page 152.
Grade	—The finished level of earth around a building.
Joists	—Horizontal members supporting floors. See Fig. 9.19, page 155.
double	—Two joints together (or close together) are mandatory under partitions.
hangers	—Metal devices forming stirrups into which the ends of joists are secured.
headers	—Short joists doubled and placed at right angles to common joists in floor openings for stairwells, chimneys or other floor openings.
stringers or trimmers	Joists doubled on the long sides —of openings as for stairwells.
tail	—Short joists between headers and wall or beam.
Keyway	—Used in concrete footings to secure the walls against lateral pressures; when waterproofed, moisture penetration is controlled.
Mudsil	—Originally it was lumber placed directly on the ground over which a building was erected. The term is used where a member is secured with anchor bolts to a masonry wall, over which a wooden sole plate is secured.
Parging or Pargetting	A coat of plaster or cement mortar applied to masonry walls and inside chimneys. Used with special mixtures to waterproof basement walls.
Pedestal	—A concrete base-pad supporting a pier in house construction. The

pier is secured to the pedestal with a metal dowel against lateral pressures.

Sill sealer —A waterproof material used between a mudsil and a sole plate. See Fig. 9.1.

Steel column —A vertical member transmitting its load to its base.

Steel girder —A main beam supporting minor beams or joists.

Stepped footings —Horizontal footings stepped or toothed into sloping ground (See Fig. 7.7 on page 102).

Subfloor —The first floor applied to joists: See floor sheathing.

Vapor barrier —Material used to arrest the passage of vapor or moisture.

Water table —The depth at which ground is saturated with water.

Weep holes —Openings left in masonry construction to allow the leaching of water from an area; prevents accumulations of water from freezing and the fracturing of walls.

The following article is reproduced through the courtesy of the Forest Products Laboratory, U.S. Department of Agriculture, Madison, Wisconsin.

9.2 INSULATION REQUIREMENTS FOR CONCRETE FLOOR SLABS ON GROUND

The use of perimeter insulation for slabs is necessary to prevent heat loss and cold floors during the heating season, except in warm climates. The proper locations for this insulation under several conditions are shown in Figs. 9.1, 9.2, 9.3.

The thickness of the insulation will depend upon requirements of the climate and upon the materials used. Some insulations have more than twice the insulating values of others. The resistance (R) per inch of thickness, as well as the heating design temperature, should govern the amount required. Perhaps two good general rules to follow are:

1. For average winter low temperatures of 0°F. and higher (moderate climates), the total R should be about 2.0 and the depth of the insulation or the width under the slab not less than 1 foot.

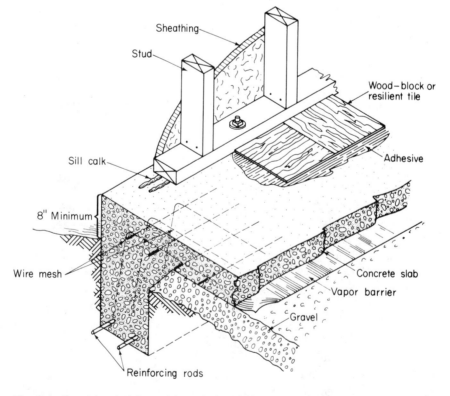

Fig. 9.1. Combined slab and foundation (thickened edge slab). *(Courtesy of the Forest Products Laboratory, U.S. Department of Agriculture.)*

2. For average winter low temperatures of −20°F. and lower (cold climates), the total *R* should be about 3.0 without floor heating and the depth or width of insulation not less than 2 feet.

Table 9.1 shows these factors in more detail. The values shown are minimum and any increase in insulation will result in lower heat losses.

Table 9.1 Resistance values used in determining minimum amount of edge insulation for concrete floor slabs on ground for various design temperatures.

Low temperatures	Depth insulation extends below grade	Resistance (R) factor	
		No floor heating	Floor heating
°F.	Ft.		
−20	2	3.0	4.0
−20	1½	2.5	3.5
0	1	2.0	3.0
+10	1	2.0	3.0
+20	1	2.0	3.0

Insulation Types

The properties desired in insulation for floor slabs are: 1) High resistance to heat transmission, 2) permanent durability when exposed to dampness and frost, and 3) high resistance to crushing due to floor loads, weight of slab, or expansion forces. The slab should also be immune to fungus and insect attack,

and should not absorb or retain moisture. Examples of materials considered to have these properties are:

1. *Cellular-glass insulation board,* available in slabs 2, 3, 4, and 5 inches thick. *R* factor, or resistivity, 1.8 to 2.2 per inch of thickness. Crushing strength, approximately 150 pounds per square inch. Easily cut and worked. The surface may spall (chip or crumble away if subjected to moisture and freezing. It should be dipped in roofing pitch or asphalt for protection. Insulation should be located above or inside the vapor barrier for protection from moisture. This type of insulation has been replaced to a large extent by the newer foamed plastics such as polystyrene and polyurethane.

2. *Glass fibers with plastic binder,* coated or uncoated, available in thicknesses of ¾, 1, 1½, and 2 inches. *R* factor, 3.3 to 3.9 per inch of thickness. Crushing strength, about 12 pounds per square inch. Water penetration into coated board is slow and inconsequential unless the board is exposed to a constant head of water, in which case this water may disintegrate the binder. Use a coated board or apply coal-tar pitch or asphalt to uncoated board. Coat all edges. Follow manufacturer's instructions for cutting. Placement of the insulation inside the vapor barrier will afford some protection.

3. *Foamed plastic* (polystyrene, polyurethane, and others) insulation in sheet form, usually available in thicknesses of ½, 1, 1½, and 2 inches. At normal

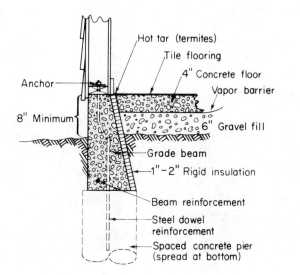

Fig. 9.2. Reinforced grade beam for concrete slab. Beam spans concrete piers located below frostline. *(Courtesy of the Forest Products Laboratory, U.S. Department of Agriculture.)*

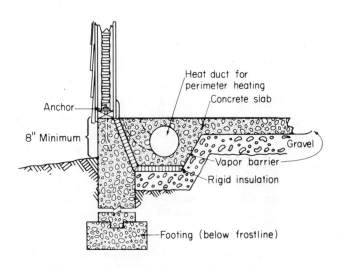

Fig. 9.3. Full foundation wall for cold climates. Perimeter heat duct insulated to reduce heat loss. *(Courtesy of Forest Products Laboratory, U.S. Department of Agriculture.)*

temperatures the R factor varies from 3.7 for polystyrenes to over 6.0 for polyurethane for a 1-inch thickness. These materials generally have low water-vapor transmission rates. Some are low in crushing strength and perhaps are best used in a vertical position and not under the slab where crushing could occur.

4. *Insulating concrete.* Expanded mica aggregate, 1 part cement to 6 parts aggregate, thickness used as required. *R* factor, about 1.1 per inch of thickness. Crushing strength, adequate. It may take up moisture when subject to dampness, and consequently its use should be limited to locations where there will be no contact with moisture from any source.

5. *Concrete made with lightweight aggregate,* such as expanded slag, burned clay, or pumice, using 1 part cement to 4 parts aggregate; thickness used as required. *R* factor, about 0.40 per inch of thickness. Crushing strength, high. This lightweight aggregate may also be used for foundation walls in place of stone or gravel aggregate.

Under service conditions there are two sources of moisture that might affect insulating materials: 1) vapor from inside the house and 2) moisture from soil. Vapor barriers and coatings may retard but not entirely prevent the penetration of moisture into the insulation. Dampness may reduce the crushing strength of insulation, which in turn may permit the edge of the slab to settle. Compression of the insulation, moreover, reduces its efficiency. Insulating materials should perform satisfactorily in any position if they do not change dimensions and if they are kept dry.

Protection Against Termites

In areas where termites are a problem, certain precautions are necessary for concrete slab floors on the ground. Leave a countersink-type opening 1-inch wide and 1-inch deep around plumbing pipes where they pass through the slab, and fill the opening with hot tar when the pipe is in place. Where insulation is used between the slab and the foundation wall, the insulation should be kept 1 inch below the top of the slab and the space should also be filled with hot tar.

9.3 GROUND SLAB REINFORCING, GRADE BEAMS AND FOUNDATION WALLS

The factors for determining the type of underpinning for residential construction are local ground conditions, the climate, and the local building code. At Fig. 9.1 is shown a combined slab and foundation which is reinforced with wire rods and wire mesh. Note the caulking, the anchor bolt, and the vapor barrier. For a definition of wire mesh see article 7.9 page 103.

The grade beam at Fig. 9.2 shows a method of providing a relatively inexpensive method of underpinning in areas where it would be too costly to excavate completely. The piers go down to solid bearing, through the frost line, water table, filled ground, and so on. They can be used for building on the brink of steep hills, thereby using land to its utmost value for spectacular views. See Fig. 7.8 page 102.

The full foundation wall for cold climates shown at Fig. 9.3 gives an excellent example of a method of avoiding creeping cold perimeters by the introduction of heat into the perimeter duct.

At Fig. 9.4 is shown a floor-joists system supported on a ledge formed at the top of the concrete wall. Note the brick veneer facing wall with an air space to the framed wall. Such veneer walls are tied to the framing with metal ties.

9.4 JOIST ENDS EMBEDDED IN CONCRETE; BALLOON FRAMING SILL

In cold climates many houses are built with a full eight feet of clearance for basements. This area affords excellent facilities for the installation of the heating system, laundry room, games room, cold storage room, workshop, and so on. During frigid cold spells such basements are a boon to parents with young children. Check with the local authority for any special building specifications for habitable basement rooms, and be aware that local taxes are greater for houses with such rooms than for those without.

In many areas it is the custom to embed (in wet concrete) the ends of treated joists at the top of the basement wall. It will be seen from Fig. 9.5 that the header and end joist (rim joists) will contain the wet concrete on the outside of the wall, but it will require formwork cut in between the joists to form the top of the concrete on the inside.

Note carefully, too, that *the concrete is never filled to the top of the joists.* If this were done, the joists would absorb water and then shrink, leaving the concrete above the surface of the joists. *This is important.* The space between the top of the concrete wall and the top of the joists is later filled with insulation, this also is important.

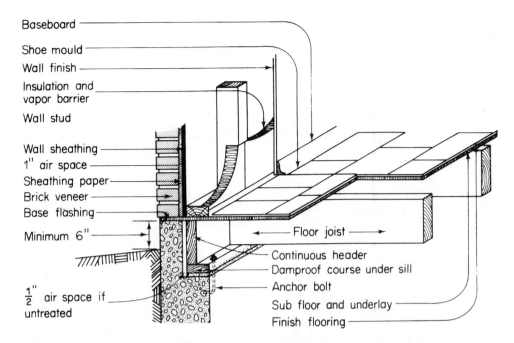

Baseboard
Shoe mould
Wall finish
Insulation and vapor barrier
Wall stud
Wall sheathing
1" air space
Sheathing paper
Brick veneer
Base flashing
Minimum 6"
$\frac{1}{2}$" air space if untreated

Floor joist
Continuous header
Damproof course under sill
Anchor bolt
Sub floor and underlay
Finish flooring

Fig. 9.4. Floor joists are supported on ledge formed in foundation wall. Joists are toe-nailed to header and sill plate. Masonry veneer supported on top of foundation wall. Wall framing supported on top of the subfloor. *(Courtesy of the Department of Forestry and Rural Development and the Central Mortgage and Housing Corporation, Ottawa.)*

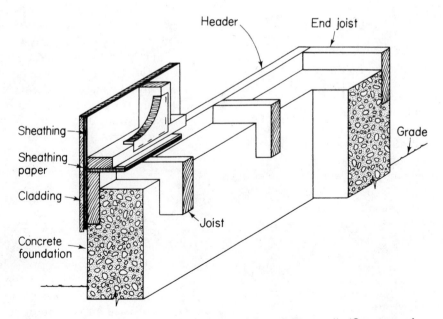

Header
End joist
Sheathing
Sheathing paper
Cladding
Joist
Concrete foundation
Grade

Fig. 9.5. Floor joists embedded in top of foundation wall. *(Courtesy of the Department of Forestry and Rural Development and the Central Mortgage and Housing Corporation, Ottawa.)*

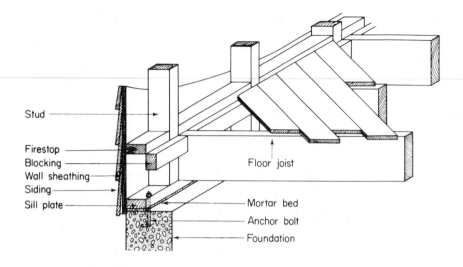

Stud

Firestop
Blocking
Wall sheathing
Siding
Sill plate

Floor joist

Mortar bed
Anchor bolt
Foundation

Fig. 9.6. Type of sill used in balloon-frame construction. *(Courtesy of the Department of Forestry and Rural Development and the Central Mortgage and Housing Corporation, Ottawa.)*

The balloon-frame sill construction shown at Fig. 9.6 should be closely studied. A great advantage of this method is in the reduction of settlement shrinkage (especially in construction of several floors). Note that the stud runs down to the sill plate, and as there is virtually no shrinkage in the wood lengthway of the grain, the shrinkage here is reduced to that of the sill plate.

In the floor assembly construction, as shown at Fig. 4.1, page 63, the shrinkage would take place in the following members: sill plate; header joists; subfloor; and the sole plate of the wall, in all, a possible shrinkage (in depth) of say 1½", against a possible shrinkage of only ⅛" in the balloon framing method.

9.5 TYPES OF CENTRALLY LOCATED SUPPORTS FOR FLOOR ASSEMBLIES

There are many methods of supporting first floor joists above a basement or cellar floor.

Note: An accepted definition of a basement is that it is more than half way out of the ground; a cellar is more than half way in the ground.

The first floor joists may be supported on masonry walls and:

(a) Brick or masonry columns supporting a wood or steel beam.

(b) Monolithic columns supporting a steel beam.

(c) Standard steel pipes supporting wood or steel beams.

(d) Telescopic metal posts supporting wood or steel beam.

(e) Lally posts, which are cylinders filled with concrete (named after their originator).

(f) A wood frame partition directly supporting wood joists without a beam.

(g) Concrete block wall with a wall plate.

Main wood beams should be metal strapped or secured by angle irons to the solid wood posts supporting them. Wood posts may also be built up by spiking together 2 x 8's or 2 x 10's and then finishing them off with a decorative plywood veneer.

Note the fire cut in the main beam, (Fig. 9.7). Such cuts are made so that in case of fire the beam may fall free without fracturing the wall. This is especially true for wood beams at their intersections with masonry walls in upper floors. Instead of a fire cut, wood beams may be anchored in metal stirrups.

Note the arrangement of the floor joists with a header on the outside of the wall, and a joist placed immediately inside the concrete wall. This is done to provide nailing for ceiling on the underside of the joists in a basement.

Standard sized 4'-0" x 8'-0" plywood subfloors may be laid in the usual manner but with a filler starter strip of plywood from the header to the first floor joist, shown at the top of the fire cut.

The disadvantage of this floor assembly is that the beam projects below the underside of the joists, a feature avoided in Figures 9.8 through 9.10.

Ledgers spiked to the sides of wood beams as at

Fig. 9.8 is a satisfactory method of constructing a floor assembly without a main beam. A further extension of this system is shown at Fig. 9.9 where the joists are secured to the built up beam with metal joist hangers (stirrups). This method prevents the ends of the joists from twisting while they are drying.

Another method of supporting wooden joists is shown at Fig. 9.10 where a metal column supports an "I"-beam; this method also gives clear headroom in the basement below, without a beam showing.

An acceptable method of supporting floor joists is shown at Fig. 9.11. With careful planning, a 2 x 6 wood framed wall may be erected to carry the joists (without a main beam) at any desirable place in the basement. Such a wall may be constructed with openings for doors, heating outlets, bathroom cabi-

nets, electric outlets, and so on. Thought must be given to the kind of finish that will be applied to the wall; for decorative fiber boards, the spacing of the studs should be arranged to minimize waste. Note the pier pad (upon which stands the wall) which is raised above the level of the floor to prevent damp penetration to the sole plate (bottom plate) of the wood framed wall.

Floors supported on telescopic legs as at Fig. 9.12 have the advantage of being adjustable, and may be raised or lowered according to the settlement of the building.

At Fig. 9.13 is shown a concrete block wall with a top wall plate and its anchor bolt. With careful planning, decorative colored concrete blocks (on one side or both) may be used in the construction of such

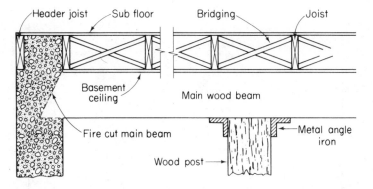

Fig. 9.7. Main wood beam supported on solid wood post.

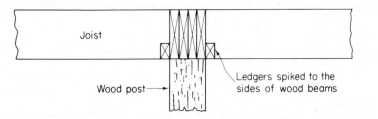

Fig. 9.8. Main wood beam with ledgers, supported on wood post.

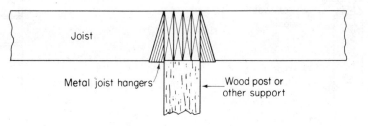

Fig. 9.9. Metal joist hangers attached to wood beam.

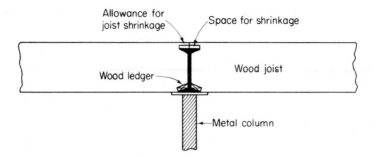

Fig. 9.10. Metal column supporting steel beam.

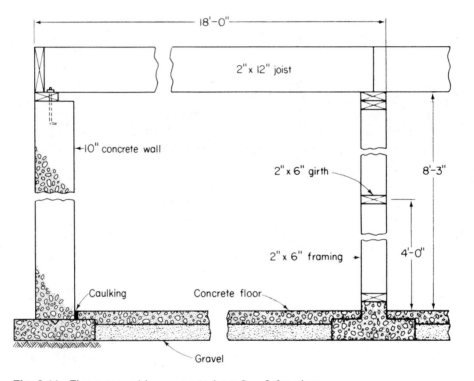

Fig. 9.11. Floor assembly supported on 2 x 6 framing.

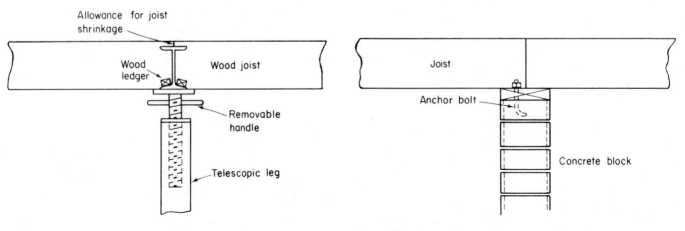

Fig. 9.12. Floor assembly supported on telescopic legs.

Fig. 9.13. Floor assembly supported on concrete block wall.

walls to form a pleasing design in the basement rooms. See Chapter 12. Here also there is no obstruction by a beam of any kind.

9.6 BRIDGING FOR FLOOR JOISTS

Three types of cross bridging are shown at Fig. 9.14. Read your local building code for maximum allowable joist spans without bridging. See also Fig. 9.19 on page 155 for in-place cross bridging.

Solid Bridging has the disadvantage of the possible shrinkage of the floor joists in thickness. Assume that a row of solid bridging is placed between sixteen consecutive joists, the total thickness of joists subject to shrinkage is $17 \times 2 = 34$ inches nominal. This could total one inch or more of shrinkage and loosen the bridgings.

When Cross Bridging is Used, the bottoms may be left without nailing until the shrinkage has taken place; later the bridgings may be hammered and nailed into place.

When Metal Strip Bridging is Used, the shrinkage of the joists may cause the metal stripping to become progressively looser.

9.7 JOIST SUPPORTS OVER WOOD BEAMS

Three methods of supporting floor joists and securing their ends from twisting are shown at Figures 9.15, 9.16, and 9.17. It is important to remember that all joists should be placed to give the maximum rigidity to the floor and to allow for the minimum waste in subfloor material.

If tongue and groove plywood in sheets of four feet by eight feet sizes are used, it is imperative to know exactly what area these sheets will actually cover. See article 9.11. If they are machine cut into exact sizes of four by eight pieces *and then tongued and grooved,* the actual floor coverage will be less and a narrower setting of the joists will be necessary. Manufacturers will produce plywood in sizes to order. *Check and be careful!*

9.8 IN-LINE JOIST SYSTEM AND FLOOR FRAMING

At Fig. 9.18 is shown the joist arrangement for a simple span. The joists should be accurately set out on each wall plate, and nailed to the header. A mason's line should then be stretched alongside each in-line joist, and the joists nailed to the center beam true to line.

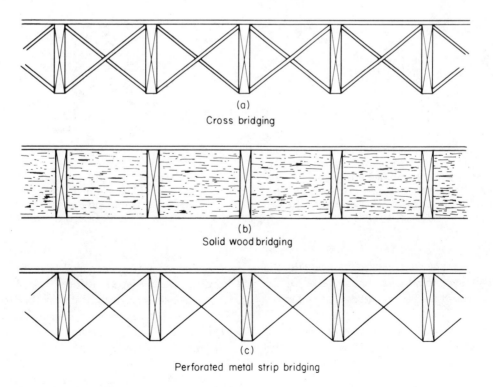

(a)
Cross bridging

(b)
Solid wood bridging

(c)
Perforated metal strip bridging

Fig. 9.14. Bridging for floor joists.

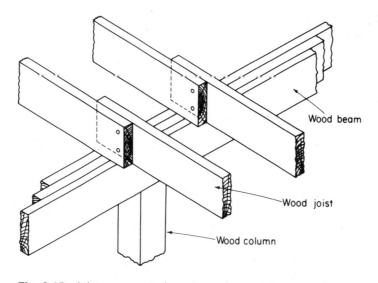

Fig. 9.15. Joists supported on top of wood beam and fastened to the beam by toe-nailing. Two $3\frac{1}{4}$ inch nails used for each joist. *(Courtesy of the Department of Forestry and Rural Development and the Central Mortgage and Housing Corporation, Ottawa.)*

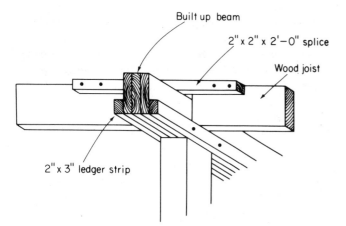

Fig. 9.16. Joists supported on ledger strip nailed to beam with two $3\frac{1}{4}$ inch nails per joist. Splice nailed to joist with two $3\frac{1}{4}$ inch nails at each end. *(Courtesy of the Department of Forestry and Rural Development and the Central Mortgage and Housing Corporation, Ottawa.)*

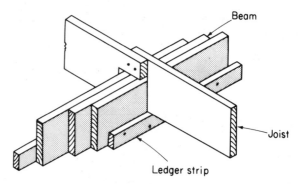

Fig. 9.17. Joists supported on ledger strip nailed to beam with two $3\frac{1}{4}$ inch nails per joist. Joists lapped and nailed together with two $3\frac{1}{4}$ inch nails. *(Courtesy of the Department of Forestry and Rural Development and the Central Mortgage and Housing Corporation, Ottawa.)*

It cannot be emphasized enough that all floor and wall framing systems should be checked thoroughly for squareness before completing the nailing. Use the 30: 40:50 system of triangulation as described in articles 6.1 and 6.2.

Study the floor assembly as shown at Fig. 9.19, and note the two systems of subflooring, plywood or diagonally boarded. The diagonally placed floorboards stiffen the whole assembly and enable other coverings such as hardwood flooring to be nailed at

any point on its surface. Floorboards placed diagonally also avoid the possibility of the edges of some longitudinally placed subfloor boards falling directly under the longitudinally placed hardwood flooring. *Remember that hardwood floors should be especially well nailed at doorways and near windows and wherever the line of traffic may be; this will prevent squeaky floors.*

Plywood subfloors are suitable for the direct application of an underfelt for wall to wall carpeting,

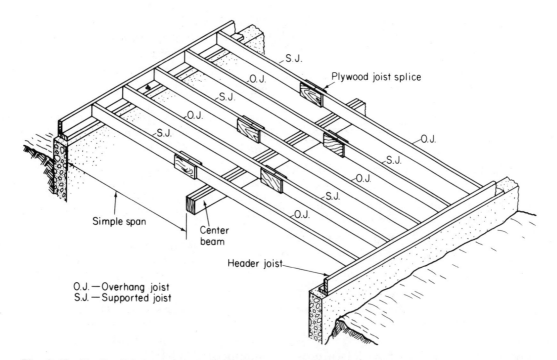

Fig. 9.18. "In-line" joist system. Alternate extension of joists over the center support with plywood gusset joint allows the use of a smaller joist size. *(Courtesy of the Forestry Products Laboratory, U.S. Department of Agriculture.)*

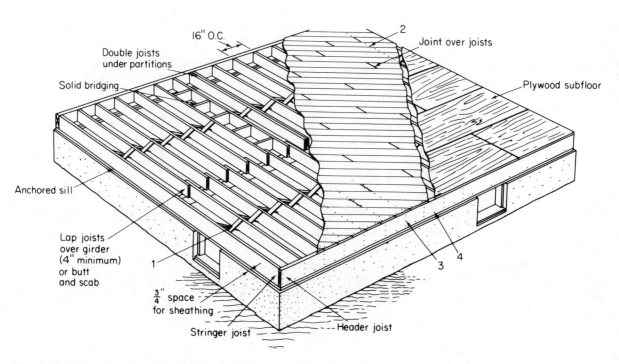

Fig. 9.19. Floor framing: (1) Nailing bridging to joists, (2) nailing board subfloor to joists, (3) nailing header to joists, and (4) toe-nailing header to sill. *(Courtesy of the Forestry Products Laboratory, U.S. Department of Agriculture.)*

or for the application of resilient or other floor coverings.

9.9 JOIST LAYOUT, 16″ ON CENTERS

It is important to select floor joists for their respective positions in the floor assembly as follows:

(a) Select the straightest joists for placing under door openings.
(b) Select straight joists for making built-up (spiked together) wood beams.
(c) Place the crown side (round side) of joists, headers, ceiling and roof members uppermost.
(d) When looking for the crown side of lumber, be sure to hold the width horizontal. If it is held in the opposite direction, its own weight may cause it to appear to have the crown side down.

When making the layout for the floor joists, proceed as follows. See Fig. 9.20.

Step 1: Nail the long header to the short one.

Step 2: From the left hand end of the header, lay off 16″ to the center line of No. 1 joist, as shown at Fig. 9.20.

Step 3: Lay off on either side of the ₵, half the thickness of the joist. **This is important!**

Step 4: Lay off 16″ from the right hand side of No. 1 joist to the right hand side of No. 2 joist. Mark this on the header with a square line and ticked to the left as shown on the drawing. Nail the right hand side of the joist to the square line.

Step 5: Lay off 16″ from the right hand side of No. 2 joist to the outside edge of No. 3 joist and mark it on the header similar to that for No. 2.

Step 6: It will be seen that from the outside edge of the header to the ₵ of joist No. 3, it is exactly 4′-0″ which is the standard size of plywood subflooring.

Step 7: To avoid 'creeping' by measuring in short 16″ steps it is better to run out a tape from the right hand mark of Joist No. 1 (not from the header), and mark 4′-0″, 8′-0″, and 16′-0″ spacings, and so on. These measures establish the certainty of the modular 4′-0″ spacings for receiving the plywood subfloor. From these spacings, lay off the 16″ ticked marks for the intermediate joists.

9.10 FRAMING FLOOR OPENINGS

Figure 9.21 illustrates the framing system for floor openings. Remember that openings in floors for stairwells and so on should in no way alter the basic layout of 16″ OCs for the joists. The same thing applies for framed walls. You should imagine the floor (or wall) as being completely covered with sheathing,

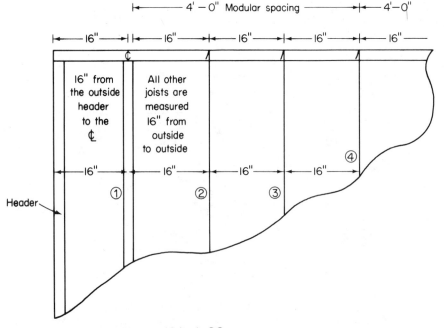

Fig. 9.20. Floor joist layout, 16 inch OCs.

and then visualize how the openings would be cut into the framing with double members surrounding the openings. *Repeat: never alter the pattern of OC joists to suit a floor opening; make the opening as if it were to 'guillotine' itself into the floor assembly.*

It is important to remember to make framed openings large enough to receive the finished stair, window or door.

Inside Chimney Openings must be large enough so that no combustible material shall be less than 2" from the masonry of the chimney. Floor openings for laundry chutes, and so on, must be large enough to accommodate any internal trim; they must not impede heating, plumbing, or air conditioning lines.

Stairwell Openings must be checked against the drawings, and, also, against the physical possibilities of installing the staircase as shown on the drawings. *Drawings are not always correct!* Assume that part of the specifications for a stairway are as follows:

The height of the stair from the finish of one floor to the finish of the one immediately above shall be 9'-0". The stair shall have fifteen equal risers and fourteen treads. The stair shall have a minimum perpendicular headroom from the angle of flight of 6'-4". Each riser shall be 7.2" (seven decimal two inches), and the treads shall be 9.8" *excluding the width of the tread nosings.* The upper floor assembly shall be 12" from finished ceiling to finished floor.

To check the stair well opening, take a piece of plywood about two feet square, and using a scale of $1\frac{1}{2}$" to 1'-0", make a sectional drawing of the staircase.

Note carefully that the scale of $1\frac{1}{2}$" to 1'-0", is ideal for on the job drawings using a carpenter's pencil, tape and framing square; where one eighth of an inch on scale represents one inch actual; and one and a half inches on scale represents one foot actual. Proceed as follows:

Step 1: Near the top of the plywood draw two faint parallel lines 12" apart and of indefinite length representing the floor assembly as at a-b and c-d in Fig. 9.22.

Step 2: From -a-, project a perpendicular line 9'-0" in length to -e- to intersect with the lower finished floor line e-f.

Step 3: Read the drawings and specifications and multiply the width of one tread by the total number of treads, thus $9.8 \times 14 = 137$" or 11'-5" nearly.

Step 4: On line e-f, from -e- measure 11'-5" to -g-, and project the perpendicular line to the first riser 7.2 to -h-.

Step 5: From -h-, project a line (angle of flight, or slope of the stair) to -a-. *It is important to remember that the line of flight starts at the top of the first riser or step to the upper floor and not from the lower floor level.* **This is very important.**

Step 6: At right angles to the *underside* of the ceiling line, project a perpendicular line 6'-4" long to intersect with the line of flight as at i-j. This is the headroom allowance. Many jurisdictions state that the

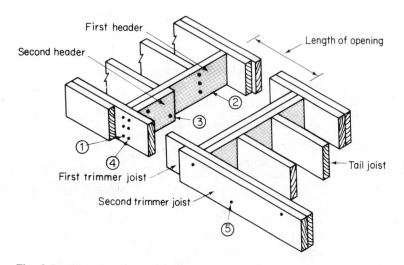

Fig. 9.21. Framing for floor openings where double headers and double trimmers are used. *(Courtesy of the Department of Forestry and Rural Development and the Central Mortgage and Housing Corporation, Ottawa.)*

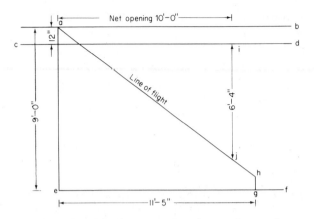

Fig. 9.22. Establishing a stairwell opening: a-b, upper floor finish; c-d, total thickness of floor assembly; e-f, lower floor level; g-h, height of the first riser; h-a, line of flight of the stairs; and i-j, minimum headroom clearance.

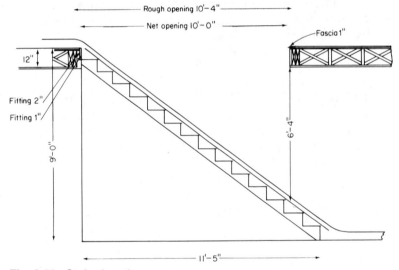

Fig. 9.23. Stair elevation.

minimum headroom clearance for houses shall not be less than 6'-4", and for all others not less than 6'-9".

Step 7: The *net* length of the **stairwell opening** is the horizontal distance between the perpendiculars a-e and i-j, which is 10'-0". Additionally, an allowance must be made for fitting the stair at its intersection with the upper floor and also for the fascia board (trim) at the other end of the opening.

Step 8: To the net length of the opening add (say) 3" for the fitting at the upper floor level, plus 1" for the fascia board at the other end, making a total of four inches. See Fig. 9.23.

Step 9: The total length of the opening will be 10'-0" in all to allow for the thickness of wall finishes. To

determine the width of the stairwell remember the thickness of the wall finish. For a fuller treatise on stairs and steps see Chapter 16.

9.11 TONGUE AND GROOVED, AND FIELD-GLUED SUBFLOORS

The following material has been excerpted from two publications titled: "Tongue and Groove Plywood" and "Field-Glued Plywood Subfloors", respectively, published by the Council of Forest Industries of British Columbia, and is here reproduced with the permission of the copyright holders.

TONGUE & GROOVE PLYWOOD

Tongue and Groove (T&G) plywood is a Sheathing or Select Sheathing grade plywood panel with a factory-machined tongue along one of the long edges and a groove along the other. T&G plywood is usually manufactured from a standard 48 by 96 in. panel blank in thicknesses of ½ in., ⅝ in., and ¾ in., although net 48 in. wide panels and other grades, sizes, and thicknesses are available on special order.

All T&G plywood manufactured by the member mills of the Plywood Manufacturers of B.C. conforms to the same thoroughly tested and approved specifications developed by PMBC. Industry-wide uniformity of the T&G profile is assured by the use of a standard test gauge thus permitting the matching of panels from different mills. T&G Douglas fir and western softwood plywoods bearing the industry-mark PMBC EXTERIOR are bonded with waterproof glue and manufactured in accordance with CSA Specifications 0121 or 0151.

STANDARD T&G PROFILE

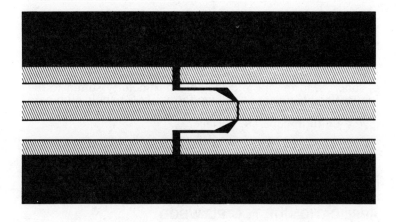

DECKING

Large, lightweight T&G plywood panels can be easily handled and quickly applied over joists or rafters. In addition to eliminating costly blocking at panel edges, T&G plywood panels save on nails and nailing time. Time studies have shown that a typical floor or roof area can be sheathed with T&G panels in approximately two-thirds the time usually required for square-edge panels and edge blocking – and fewer nails are required.

T&G plywood panels interlock to ensure the effective transmission of loads across panel joints, eliminating differential deflection between panel edges and providing a strong, uniform deck suitable for any finish flooring or roofing material.

COMBINED SUBFLOORING AND UNDERLAYMENT

T&G plywood can be applied directly over floor joists to form a combined subfloor and underlayment. This economical method of floor construction saves labour and materials by eliminating the need for a separately-installed underlayment. The smooth uniform surface is ideal for laying tile, linoleum and carpet. Recommended plywood thicknesses and nailing schedules are shown in the table on page 4. Annularly grooved nails should be used in preference to common nails for combined subflooring and underlayment.

WALL CLADDING

One-half inch thick T&G plywood may be used as wall cladding on vertical supports spaced up to 48 in. oc. Plywood should be placed with face grain perpendicular to supports and nailed with 2 in. common nails spaced at intervals not exceeding 6 in. along all supporting members. This application of T&G plywood is suitable as a single-finish combined wall sheathing and cladding on industrial, agricultural and similar utility buildings provided all square edge joints are caulked or covered and corrosion-resistant nails used to fasten the T&G plywood to the supporting members.

CONCRETE FORMING

Large T&G panels can be used in the construction of all types of concrete formwork and they are particularly suited to the fabrication of gang or panelized forms. T&G panels eliminate differential deflection between adjoining panel edges and give a uniform concrete surface.

INSTALLATION NOTES

APPLICATION OF T&G PLYWOOD

It is advisable to plan placement of T&G panels to minimize cutting, waste, and labour. T&G panels should be placed with the face grain at right angles to the supports. End joints which must occur over supports should be staggered. When placing the tongued edge of a panel into the groove of a fastened panel care should be taken to ensure that tongues and grooves are not damaged.

The recommended application sequence for T&G plywood is:

1. It is important that the first row of panels be aligned straight and true. To ensure this, measure 48 inches in from the ends of the two outside joists or rafters and snap a chalk line between these points.

2. Lay the first row of T&G plywood panels with the grooved edge along the chalk line and tack in position with four to six nails per panel.

   NOTE: Each T&G panel is stamped 'THIS SIDE DOWN' to ensure correct installation.

3. Place a second row of T&G plywood panels along the other side of the chalkline with the tongues fitting into the grooves of the panels in the first row and having the end joints between plywood panels staggered relative to the first row. Do not attempt to force a tight joint between panel faces because the T&G joint is designed to butt at the tip of the tongue leaving a $\frac{1}{32}$ inch gap on the face and underside.

4. Sheathe the remaining floor or roof area, tacking all panels in place. If the T&G plywood is being used as roof sheathing or subflooring, the panels can be nailed securely to the rafters or joists at this stage. If the plywood is being used for combined subflooring and underlayment as a base for resilient flooring, panels should be left until the finish flooring is to be laid, at which time all nailing should be completed and the nail heads set.

5. Protect T&G plywood combined subflooring and under-layment from excessive moisture to obtain best results with resilient flooring.

NAILING

Nailing schedules for T&G plywood are shown in the application tables on page 4. To achieve the maximum possible withdrawal strength and resistance to nail popping, annularly grooved nails should be used when laying floor decking. T&G plywood can also be fastened to supporting members with power-driven nails or staples.

HINTS FOR REDUCING NAIL POPPING

1. Select joists so that members with similar crowns will be placed side by side.

2. Use dry lumber whenever possible. If green lumber is used do not fasten T&G plywood permanently until the joists have attained equilibrium moisture content. The optimum moisture content is 10-12%.

3. Use the shortest possible annularly grooved nails consistent with the recommendations given on page 4. Drive nails at a slight angle and set all nail heads just prior to laying resilient flooring.

PREPARING SURFACE FOR FINISH FLOORING

To obtain the best results with light gauge resilient flooring over T&G plywood, it is important that the plywood floor be treated with care throughout the building's construction. Reasonable precautions should be taken to keep the plywood clean, dry and undamaged. The following procedure is recommended to prepare the floor for tile or linoleum:

1. Thoroughly clean the floor surface.

2. Apply hard-setting filler with a wide trowel over the entire area, filling all joints and irregularities. Many builders have reported that trowelling filler over the entire surface is more efficient than spot filling and touch sanding.

3. An alternative method that should be considered when the joists are very wet is to fill only the joints and damaged areas, leaving the nail holes unfilled to allow for any lifting of the nail heads as the joists dry. This will prevent nail heads dimpling the resilient flooring.

RECOMMENDED PLYWOOD GRADES AND THICKNESSES

Sheathing or Select Sheathing grades of T&G Douglas fir or western softwood plywood are recommended for use as combined subflooring and underlayment for field-glued floor systems when strength and economy are required and a sanded finish is unnecessary. Recommended thicknesses for various joist spacings and finish floorings are given in the table opposite.

All T&G plywood produced by the plywood manufacturing members of the Council of the Forest Industries of British Columbia conforms to a standard, thoroughly tested and approved specification. Industry-wide uniformity of the T&G profile is assured by the use of a standard test guage, thus permitting the matching of panels from different CFI member mills.

T&G Douglas fir and western softwood plywoods bearing the registered certification mark PMBC EXTERIOR stamped on panel ends are bonded with waterproof glue and manufactured in accordance with Canadian Standards Association Specifications 0121 or 0151.

AVAILABLE GLUES

Several brands of elastomeric glue are available from building supply dealers. Most are conveniently packaged in spouted, ready-to-use cartridges for hand or pneumatic caulking guns. Some brands are available in bulk for use with portable pneumatic systems. These special construction adhesives are not to be confused with ordinary drywall glues. Use only those glues specified for construction purposes and apply according to manufacturer's instructions.

GLUING CONDITIONS

Gluing can be accomplished with unseasoned lumber because the resulting adhesive strength is only about 20% lower than with dry lumber. Gluing can also be accomplished in below freezing weather with several of the adhesives available, even using lumber with frosty surfaces, provided that the temperature rises above freezing reasonably soon after the floor is applied. Gluing should be avoided during continuous freezing weather because none of the adhesives develop adequate shear strength under such conditions.

APPLICATION SEQUENCE FOR FIELD-GLUED T&G PLYWOOD SUBFLOORS

It is advisable to plan placement of combined subfloor and underlayment panels to minimize cutting, waste, and labour. Panels should be placed with the face grain at right angles to the joists. End joints which must occur over joists should be staggered. When placing the tongued edge of a panel into the groove of a fastened panel care should be taken to ensure that tongues and grooves are not damaged.

For best results, both floor joists and plywood should be dry, the temperature above freezing, and in all cases the glue manufacturer's instructions should be followed.

1 It is important that the first row of panels be aligned straight and true. To ensure this, measure 48 inches in from the ends of the two outside joists and snap a chalk line between these points. This line will also act as a boundary when applying glue for the first row.

2 Lay a $\frac{1}{8}''$ to $\frac{1}{4}''$ diameter bead of glue along the joists and sill. On wide areas apply glue in a serpentine pattern. Where panel ends will butt on the joists, lay two beads of glue to ensure that each panel end will be glued. Apply only enough glue to lay one or two panels at a time (unless the glue manufacturer's instructions permit covering a larger area in advance of laying the panels).

3 Lay the first row of T&G plywood panels with the grooved edge along the chalk line and nail in position with nails spaced according to the provisions of the table opposite. Note that each T&G panel is stamped "THIS SIDE DOWN" to ensure correct application.

4 Place a second row of T&G plywood panels along the other side of the chalkline with the tongues fitting into the grooves of the panels in the first row and having the end joints between plywood panels staggered relative to the first row. Do not attempt to force a tight joint between panel faces because the T&G joint is designed to butt at the tip of the tongue leaving a $\frac{1}{32}''$ gap on the face and underside. A $\frac{1}{16}''$ gap should also be left between panel ends to allow for expansion. For extra stiffness, a bead of glue, either continuous or intermittent, may be squeezed into the grooves before mating the panels.

5 Sheathe the remaining floor area, applying glue in advance of each panel and nailing securely in place within the glue manufacturer's specified assembly time.

6 Protect the plywood from excessive moisture and damage during construction to obtain best results with resilient flooring.

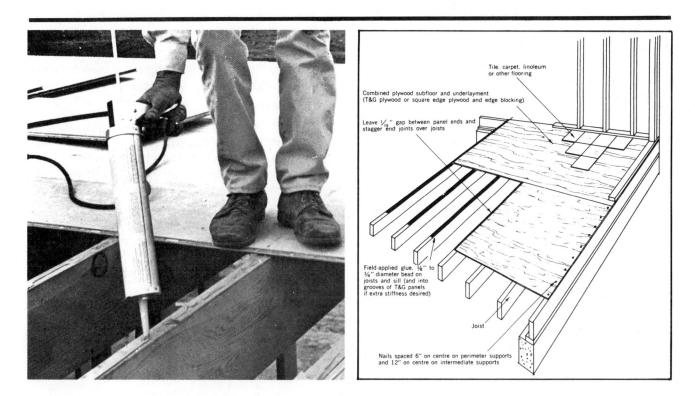

NATIONAL BUILDING CODE REQUIREMENTS FOR DOUGLAS FIR AND WESTERN SOFTWOOD PLYWOOD SUBFLOORING OR COMBINED SUBFLOOR AND UNDERLAYMENT FOR VARIOUS JOIST SPACINGS AND FINISH FLOORINGS

BUILDING ELEMENT	Maximum Support Spacing centre to centre, inches	SPECIAL CONDITIONS	Minimum Plywood Thickness, inches	Minimum Nail Length, inches C—common AG—annularly grooved	Maximum Nail Spacing, inches		NOTES
					Perimeter Supports	Intermediate Supports	
SUBFLOORING AND COMBINED SUBFLOOR AND UNDERLAYMENT	16	T&G Plywood or 2" x 2" blocking at joints	1/2	1 3/4 AG	6	12	For resilient type finish floor without separate underlay. Notes 1 and 2 apply.
	16	No joint support required	1/2	2 C 1 3/4 AG	6	12	For other than resilient finish floor. Notes 1 and 2 apply.
	16	Phenolic bonded exterior type plywood equivalent to nominal 1" finish floor or T&G subfloor	5/8	2 C 1 3/4 AG	6	12	For fire-resistive construction. Note 2 applies.
	20	T&G plywood or 2" x 2" blocking at joints	5/8	1 3/4 AG	6	12	For resilient type finish floor without separate underlay. Notes 1 and 2 apply.
	20	No joint support required	5/8	2 C 1 3/4 AG	6	12	For other than resilient finish floor. Notes 1 and 2 apply.
	24	No joint support required	1/2	2 C 1 3/4 AG	6	12	For matched hardwood floor at least 3/4" thick at right angles to joists. Notes 1 and 2 apply.
	24	T&G plywood or 2" x 2" blocking at joints	3/4	1 3/4 AG	6	12	For resilient type finish floor without separate underlay. Notes 1 and 2 apply.
	24	No joint support required	3/4	2 C 1 3/4 AG	6	12	For other than resilient finish floor. Notes 1 and 2 apply.
UNDERLAYMENT	Continuous		1/4	3/4 AG	6	8 in. grid	Staples in lieu of nails: For 1/4" plywood 18 ga., 7/8" long, 3/8" crown. For 5/16" and 3/8" plywood 18 ga., 1 1/8" long, 3/8" crown.
	Continuous		5/16	7/8 AG	6	8 in. grid	Underlay beneath resilient or ceramic floors applied with an adhesive shall have all holes or open defects on the surface patched in such a manner that the defects will not be transmitted through to the finished surface. (A sanded surface is recommended under 0.080" or thinner tile).

NOTE 1: Plywood shall be applied with face grain at right angles to supports.

NOTE 2: 16 ga. 3/8" crown staples permitted in lieu of nails: 1 1/2" long for 5/16" and 3/8" plywood, 2" long for 1/2" to 3/4" plywood.

CHAPTER 9 REVIEW QUESTIONS

1. Make a sketch of a foundation (no basement) wall with an insulated perimeter heat duct suitable for use in frigid climatic areas such as North America.

2. Make a sketch of the top of a concrete wall supporting cast-in-place joist endings.

3. Define the difference between a basement and a cellar.

4. Sketch and define a fire-cut in a wood beam.

5. Sketch six methods of supporting first floor joists between supporting concrete walls.

6. Sketch a plan, and arrow-name the different members of a floor opening suitable for a stairwell.

7. What is the minimum allowable headroom in your area for a single residential stair, and for a stairway from the first to the second floor?

8. What is the minimum headroom allowance and width of stair for an apartment building in your area?

9. What is the maximum height of a riser for a two story single house in your area?

10. Define the soffit of a stair.

11. Make a neat sketch of the following: (a) a fire cut in a beam at its intersection with a masonry wall; (b) a cavity wall of brick and concrete block; (c) a keyway in a concrete footing; (d) flashing where wood framing adjoins a masonry wall; and (e) a stepped footing.

12. Make a neat freehand sketch of balloon framing at its intersection with the concrete foundation, and show: (a) the sill plate; (b) the fire stop, (c) the floor joist, and (d) the blocking which supports the subfloor.

10

Wall Framing, Anchors, Lumber Standards, Fasteners, and Scaffolding

In this chapter we shall present a report on lumber standards; wall anchors; conventional nailing methods, and metal wood fasteners in wood framing; steel studs; floor decks; and an actual case study of a housing development; and scaffolding.

10.1 NEW LUMBER STANDARDS

The following two pages have been excerpted from a special report on **new lumber standards which were developed, simultaneously, in the United States and Canada.** The material is here presented through the courtesy of the Canadian Wood Council. **The following two pages are very important.**

SPECIAL REPORT ON

NEW LUMBER STANDARDS

GENERAL INFORMATION

CANADIAN WOOD COUNCIL ■ CONSEIL CANADIEN DU BOIS

300 COMMONWEALTH BUILDING   77 METCALFE STREET   OTTAWA , CANADA K1P 5L6

SR2-71 (Rev.)   MAY, 1971

New standards for softwood lumber have been developed simultaneously in Canada and the United States, as a result of development extending over nearly ten years. The essential changes are
● new standard dressed sizes, for green and dry lumber, and
● new uniform grading rules.

These changes in lumber standards result in a better-regulated range of products. The changes benefit producers, builders, designers, dealers and wholesalers, building officials, and ultimate consumers, by
● relating manufactured sizes to moisture content so that end-use sizes are the same for lumber manufactured dry or green;
● simplifying and grouping grades and species for most efficient end use and better product identification;
● combining and simplifying grading rules for all softwood lumber throughout Canada;
● making sizes, grades, and grade names for dimension lumber and most other lumber products uniform throughout North America, thus unifying Canada's domestic and important U.S. export markets;
● conserving forest resources through more efficient use, better recovery of raw materials, and less waste in remanufacture; and
● providing a product geared to modern modular methods and precise engineering.

Canadian Lumber Standards

The new grading rule is published by the National Lumber Grades Authority, a Canadian agency that replaces all former Canadian rules-writing agencies. It conforms to the requirements of the 1970 edition of CSA Standard 0141, Softwood Lumber, and has been certified by the Canadian Lumber Standards Administrative Board to apply throughout Canada.

Lumber Grade Categories and Species

The new rule classifies dimension lumber (2" to 4" in nominal thickness) into two width categories and five use categories. Dimension lumber up to 4" wide is classified as *Structural Light Framing, Light Framing,* or *Stud.* Dimension Lumber 6" and wider is classified as *Structural Joists and Planks.* In addition, a single *Appearance Framing* category of 2" and wider dimension is designed for special uses where appearance is of prime importance. The NLGA rule also covers larger timbers, decking, and alternative grades for boards. Unlike earlier grading rules, the NLGA rule provides for certain commercial groups of species of similar characteristics that are normally manufactured and marketed together.

Lumber Shrinkage and Size

Standard sizes prior to this year applied to all lumber regardless of the moisture content at the time of surfacing, with unseasoned lumber dressed to the same size as dry lumber. Unseasoned lumber shrinks to a smaller size when used in dry locations. The new CSA standard provides for two sets of sizes to overcome this discrepancy; unseasoned lumber is dressed to a larger size than dry lumber. The difference in size is just enough so that when used in dry conditions, unseasoned lumber will shrink to be equivalent in size and load capacity to lumber that was surfaced dry.

Transition from "Old" to "New"

After August 1, 1971, all CLS lumber is to be manufactured to the new standards. Because of stocks on hand and the complex distribution chain, there must continue to be a period of overlap during which either standard is acceptable. Amendments to the National Building Code of Canada 1970 through revisions to be issued in July 1971 provide for alternatives, as will the Canadian Code for Residential Construction, (formerly Residential Standards).

Span tables for both old and new grades and sizes will be available, and used by Central Mortgage and Housing Corporation for NHA-financed construction. The lumber buyer or inspector can always tell lumber of the "new" standard by the presence of either "S-DRY" or "S-GRN" on the grade stamp (see Moisture Content Classifications, page 3).

Authority for Change

● **Sizes, grading requirements, inspection:** CSA Standard 0141-1970, Softwood Lumber; CSA[1], $2.00
● **Grades and commercial species groups:** Standard Grading Rules for Canadian Lumber; NLGA[2], $1.50
● **Stresses and design data:** CSA Standard 086-1970, Engineering Design in Timber; CSA[1], $6.00, and Canadian Structural Design Manual, Supplement No. 4 to the National Building Code of Canada 1970; NRC[3], $5.00
   Span tables for joists and rafters: National Building Code of Canada 1970 (NRC No. 11246, revised July, 1971); NRC[3], $5.00 casebound, $1.50 paperback, and Canadian Code for Residential Construction 1970 (used by CMHC) (NRC No. 11562, July, 1971); NRC[3], free; also spans for "new" rules and sizes in a separate booklet (NRC No. 11862).
● **Grade stamping regulation and facsimiles:** Available from Canadian Lumber Standards Administrative Board of CSA[4]

---

1 Canadian Standards Association, 178 Rexdale Blvd., Rexdale 603, Ontario.
2 National Lumber Grades Authority, 1500-1055 West Hastings St., Vancouver 1, B.C.
3 ACNBC, National Research Council, Ottawa, Ont. K1A OR6.
4 Canadian Lumber Standards Administrative Board, 15th floor, Guinness Tower, Vancouver 1, B.C.

SIZES OF LUMBER AND TIMBERS

Moisture Content Classifications

Characteristics	Moisture Content		
	Unseasoned (Green)	Dry	Especially Dried to Low (15%) Moisture Content
Grade Stamp Designation	S - GRN	S - DRY	MC - 15
Size When Surfaced	See table for "Surfaced Green"	See table for "Surfaced Dry"	See table for "Surfaced Dry"
Required Maximum Moisture Content	None	19%	15%
Average Moisture Content	--	About 15%	Less than 15% about 10-12%
Size to be Used for Section Properties and Working Stress	Dry	Dry	Dry

"Old" and "New" Sizes of Dimension Lumber and Boards Compared

Nominal Sizes	Old Sizes (in.) (Dry or Unseasoned)	New Sizes (in.)	
		Unseasoned	Dry
Dimension			
2 x 4	1 ⅝ x 3 ⅝	1 ⁹⁄₁₆ x 3 ⁹⁄₁₆	1 ½ x 3 ½
2 x 6	1 ⅝ x 5 ½	1 ⁹⁄₁₆ x 5 ⅝	1 ½ x 5 ½
2 x 8	1 ⅝ x 7 ½	1 ⁹⁄₁₆ x 7 ½	1 ½ x 7 ¼
2 x 10	1 ⅝ x 9 ½	1 ⁹⁄₁₆ x 9 ½	1 ½ x 9 ¼
2 x 12	1 ⅝ x 11 ½	1 ⁹⁄₁₆ x 11 ½	1 ½ x 11 ¼
Boards			
1 x 4	¾ x 3 ⅝	¾ x 3 ⁹⁄₁₆	¹¹⁄₁₆ x 3 ½
1 x 6	¾ x 5 ½	¾ x 5 ⅝	¹¹⁄₁₆ x 5 ½
1 x 8	¾ x 7 ½	¾ x 7 ½	¹¹⁄₁₆ x 7 ¼
1 x 10	¾ x 9 ½	¾ x 9 ½	¹¹⁄₁₆ x 9 ¼
1 x 12	¾ x 11 ½	¾ x 11 ½	¹¹⁄₁₆ x 11 ¼

New Finish, Flooring, Ceiling, Partition and Stepping Lumber Sizes

Product	Thickness		Width	
	Nominal	Actual Surfaced Dry (in.)	Nominal	Actual Surfaced Dry (in.)
Finish	⅜	⁵⁄₁₆	2	1 ½
	½	⁷⁄₁₆	3	2 ½
	⅝	⁹⁄₁₆	4	3 ½
	¾	⅝	5	4 ½
	1	¾	6	5 ½
	1 ¼	1	7	6 ½
	1 ½	1 ¼	8	7 ¼
	1 ¾	1 ⅜	9	8 ¼
	2	1 ½	10	9 ¼
	2 ½	2	11	10 ¼
	3	2 ½	12	11 ¼
	3 ½	3	14	13 ¼
	4	3 ½	16	15 ¼
Flooring	⅜	⁵⁄₁₆	2	1 ⅛
	½	⁷⁄₁₆	3	2 ⅛
	⅝	⁹⁄₁₆	4	3 ⅛
	1	¾	5	4 ⅛
	1 ¼	1	6	5 ⅛
	1 ½	1 ¼		
Ceiling	⅜	⁵⁄₁₆	3	2 ⅛
	½	⁷⁄₁₆	4	3 ⅛
	⅝	⁹⁄₁₆	5	4 ⅛
	¾	¹¹⁄₁₆	6	5 ⅛
Partition	1	²³⁄₃₂	3	2 ⅛
			4	3 ⅛
			5	4 ⅛
			6	5 ⅛
Stepping	1	¾	8	7 ¼
	1 ¼	1	10	9 ¼
	1 ½	1 ¼	12	11 ¼
	2	1 ½		

Size Differential Between Unseasoned and Dry Lumber

Nominal Dimension (in.)	Size Differential at Time of Surfacing between Unseasoned and Dry Lumber (in.)
2 to 4 ½	¹⁄₁₆
5 to 7	⅛
8 or more	¼

Dimension Lumber Sizes and Section Properties

Nominal Size (in.)	Surfaced Green Net Size (in.)	Surfaced Dry			
		Net Size (in.)	Area (in.²)	Section Modulus (in.³)	Moment of Inertia (in.⁴)
2 x 2	1 ⁹⁄₁₆ x 1 ⁹⁄₁₆	1 ½ x 1 ½	2.25	0.56	0.42
2 x 3	1 ⁹⁄₁₆ x 2 ⁹⁄₁₆	1 ½ x 2 ½	3.75	1.56	1.95
2 x 4	1 ⁹⁄₁₆ x 3 ⁹⁄₁₆	1 ½ x 3 ½	5.25	3.06	5.36
2 x 6	1 ⁹⁄₁₆ x 5 ⅝	1 ½ x 5 ½	8.25	7.56	20.80
2 x 8	1 ⁹⁄₁₆ x 7 ½	1 ½ x 7 ¼	10.87	13.14	47.63
2 x 10	1 ⁹⁄₁₆ x 9 ½	1 ½ x 9 ¼	13.87	21.39	98.93
2 x 12	1 ⁹⁄₁₆ x 11 ½	1 ½ x 11 ¼	16.87	31.64	177.98
3 x 6	2 ⁹⁄₁₆ x 5 ⅝	2 ½ x 5 ½	13.75	12.60	34.66
3 x 8	2 ⁹⁄₁₆ x 7 ½	2 ½ x 7 ¼	18.12	21.90	79.39
3 x 10	2 ⁹⁄₁₆ x 9 ½	2 ½ x 9 ¼	23.12	35.65	164.89
3 x 12	2 ⁹⁄₁₆ x 11 ½	2 ½ x 11 ¼	28.12	52.73	296.63
4 x 4	3 ⁹⁄₁₆ x 3 ⁹⁄₁₆	3 ½ x 3 ½	12.25	7.15	12.50
4 x 6	3 ⁹⁄₁₆ x 5 ⅝	3 ½ x 5 ½	19.25	17.65	48.53
4 x 8	3 ⁹⁄₁₆ x 7 ½	3 ½ x 7 ¼	25.37	30.66	111.15
4 x 10	3 ⁹⁄₁₆ x 9 ½	3 ½ x 9 ¼	32.37	49.91	230.84
4 x 12	3 ⁹⁄₁₆ x 11 ½	3 ½ x 11 ¼	39.37	73.83	415.28

10.2 CONVENTIONAL WALL FRAMING

Before making any wall frame layout, the openings in the subfloor should be checked for the correctness of their locations and sizes. At this stage, it is relatively inexpensive to correct any fault before further construction begins. See article 9.10 for stairwell openings. When everything is satisfactory, the floor should be swept clean, and the following suggested steps taken to lay out the floor plates (sole plates).

Step 1: Study Figs. 10.1 and 10.2 and decide what type of stud assembly is to be used for corners and intersecting internal partitions. In this discussion, we will use the stud assembly as at Fig. 10.1A.

Step 2: Examine the drawings. (Some contractors glue them onto pieces of plywood, and clear-varnish them as a protection against the weather and to prevent persons defacing them with pencil sketches and calculations.) Check to see if partition and wall openings, such as doors and windows, are dimensioned to center lines; if so, make an allowance on either side for half their widths, and that of the sole plate of wall partitions.

Step 3: Strike chalk lines on the subfloor for partitions, and check the diagonals for square.

Step 4: From the lumber pile, select the straightest and best material for corner posts, window and door openings, and sole plates. See Fig. 4.1 on page 63 for the named parts of a typical wood framed wall section.

Step 5: Square-cut the ends of two pieces of sole plate, and loosely place them on the subfloor at right angles to each other as at Fig. 10.3. Using a square-ended piece of 2 x 4, about a foot long (called a short end), stand it square to the end of the plate; with a sharp pencil, draw the position of the corner stud and mark it as at Fig. 10.3a. *You will never waste time by sharpening a pencil or any other tool.*

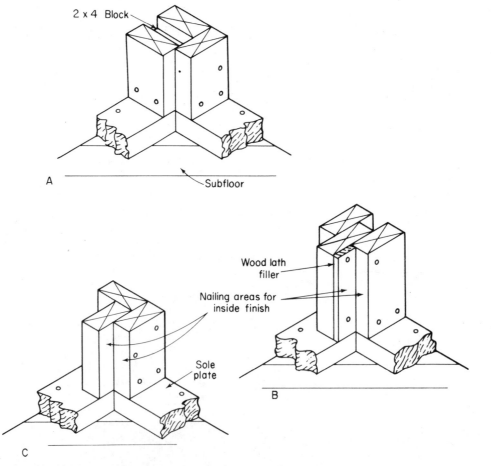

Fig. 10.1. Examples of corner stud assembly: (a) standard outside corner; (b) special corner with lath filler; (c) special corner without lath filler. *(Courtesy of Forest Products Laboratory, U.S. Department of Agriculture.)*

Step 6: From the corner post thus marked, again stand the short end and mark its width as at Fig. 10.3b. which represents the thickness of the blocking for the built up corner post. See also Fig. 10.1a.

Step 7: From the blocking mark, draw the inside starter stud as at Fig. 10.3c.

Step 8: Place the other wall plate at right angles snug to the long one. Stand the short end as shown on the drawing and mark the inside starter stud of the short wall plate as at Fig. 10.3d. It is recommended the actual material be used (instead of a ruler) for layout wherever practicable.

Step 9: From the inside intersection of the sole plates, measure 16″ each way as at Fig. 10.4. These are the center lines of the No. 2 studs, respectively. **This is important.** On either side of the center line, mark $\frac{3}{4}$″ as shown. This is the location of the No. 2 stud for each wall plate.

Step 10: From the outside edge of each **No. 2 stud** (the edge furthest away from the intersection of the plates), run out a tape and mark 4′-0″ intervals (with ticked marks) as shown. The tick will be covered later by the nailed studs. Then lay off the intermediate studs with the 16″ tongue of a carpenter's framing square as shown. The reason for laying off the 4′-0″ intervals

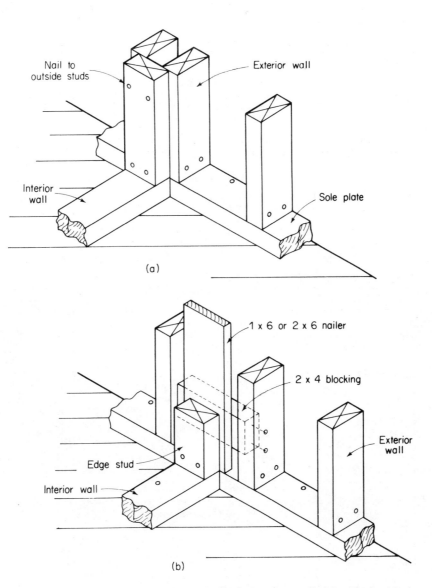

(a)

(b)

Fig. 10.2. Intersection of interior wall with exterior wall: (a) with doubled studs on outside wall, (b) partition between outside studs. *(Courtesy of Forest Products Laboratory, U.S. Department of Agriculture.)*

with a tape is to avoid the creeping which may occur if all the studs are stepped off with the framing square. This way it is certain that all 4'-0" wide fiber boards will fall exactly on the center line of a stud. See Fig. 10.5 for the named members of a rough frame opening.

The Carpenter's Framing Square

(a) The body of the framing square is 2 inches wide and 24 inches long.

(b) The tongue is 1½ inches wide and 16 inches long. In some countries the tongue is 18 inches long.

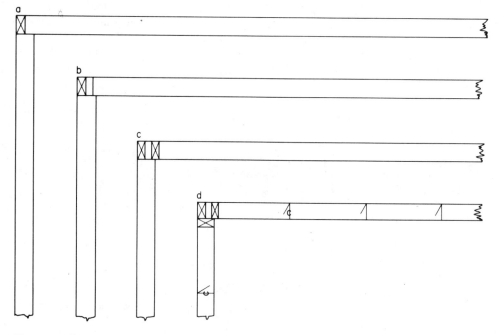

Fig. 10.3. Corner post layout.

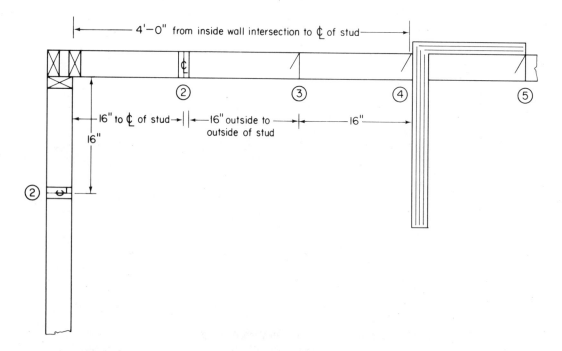

Fig. 10.4. Plate layout.

(c) The tool has two rectangular parts at right angles to each other. It usually has the rafter tables and the name of the manufacturer on the face side.

(d) There are eight edges on the framing square. Each edge is divided into inches and fractions of an inch. Readings of $\frac{1}{16}''$, $\frac{1}{8}''$, $\frac{1}{4}''$, $\frac{1}{2}''$, $\frac{1}{10}''$, $\frac{1}{5}''$, or $\frac{1}{3}''$ may be made on one or more edges. The use to which such graduations may be put in carpentry layout is indicated in the following example.

A one-twelfth scale drawing is to be made of a rectangular building measuring 12'-3" by 20'-6". By use of the framing square edge having $\frac{1}{12}''$ graduations, these measurements can be scaled down immediately to read $12\frac{3}{12}''$ by $20\frac{6}{12}''$. The drawing may be done quickly and may just as easily be read. See Fig. 10.6.

10.3 ROUGH OPENINGS FOR DOORS AND WINDOWS

Irrespective of the sizes or numbers of doors or window openings in a wood frame wall, the 16" OCs of all studs must be constant; the framing for the openings is imposed on the main framing at the place desired.

It is important to read the manufacturer's catalog, which gives the actual sizes of door and window frame units and the **relative sizes for the rough openings.** Assuming that a newly framed wall has been

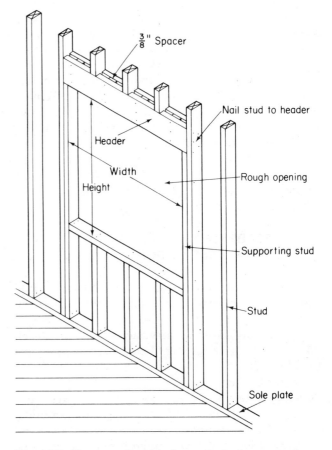

Fig. 10.5. Headers for windows and door openings. *(Courtesy of Forest Products Laboratory, U.S. Department of Agriculture.)*

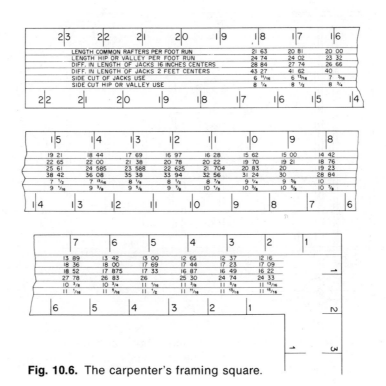

Fig. 10.6. The carpenter's framing square.

sheathed, insulated, and finished with hardboard before it was discovered that the opening was too small for the unit, this would be a major (and expensive) mistake to correct. If units are delivered onto the job before the wall framing has begun, take the widths and heights on a piece of lath and allow sufficient clearance for fitting them into the rough openings. From my personal experience, I would never use a tape if I could manage without it. Similarly, it is better to take the height or width of an opening on a pair of slip sticks, transfer this length to the lumber to be fitted into the opening, and cut!

10.4 PREFABRICATION AND SEMI-PREFABRICATION

If an area is indiscriminately subdivided into building lots and uses a basic plan with variations of doors, windows and roof treatments, a number of houses in any given district may complement each other. Such a development lends itself to intensive prefabrication methods such as prefabricated basement walls; floors; walls; pre-fitted windows and doors complete with hardware; meter boxes; ceilings; roofs; chimneys; hearths; firegrate surrounds; clothes closets; complete bathrooms with plumbing units, vanities, towel rails, and mirrors; kitchen cabinets; ranges, refrigerators.

In addition, the construction industry lends itself to an ever increasing number of specializing subtradesmen. This new age of house prefabrication requires administrators, with special organizing ability, who can coordinate the different assembly operations, so that a whole complex of houses may be given consecutive treatment by consecutive specializing crews of workmen with their own special tools, equipment, and techniques.

Semi-Prefabrication has been practiced, with continuing success in the saving of material and labor costs, for many years. As an example, it is particularly important when working in frigid or very hot temperatures to afford men some cover as soon as possible. When the underpinning has been completed the following method (with your own variations) may be adopted:

Step 1: On the completed subfloor, prefabricate one long wall (complete with its sheathing, small glazed windows, and hardware-fitted prehung doors secured in their respective positions) and complete with outside trim. Meanwhile on the floor, the top plates should be marked off for the positions of the prefabricated roof trusses spaced that will be spaced at 16" or 24" OCs. *Remember that the less work that has to be done in space, the safer, faster, and more profitable will be the enterprise.*

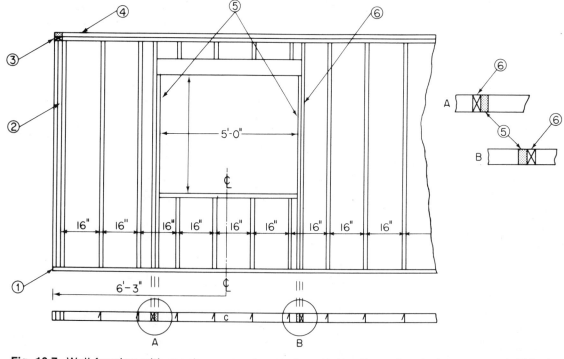

Fig. 10.7. Wall framing with rough opening for window. Note: all regular studs are spaced 16 inches OCs. (1) Bottom or sole plate, (2) built-up corner post, (3) top plate, (4) cap plate, (5) trimmer stud, and (6) window framing through stud.

Step 2: Raise the wall into position and brace it well, *especially if it has to be left overnight.* Nail the sole plate through the subfloor to the joist header, but leave four or five feet unnailed at each end of the wall ready for jockeying the end walls into position. See a case study of housing development, page 185.

Step 3: If the door and window units are not placed before the wall is raised, cover the rough opening with 4 mil polyethylene.

Step 4: Prefabricate the opposite long wall and raise it into its position as in Step 2.

Step 5: Prefabricate one short wall, raise it into position by jockeying it between the unnailed end plates of the long walls, position it carefully, nail it through to the floor joists. Then spike its end studs to the long wall corner studs.

Step 6: Prefabricate the other short wall and give it the same treatment as in Step 5.

Step 7: The roof trusses may be made on the subfloor, or bought from, and delivered to the job by the manufacturer. Be sure to have them on hand when required. The delivery truck should place sufficient trusses for each house and as near to each as possible. If you are making your own trusses, you could use one subfloor upon which to lay out a jig and make a great number, on the job, with a steady work crew. See article 14.18, page 253.

Step 8: Raise the roof trusses and spike them to their premarked positions on the top plate of the walls.

Step 9: Sheath the roof and apply shakes, shingles or other roofing material, and the place is virtually stormproof. Notice at this stage that no internal walls have been made or positioned. At this stage, the four walls with the roof is a free standing building, which should be well braced against sudden gusts of wind.

Step 10: Neatly install the insulation around the inside perimeter walls. Note that this is another reason why all wall studs should be placed 16" OCs irrespective of any wall openings.

Step 11: As soon as the internal walls are secured in position, the structure is ready for the following trades: electrician, plumber, tinsmith for heating ducts, air conditioning, telephone, cable vision, and the building inspectors.

With careful planning, an excellent field-type of production line assembly can be achieved with excellent worker participation. Men like to see fast physical production, especially when their work is quickly shaded from extremes of climate. Remember, too, that subtrades will have prefabricated some of their plumbing lines, ducts, and so on, in their own shops.

10.5 WOOD BACKING FOR SUBTRADES

Before the carpenters' rough wood framing is completed and handed over to other trades, inspect thoroughly each room to be sure that wood backing has been provided for affixing units. See Fig. 10.8. It will be seen from an inspection of the intersection of any wall frame with another, that without a short end of 2 x 4 nailed at the foot of all intersecting wall studs, that the carpenter will not have sufficient bearing to readily nail the baseboard. The time taken in spiking these nogs in place will be well repaid in the time saved by the finishing carpenter.

A list of other units, or items requiring backing is as follows:

Towel rail	Bathroom:
Plumbing units	toilet, wash basin,
Hot air registers	vanity towel bars,
Girts or fire stops	coat hangers, cabinets, soap and toilet tissue
Chimney hearths and fireplaces	holders, shower faucets and doors, curtain rods, facilities for
Undercarriage for stairs	repairing—plumbing
Stair handrail brackets	services behind all faucets (may be a loose
Drywall backing at ceiling intersections with walls	door in the adjoining room).
Ceiling vents, fans, and chandeliers	Milk and clothes chutes
	Ceiling access door
Plaster grounds	Ironing board
Closet shelves and rails	Backing for telephone, cable vision, electric and
Window valance	vacuum cleaning outlets.

This list is not exhaustive and it should be added to with each new experience. It is expensive to have to call a carpenter from another job to provide backing, especially after the hardboard to all the walls has been completed. Check and be careful!

See Figs. 10.9, 10.10, and 10.11, and pages 174 and 175.

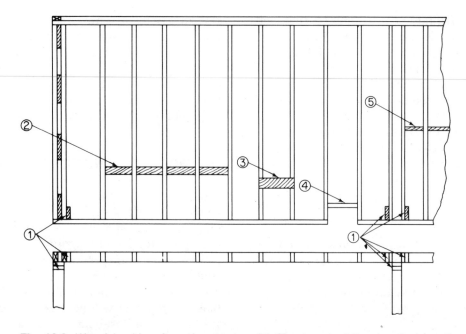

Fig. 10.8. Wood backing for other trades. (1) Short end of 2 x 4 nailed to all framing intersections for fixing baseboards, (2) backing for towel rails, (3) backing for plumbers unit, let in flush, (4) framing for tinsmiths hot air register, and (5) girts, or fire stops.

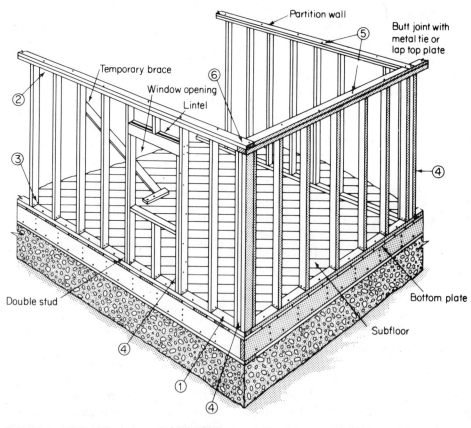

Fig. 10.9. Wall framing used with platform construction. *(Courtesy of the Department of Forestry and Rural Development and the Central Mortgage and Housing Corporation, Ottawa.)*

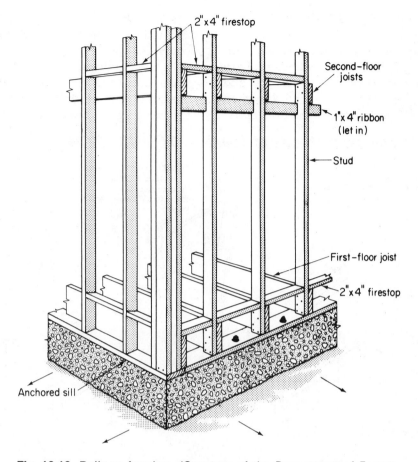

Fig. 10.10. Balloon framing. *(Courtesy of the Department of Forestry and Rural Development and the Central Mortgage and Housing Corporation, Ottawa.)*

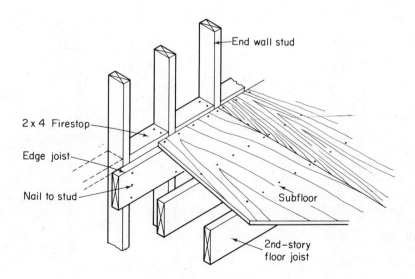

Fig. 10.11. End wall framing for balloon construction (junction of first-floor ceiling and upper-story floor framing). *(Courtesy of Forest Products Laboratory, U.S. Department of Agriculture.)*

10.6 STRUCTURAL WOOD FASTENERS

Wood fasteners for wood framing are finding increasing favor among builders and contractors. They are comparatively inexpensive for original cost and especially for labor costs. The information shown on pages 176 through 183 has been excerpted from booklets published by the "Teco Engineering Company", and is here reproduced with the permission of the copyright holder.

Note carefully the types of anchors used in the walls. Check the local by-laws; the stronger the wind pressure the more stringent will be the requirements for anchorage of buildings to their foundations.

Several methods of framing: FOUNDATION LEVEL

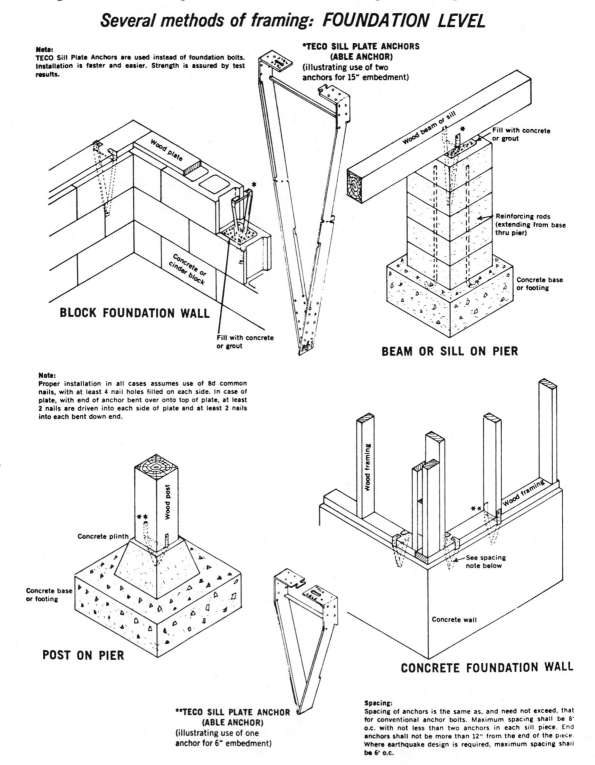

Note:
TECO Sill Plate Anchors are used instead of foundation bolts. Installation is faster and easier. Strength is assured by test results.

*TECO SILL PLATE ANCHORS (ABLE ANCHOR)
(illustrating use of two anchors for 15" embedment)

BLOCK FOUNDATION WALL

BEAM OR SILL ON PIER

Note:
Proper installation in all cases assumes use of 8d common nails, with at least 4 nail holes filled on each side. In case of plate, with end of anchor bent over onto top of plate, at least 2 nails are driven into each side of plate and at least 2 nails into each bent down end.

POST ON PIER

**TECO SILL PLATE ANCHOR (ABLE ANCHOR)
(illustrating use of one anchor for 6" embedment)

CONCRETE FOUNDATION WALL

Spacing:
Spacing of anchors is the same as, and need not exceed, that for conventional anchor bolts. Maximum spacing shall be 8' o.c. with not less than two anchors in each sill piece. End anchors shall not be more than 12" from the end of the piece. Where earthquake design is required, maximum spacing shall be 6' o.c.

Several methods of framing: FLOOR LEVEL

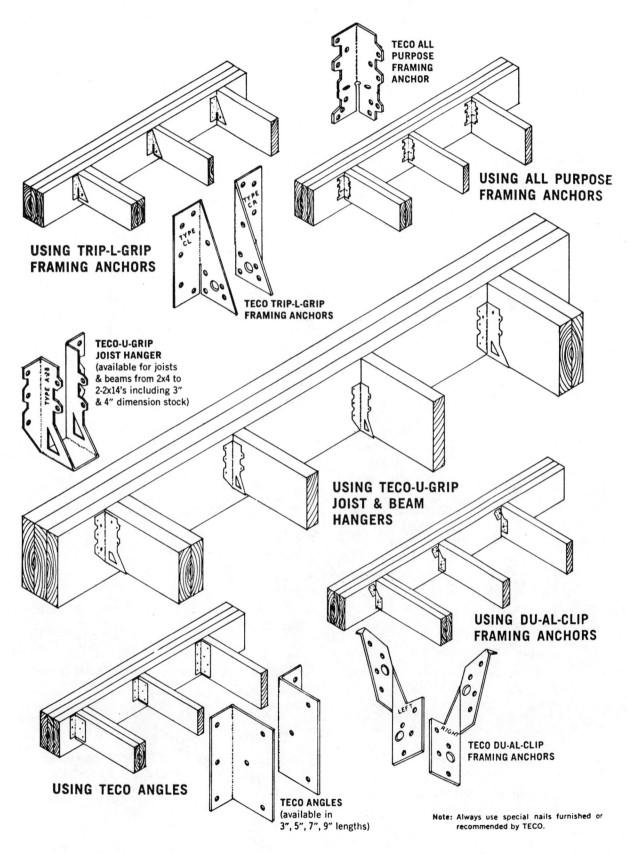

TECO ALL PURPOSE FRAMING ANCHOR

USING ALL PURPOSE FRAMING ANCHORS

USING TRIP-L-GRIP FRAMING ANCHORS

TYPE CL

TYPE CR

TECO TRIP-L-GRIP FRAMING ANCHORS

TYPE A28

TECO-U-GRIP JOIST HANGER (available for joists & beams from 2x4 to 2-2x14's including 3" & 4" dimension stock)

USING TECO-U-GRIP JOIST & BEAM HANGERS

USING DU-AL-CLIP FRAMING ANCHORS

LEFT

RIGHT

TECO DU-AL-CLIP FRAMING ANCHORS

USING TECO ANGLES

TECO ANGLES (available in 3", 5", 7", 9" lengths)

Note: Always use special nails furnished or recommended by TECO.

Several methods of framing: WALL CONNECTIONS

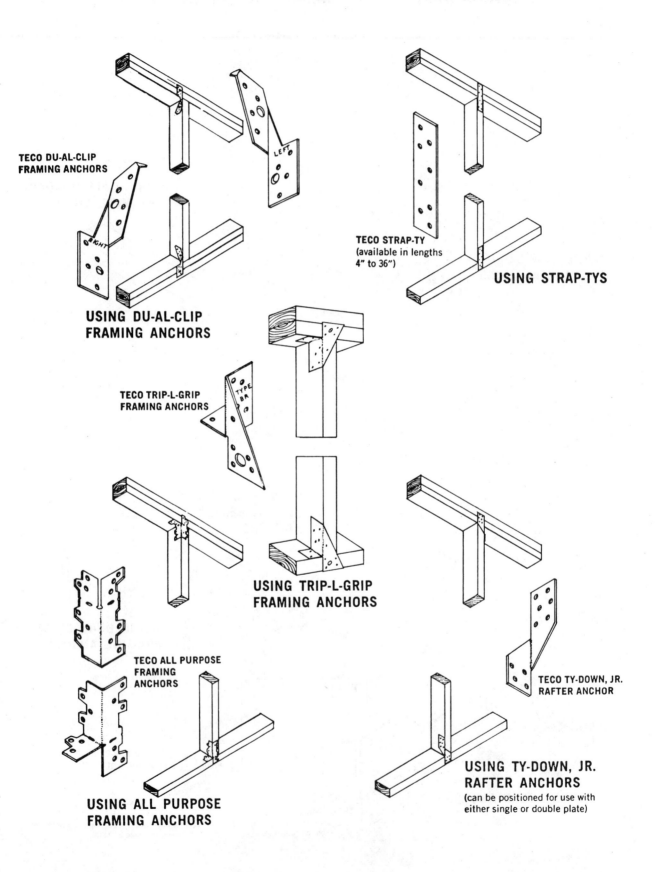

TECO DU-AL-CLIP FRAMING ANCHORS

USING DU-AL-CLIP FRAMING ANCHORS

TECO STRAP-TY
(available in lengths 4" to 36")

USING STRAP-TYS

TECO TRIP-L-GRIP FRAMING ANCHORS

USING TRIP-L-GRIP FRAMING ANCHORS

TECO ALL PURPOSE FRAMING ANCHORS

USING ALL PURPOSE FRAMING ANCHORS

TECO TY-DOWN, JR. RAFTER ANCHOR

USING TY-DOWN, JR. RAFTER ANCHORS
(can be positioned for use with either single or double plate)

Several methods of framing: MISCELLANEOUS APPLICATIONS

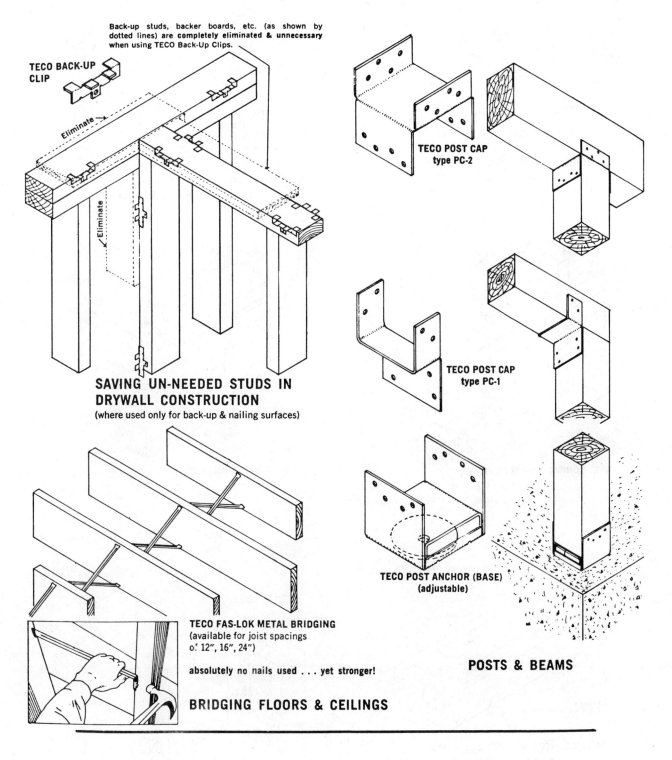

Back-up studs, backer boards, etc. (as shown by dotted lines) are **completely eliminated & unnecessary** when using TECO Back-Up Clips.

TECO BACK-UP CLIP

Eliminate

Eliminate

SAVING UN-NEEDED STUDS IN DRYWALL CONSTRUCTION
(where used only for back-up & nailing surfaces)

TECO POST CAP
type PC-2

TECO POST CAP
type PC-1

TECO POST ANCHOR (BASE)
(adjustable)

TECO FAS-LOK METAL BRIDGING
(available for joist spacings
of 12", 16", 24")

absolutely no nails used . . . yet stronger!

BRIDGING FLOORS & CEILINGS

POSTS & BEAMS

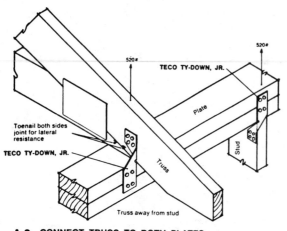

A-3   CONNECT TRUSS TO BOTH PLATES
   A-4   CONNECT STUD TO BOTH PLATES

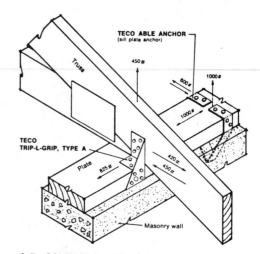

A-5   CONNECT TRUSS TO WALL PLATE
   A-6 CONNECT SILL PLATE TO MASONRY

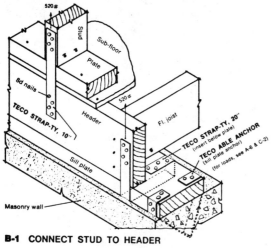

B-1   CONNECT STUD TO HEADER
   B-2   CONNECT HEADER TO SILL PLATE
      B-4 CONNECT SILL PLATE TO MASONRY

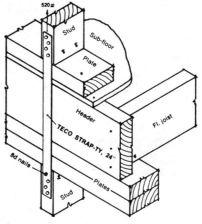

B-3   CONNECT STUD TO STUD

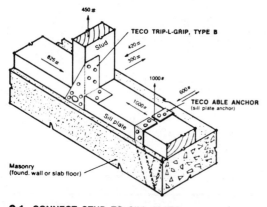

C-1   CONNECT STUD TO SILL PLATE
   C-2 CONNECT SILL PLATE TO MASONRY

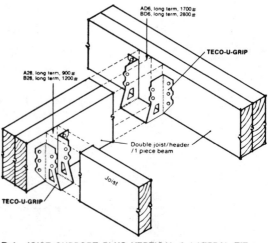

D-1   JOIST SUPPORT PLUS VERTICAL & LATERAL TIE

Several methods of framing: ROOF ANCHORAGE

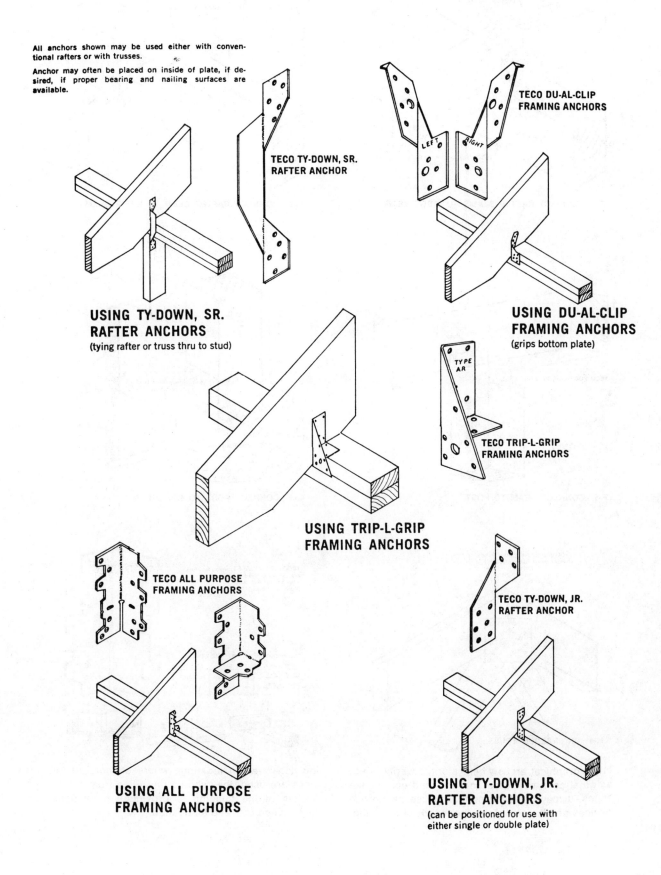

All anchors shown may be used either with conventional rafters or with trusses.

Anchor may often be placed on inside of plate, if desired, if proper bearing and nailing surfaces are available.

TECO TY-DOWN, SR. RAFTER ANCHOR

TECO DU-AL-CLIP FRAMING ANCHORS

USING TY-DOWN, SR. RAFTER ANCHORS
(tying rafter or truss thru to stud)

USING DU-AL-CLIP FRAMING ANCHORS
(grips bottom plate)

TECO TRIP-L-GRIP FRAMING ANCHORS

USING TRIP-L-GRIP FRAMING ANCHORS

TECO ALL PURPOSE FRAMING ANCHORS

TECO TY-DOWN, JR. RAFTER ANCHOR

USING ALL PURPOSE FRAMING ANCHORS

USING TY-DOWN, JR. RAFTER ANCHORS
(can be positioned for use with either single or double plate)

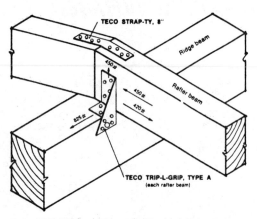

E-1 CONNECT RAFTER BEAMS TO RIDGE BEAM

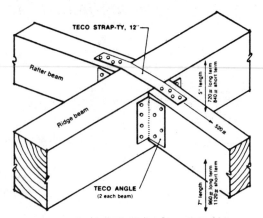

E-2 CONNECT RAFTER BEAM TO RIDGE BEAM

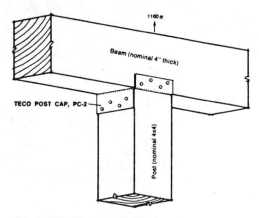

F-1 CONNECT BEAM TO POST

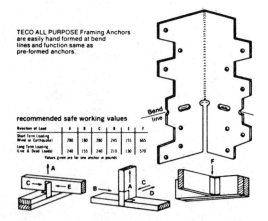

G-1 CONNECT POST TO CONCRETE

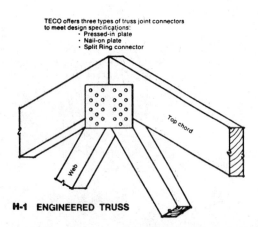

H-1 ENGINEERED TRUSS

In geographical areas where storm severity is not as great as in hurricane, tornado, and earthquake zones, it may not be essential to use anchorage devices on **all** joints along the length of the build-

ing. However, in locations where anchorage devices are used, they should be in-line from top to bottom of the building to provide a continuous tie between roof and foundation.

Several methods of framing: ROOF CONSTRUCTION

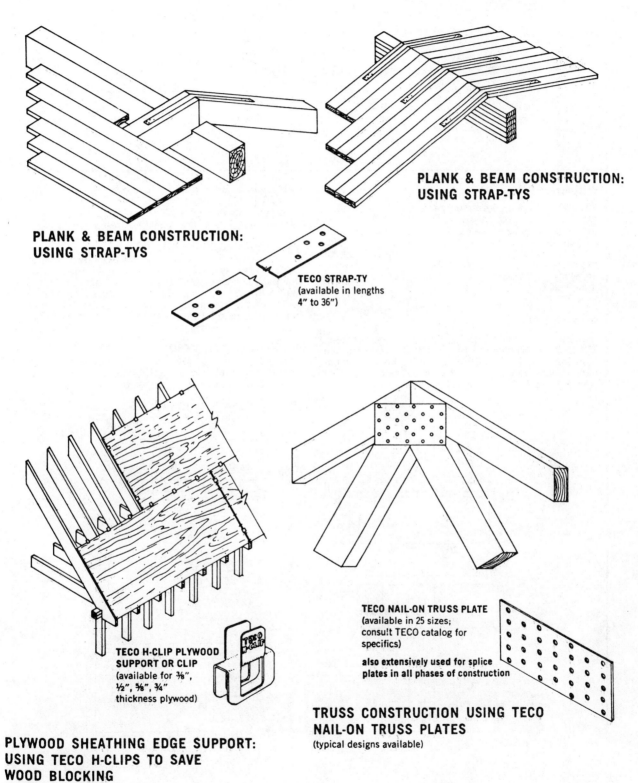

PLANK & BEAM CONSTRUCTION: USING STRAP-TYS

PLANK & BEAM CONSTRUCTION: USING STRAP-TYS

TECO STRAP-TY (available in lengths 4" to 36")

TECO H-CLIP PLYWOOD SUPPORT OR CLIP (available for ⅜", ½", ⅝", ¾" thickness plywood)

PLYWOOD SHEATHING EDGE SUPPORT: USING TECO H-CLIPS TO SAVE WOOD BLOCKING (also provides automatic end spacing for expansion & contraction of plywood)

TECO NAIL-ON TRUSS PLATE (available in 25 sizes; consult TECO catalog for specifics)

also extensively used for splice plates in all phases of construction

TRUSS CONSTRUCTION USING TECO NAIL-ON TRUSS PLATES (typical designs available)

TRUSS PLATES (nailed type)

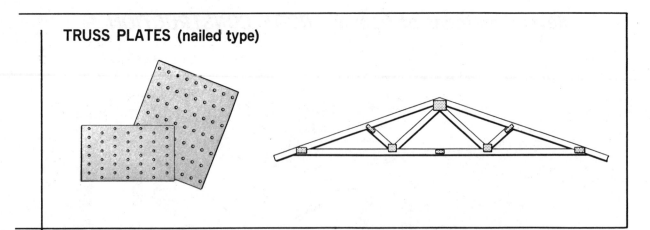

where used

Where single plane assembly of residential and other types of light roof trusses is desired, TECO Truss Plates provide economical clear span construction without the requirement for costly fabricating equipment. No special presses or rollers are needed for the proper assembly of members. In most situations an efficient fabricating line can be set up for less than $300.00.

description

Manufactured from 20 gauge zinc coated sheet steel, TECO Truss Plates are pre-punched to receive 8d 1½″ length nails. The TECO plate system accommodates spans ranging from 16′ to 32′ and slopes of from 3:12 to 7:12. Designed in accordance with HUD

(FHA) Construction Standards, the system is covered by HUD (FHA) Bulletin SE 297.
Complete details on design and fabrication with TECO Truss Plates are available free upon request.

suggested specification

Trusses shall be built using TECO Truss Plates (nailed type), as manufactured by Timber Engineering Company, Washington, D.C. Nails shall be either 1½″ x 0.133″ diameter (square barbed) or 1½″ x 0.131″ diameter (round barbed) and shall be placed in requisite number of nail holes as shown on TECO designs.

shipping information

Packed: in varying quantities, dependent on sizes. Contact nearest TECO distributor (or TECO) for standard packing. Nails are packed in 50# cartons.

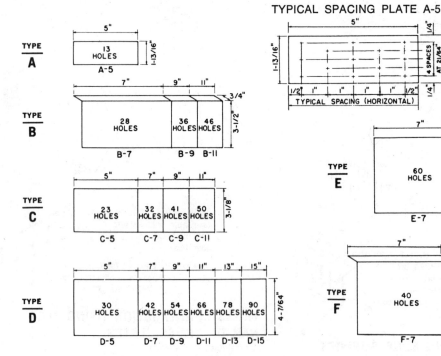

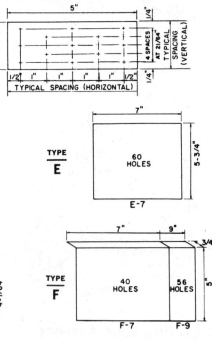

TIMBER ENGINEERING COMPANY

10.7 CASE STUDY OF A HOUSING DEVELOPMENT

The following case study of Acadia Park Housing Development project is here reproduced through the courtesy of the Forest Industries of B.C.

Acadia Park Housing Development

Over 180,000 sq. ft. of ⅝ inch Tongue-and-Groove fir plywood was used for combined subfloor and underlayment at Acadia Park, a 175-unit row-housing development for married students at the University of British Columbia. The project's architect, Vladimir Plavsic, specified T&G fir plywood flooring throughout Acadia Park for two reasons. "It goes down faster," he said, and "it gives a better surface for the application of the finished flooring."

Speed was important in the construction of Acadia Park for the 25-acre site had to go from wilderness to finished housing in just ten months. The project is the university's first venture into housing specifically designed for married students, which each year make up an increasingly large percentage of UBC's student body.

The two-storey houses are grouped in a series of nine U-shaped clusters, each enclosing a parking and play area. Kitchen windows face this area so mothers can keep an eye on children playing outside. The living rooms and bedrooms are located at the rear, to provide more peace and privacy.

The development consists of 1- and 2-bedroom suites, with a small number of 3-bedroom units. In addition to the 175-unit housing development, there is a 100-suite high-rise apartment building, the first on the university campus.

Families with several small children rent the houses, while the

Tongue and groove fir plywood was used for combined subflooring and underlayment throughout Acadia Park. Wall sections were pre-assembled at the site, then lifted into place.

apartments are for couples with no children or with babies. The 15-storey apartment building has study rooms on the top floor and students who use them are issued special keys. Other specialized features of Acadia Park include an enclosed play area, communal laundry facilities and a social area as part of the study facilities on the top floor of the highrise.

Timber frame construction was used throughout the housing project, which was built by Laing Construction and Equipment Ltd. Prefabrication was called for wherever possible, in an effort to speed construction. A total of 1600 pre-assembled roof components, including more than 1400 roof trusses, were prefabricated by Fiscus Building Components of North Vancouver. The trusses, each spanning 27 ft, have machine-nailed plywood gussets and 2x4 hemlock framing members.

Architect Plavsic called for plywood wall sheathing to speed construction and provide a good base

Roof trusses and gable ends were prefabricated by Fiscus Components of North Vancouver, B.C.

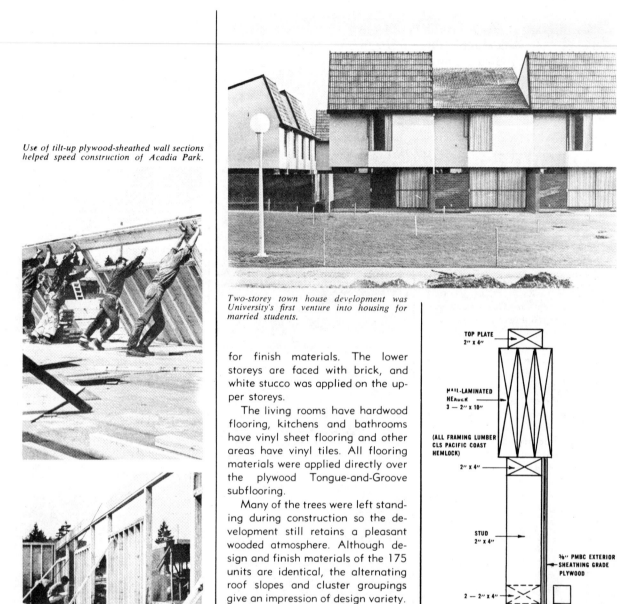

Use of tilt-up plywood-sheathed wall sections helped speed construction of Acadia Park.

Two-storey town house development was University's first venture into housing for married students.

Erected wall unit is held in place with lumber while crew completes nailing.

for finish materials. The lower storeys are faced with brick, and white stucco was applied on the upper storeys.

The living rooms have hardwood flooring, kitchens and bathrooms have vinyl sheet flooring and other areas have vinyl tiles. All flooring materials were applied directly over the plywood Tongue-and-Groove subflooring.

Many of the trees were left standing during construction so the development still retains a pleasant wooded atmosphere. Although design and finish materials of the 175 units are identical, the alternating roof slopes and cluster groupings give an impression of design variety.

The $4.5 million project is financed through Central Mortgage and Housing Corporation, which provides 50-year mortgages at 6⅛ percent for student residences on university campuses. The development is planned to be a self-sustaining, self-liquidating operation with rents covering operating costs, principal and interest.

TOP PLATE
2″ x 4″

NAIL-LAMINATED HEADER
3 — 2″ x 10″

(ALL FRAMING LUMBER CLS PACIFIC COAST HEMLOCK)

2″ x 4″

STUD
2″ x 4″

⅜″ PMBC EXTERIOR SHEATHING GRADE PLYWOOD

2 — 2″ x 4″

BRICK VENEER

BOTTOM PLATE
2″ x 4″

AIR SPACE

Cross section of lower tilt-up wall.

10.8 STEEL STUDS AND FLOOR DECKS

The material on pages 187 through 190 has been excerpted from the floor and steel catalogs on the Inland-Ryerson Construction Products Company, and is here reproduced with their permission.

MILCOR WIDE FLANGE STEEL STUDS

The 1⅜″ wide flanges of Milcor Wide Flange Steel Studs provide ample bearing and attachment surfaces for butt joints of sheet materials. The designer is thus offered wide latitude in his selection of interior and exterior facings. Self-drilling, self-tapping Milcor Tite-lock screws provide a quick means of attachment. Use of these studs eliminates the need for costly nailable studs.

product data

MATERIAL: 16 ga., 18 ga., 20 ga. — Steel, prime painted with red oxide paint, or Galvanized Steel.

SIZES: 2½″, 3¼″, 3⅝″, 4″ and 6″.

LENGTHS:

STUDS — 7-ft. to 32-ft. in increments of one inch.

TRACK — 10-ft. and 20-ft.

BRIDGING — 12-ft. only. (Same bridging as for structural studs. See pg. 5 for properties.)

PACKING: 2½″, 3¼″, 3⅝″ and 4″ — 6 pcs. per bundle; 6″ — 4 pcs. per bundle.

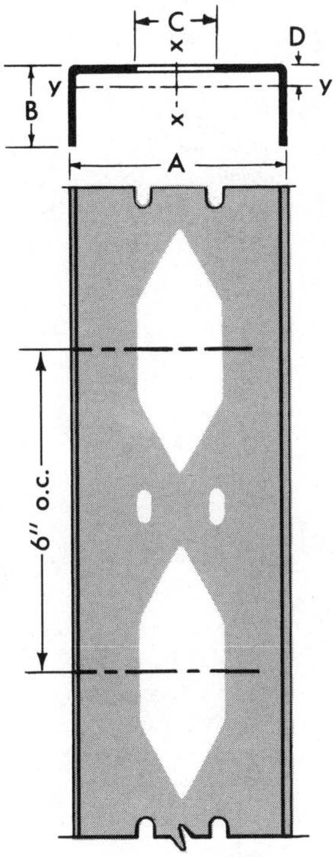

Punched stud

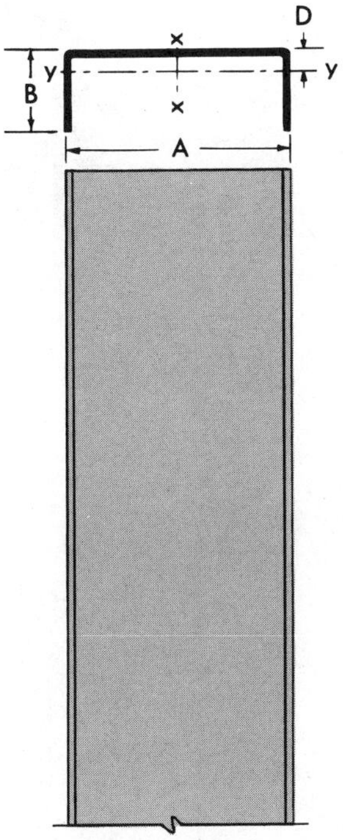

Unpunched track (Punched track furnished on special order)

physical and structural properties:

Unit	Nominal Size	Dimensions (Inches) A	B	C	Net Weight (Lbs. per Ft.)	Cross Sectional Area (Sq. In.)	Allowable Compression Stress (Lbs. per Sq. In.)	Column Factor (Q)	About Major Axis I_x (In.⁴)	S_x (In.³)	r_x (In.)	About Minor Axis I_y (In.⁴)	r_y (In.)	D (In.)	Resistance Moment
16 GA. WIDE FLANGE STUD	2½″	2½	1⅜	1¼	0.923	0.226	16,811	0.623	0.293	0.235	1.139	0.045	0.447	0.510	3,951
	3¼″	3¼	1⅜	1¼	1.079	0.271	16,811	0.623	0.546	0.336	1.419	0.054	0.446	0.430	5,648
	3⅝″	3⅝	1⅜	1¼	1.158	0.294	16,811	0.623	0.708	0.391	1.553	0.057	0.441	0.400	6,573
	4″	4	1⅜	1½	1.174	0.302	16,811	0.623	0.890	0.445	1.716	0.058	0.439	0.389	7,481
	6″	6	1⅜	2½	1.456	0.361	16,811	0.606	2.334	0.778	2.543	0.065	0.423	0.331	13,079
18 GA. WIDE FLANGE STUD	2½″	2½	1⅜	1¼	0.694	0.183	12,572	0.629	0.239	0.191	1.144	0.037	0.449	0.504	2,401
	3¼″	3¼	1⅜	1¼	0.811	0.219	12,572	0.629	0.444	0.273	1.425	0.044	0.448	0.425	3,432
	3⅝″	3⅝	1⅜	1¼	0.870	0.237	12,572	0.629	0.575	0.317	1.559	0.046	0.443	0.394	3,985
	4″	4	1⅜	1½	0.882	0.243	12,572	0.629	0.721	0.361	1.722	0.047	0.441	0.384	4,538
	6″	6	1⅜	2½	1.094	0.291	12,572	0.592	1.887	0.629	2.548	0.053	0.425	0.326	7,908
20 GA. WIDE FLANGE STUD	2½″	2½	1⅜	1¼	0.531	0.138	10,000	0.370	0.182	0.146	1.149	0.028	0.451	0.498	1,460
	3¼″	3¼	1⅜	1¼	0.620	0.165	10,000	0.370	0.338	0.208	1.430	0.033	0.449	0.419	2,080
	3⅝″	3⅝	1⅜	1¼	0.665	0.179	10,000	0.370	0.437	0.241	1.564	0.035	0.445	0.389	2,410
	4″	4	1⅜	1½	0.675	0.184	10,000	0.370	0.548	0.274	1.727	0.036	0.443	0.379	2,740
	6″	6	1⅜	2½	0.837	0.219	10,000	0.324	1.430	0.477	2.554	0.040	0.427	0.321	4,770
16, 18 or 20 Ga. UNPUNCHED DEEP LEG TRACK	2½″	2 11/16	1 5/16	Net Weights Unpunched Track (Lbs. per lin. ft.):											
	3¼″	3 7/16	1 5/16	16 GA.—2½″—1.048 lbs.; 3¼″—1.204 lbs.; 3⅝″—1.283 lbs.; 4″—1.361 lbs.; 6″—1.777 lbs.											
	3⅝″	3 13/16	1 5/16	18 GA.—2½″—0.788 lbs.; 3¼″—0.905 lbs.; 3⅝″—0.964 lbs.; 4″—1.023 lbs.; 6″—1.336 lbs.											
	4″	4 3/16	1 5/16	20 GA.—2½″—0.603 lbs.; 3¼″—0.692 lbs.; 3⅝″—0.737 lbs.; 4″—0.782 lbs.; 6″—1.022 lbs.											
	6″	6 3/16	1 5/16												

NOTES: 1. All 16 gage Studs are formed from steel meeting the requirements of ASTM A570-66T, except that the steel shall be 50,000 psi. The 16 gage Track and all 18 and 20 gage members are 33,000 psi steel of standard commercial quality.
2. Q values listed are to be used when computing axial loads, and are based on the net areas of the sections.
3. All structural properties are computed in accordance with A.I.S.I. "Specifications for the Design of Cold Formed Steel Structural Members."

CONSTRUCTION DETAILS

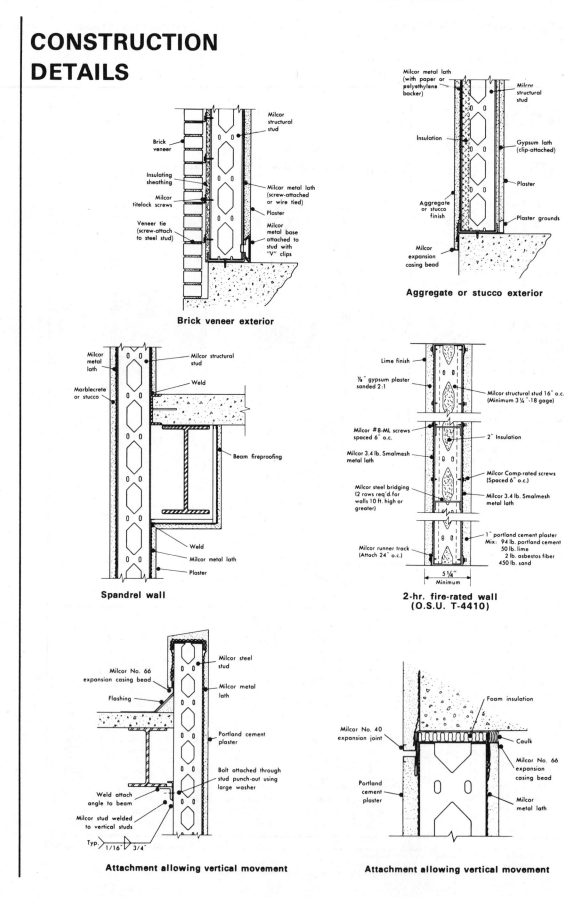

Brick veneer exterior

Aggregate or stucco exterior

Spandrel wall

2-hr. fire-rated wall
(O.S.U. T-4410)

Attachment allowing vertical movement

Attachment allowing vertical movement

MILCOR STEEL STUDS

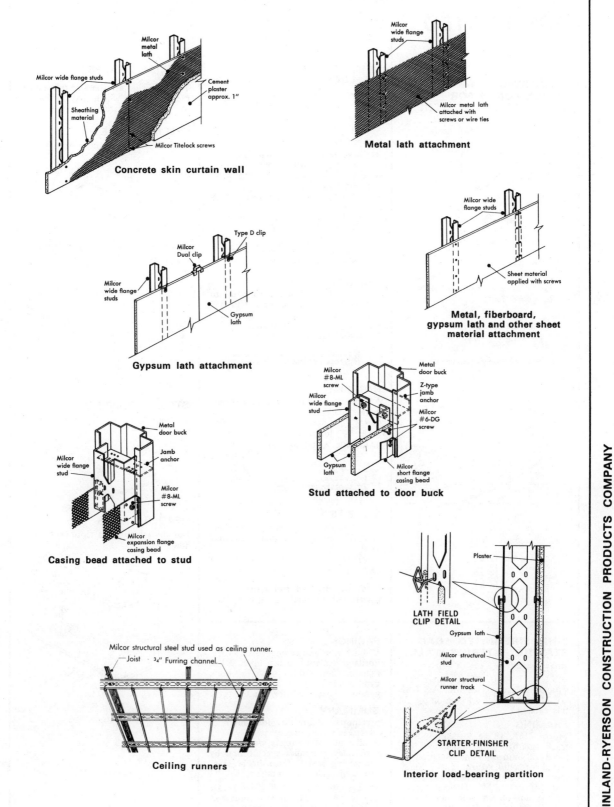

Concrete skin curtain wall

- Milcor wide flange studs
- Milcor metal lath
- Sheathing material
- Cement plaster approx. 1"
- Milcor Titelock screws

Metal lath attachment

- Milcor wide flange studs
- Milcor metal lath attached with screws or wire ties

Gypsum lath attachment

- Type D clip
- Milcor Dual clip
- Milcor wide flange studs
- Gypsum lath

Metal, fiberboard, gypsum lath and other sheet material attachment

- Milcor wide flange studs
- Sheet material applied with screws

Casing bead attached to stud

- Metal door buck
- Jamb anchor
- Milcor wide flange stud
- Milcor #8-ML screw
- Milcor expansion flange casing bead

Stud attached to door buck

- Milcor #8-ML screw
- Milcor wide flange stud
- Metal door buck
- Z-type jamb anchor
- Milcor #6-DG screw
- Gypsum lath
- Milcor short flange casing bead

Ceiling runners

- Milcor structural steel stud used as ceiling runner.
- Joist
- ¾" Furring channel.

Interior load-bearing partition

- Plaster
- LATH FIELD CLIP DETAIL
- Gypsum lath
- Milcor structural stud
- Milcor structural runner track
- STARTER-FINISHER CLIP DETAIL

For additional details, contact Inland-Ryerson Construction Products Company.

INLAND-RYERSON CONSTRUCTION PRODUCTS COMPANY

MILCOR STEEL STUDS

ASSOCIATED DATA

SUGGESTED DESIGN LOADS FOR POWER DRIVEN FASTENERS

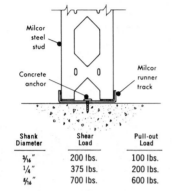

Milcor steel stud

Concrete anchor

Milcor runner track

Shank Diameter	Shear Load	Pull-out Load
$3/16''$	200 lbs.	100 lbs.
$1/4''$	375 lbs.	200 lbs.
$5/16''$	700 lbs.	600 lbs.

Above design loads assume minimum 2,500 psi concrete and fastener driven into concrete a distance of six times its diameter.

Anchor Size and Spacing Selection Graph

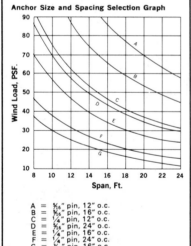

Wind Load, PSF.

Span, Ft.

A = $5/16''$ pin, 12" o.c.
B = $5/16''$ pin, 16" o.c.
C = $1/4''$ pin, 12" o.c.
D = $5/16''$ pin, 24" o.c.
E = $1/4''$ pin, 16" o.c.
F = $1/4''$ pin, 24" o.c.
G = $3/16''$ pin, 16" o.c.

MILCOR TITELOCK SCREWS

Self-drilling, self-tapping screws provide a fast, easy method of fastening materials to steel studs or other metal components as thick as $1/16''$.

NO. 6-DG

Attaches drywall or gypsum lath. The $3/16''$ flared head with slightly rounded edge produces excellent seating without rupturing paper skins.
SIZES: $3/4''$, 1", $1 5/8''$ long — #6 shank.
PACKING (per ctn.): $3/4''$ — 2,500; 1" — 2,000; $1 5/8''$ — 1,000.

No. 6-DG — $1 5/8''$

NO. 6-PL

Attaches plywood sub-flooring up to $3/4''$ thick to metal members. Has $5/16''$ diameter flared head.
SIZE: $1 7/8''$ long — #6 shank.
PACKING: 2,000 per ctn.

No. 6-PL

NO. 8-ML

Attaches metal lath, lathing accessories, metal panels, etc. The $7/16''$ diameter pan washer head spans lath openings.
SIZE: $1/2''$ long — #8 Shank.
PACKING: 2,500 per ctn.

No. 8-ML

No. 8-FS

NO. 8-FS

Attaches wood fiber sheathing, rigid insulation, etc. Has $7/16''$ diameter flared head.
SIZE: $1 1/4''$ long — #8 shank.
PACKING: 1,000 per ctn.

One #2 Phillips head bit furnished in each carton.

SPECIAL SHAPES

The following are some of the supplementary items which may be needed for a complete framing system.

SPECIAL ANGLE

For attachment of stud panels.

SPECIAL DEEP LEG TRACK

For walls designed to allow vertical movement.

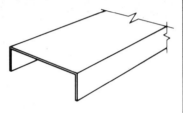

STEEL STRAPPING

For racking resistance and wind shear bracing.

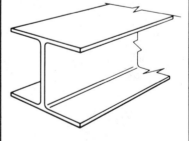

HEADER BEAMS

For distribution of floor/roof joist loads.

2-HR. FIRE RATED, LOAD BEARING EXTERIOR WALL

FIRE TEST

O.S.U. #T-4410. (Test panels constructed in accordance with A.S.T.M. #E-119. See detail drawing on pg. 22 this catalog.)

FRAMING

Milcor $3 1/4''$, 18 ga. structural studs, track and bridging fabricated into panels. 2" mineral wool insulation inserted in cavity. Milcor 3.4 lb. Smalmesh metal lath attached both sides: with Milcor #8-ML Titelock Screws on gypsum plaster side and Milcor #8-CR Comp-Rated on Portland cement side.

FACINGS

1" thick Portland cement mix: 94 lbs. Portland cement; 50 lbs. lime; 2 lbs. asbestos fiber; 450 lbs. sand. $7/8''$ thick gypsum plaster mix: 100 lbs. fibered gypsum plaster; 200 lbs. sand.

SUMMARY

System passed 2-hr. fire, load and hose stream test both sides.
Availability of complete wall, tested both sides, simplifies specification by eliminating search for suitable separate 2-hr. rated assemblies.

Ratings also apply to all wider 18 ga. and/or 16 ga. Milcor structural and wide flange steel studs at their respective design loads.

INLAND-RYERSON CONSTRUCTION PRODUCTS COMPANY

10.9 SCAFFOLDING

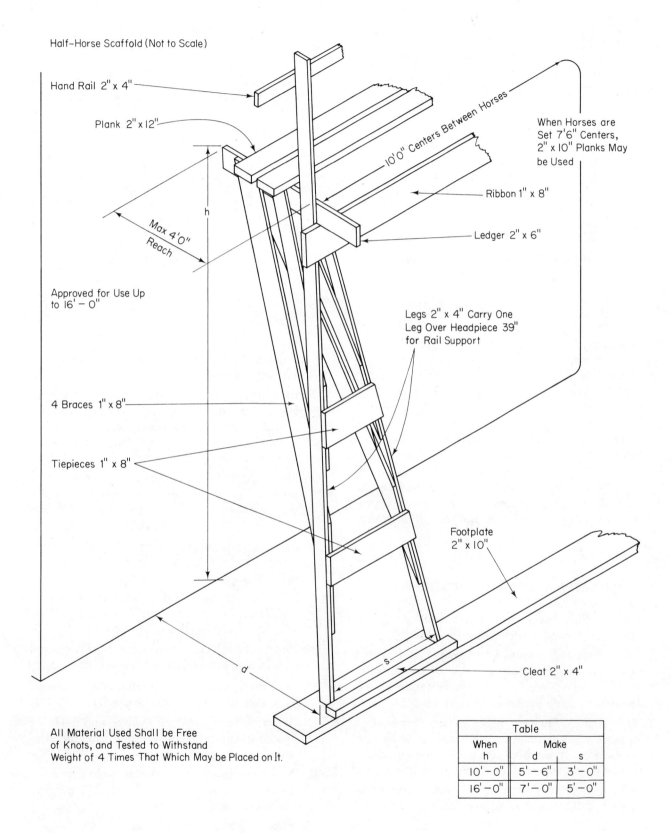

Half–Horse Scaffold (Not to Scale)

Hand Rail 2" x 4"

Plank 2" x 12"

10'0" Centers Between Horses

When Horses are
Set 7'6" Centers,
2" x 10" Planks May
be Used

Max 4'0"
Reach

h

Ribbon 1" x 8"

Ledger 2" x 6"

Approved for Use Up
to 16' – 0"

Legs 2" x 4" Carry One
Leg Over Headpiece 39"
for Rail Support

4 Braces 1" x 8"

Tiepieces 1" x 8"

Footplate
2" x 10"

d

s

Cleat 2" x 4"

All Material Used Shall be Free
of Knots, and Tested to Withstand
Weight of 4 Times That Which May be Placed on It.

Table		
When	Make	
h	d	s
10'–0"	5'–6"	3'–0"
16'–0"	7'–0"	5'–0"

Fig. 10.12.　Half horse scaffold (not to scale).

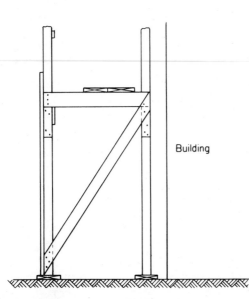

Fig. 10.13. Double pole scaffold.

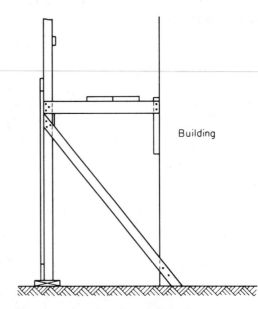

Fig. 10.14. Single pole scaffold.

An excellent source of information regarding all types of scaffolding is your local Workmen's Compensation Board. As with all other aspects of building construction; better and safer general working conditions are being developed. It must be kept in mind that scaffolding may have to withstand the rigors of frigid temperatures, occasional unexpected gusts of wind, extremes of heat, hail, snow, or torrential rain. The material on pages 193 through 194 has been excerpted from literature supplied by the Morgen Manufacturing Company and is here reproduced with the permission of the copyright holder.

CHAPTER 10 REVIEW QUESTIONS

1. Explain how the new dimension standards for softwood lumber differs from the old.

2. The old size of a piece of 2 x 4 was $1\frac{5}{8}''$ x $3\frac{5}{8}''$. What is the new size of a piece of *seasoned* 2 x 4 and the new size of a piece of 2 x 4 in its dry state? *This is important; study the tables on pages 166–167.*

3. Sketch the following types of structural wood fasteners: (a) joist and beam hangers; (b) all purpose framing anchors; (c) post anchors (base); (d) post cap; (e) metal bridging; (f) wall anchors.

4. Which method of securing wood framing members together do you prefer between (a) struc-

tural wood fasteners, and (b) nailing the members together? Support your choice in not more than 120 words.

5. Inquire at your local workmen's compensation board for drawings of wooden scaffolding, wooden shoring for trench work, types of ladders permissible, and the width of scaffold decks allowable.

6. Study the three examples of corner stud assemblies at Fig. 10.1. State your preference and support it in writing of not more than 100 words.

7. Study Fig. 10.2, page 169 and state which is the most economical on labor costs. *Remember that labor costs cover other trades too.*

8. Study Fig. 10.5, page 191. What size spacer should be used (on the header) with the new lumber standards for the drawing shown on pages 193–194. *This is an important question.*

9. List twelve places where wood backing is necessary in wood construction. See Fig. 10.8, page 174.

10. In which types of residential construction is it desirable that steel studs be used?

11. Research your Building Bylaws, and make a neat sketch of the types of wall and roof anchors that are mandatory in your area.

12. Name six different types of scaffold that are mandatory in your area. (See article 1.7, page 11 for class research.)

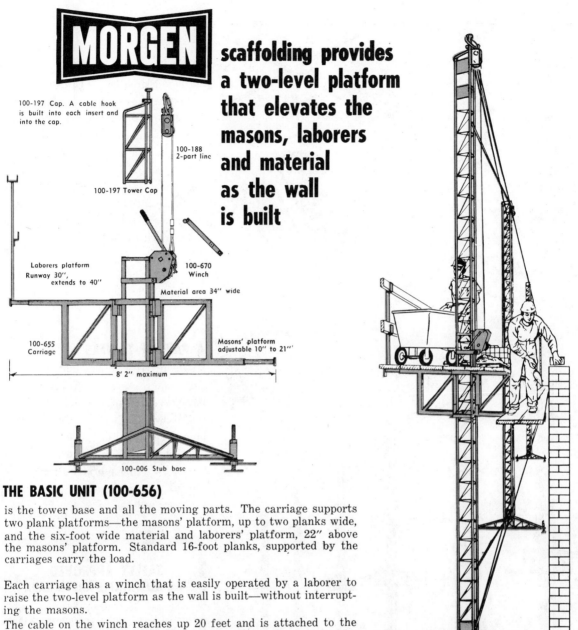

MORGEN scaffolding provides a two-level platform that elevates the masons, laborers and material as the wall is built

100-197 Cap. A cable hook is built into each insert and into the cap.

100-188 2-part line

100-197 Tower Cap

Laborers platform Runway 30″, extends to 40″

100-670 Winch

Material area 34″ wide

100-655 Carriage

Masons' platform adjustable 10″ to 21″

8′ 2″ maximum

100-006 Stub base

THE BASIC UNIT (100-656)

is the tower base and all the moving parts. The carriage supports two plank platforms—the masons' platform, up to two planks wide, and the six-foot wide material and laborers' platform, 22″ above the masons' platform. Standard 16-foot planks, supported by the carriages carry the load.

Each carriage has a winch that is easily operated by a laborer to raise the two-level platform as the wall is built—without interrupting the masons.

The cable on the winch reaches up 20 feet and is attached to the highest cable hook it will reach. The carriage can be locked to the tower at any level to permit attaching the cable to a higher hook.

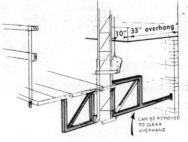

10″ 33″ overhang

CAN BE REMOVED TO CLEAR OVERHANG

Normal Two-Level Planking

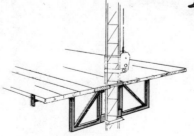

Planked Flat for Inside Work or Overhead Obstructions

Three-Level Planking for Stairwells, Elevator Shafts or Special Applications

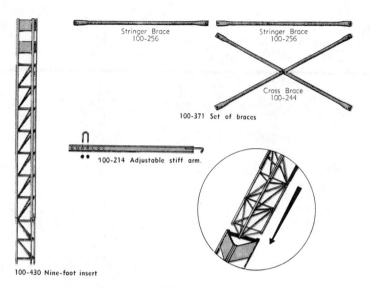

Stringer Brace
100-256

Stringer Brace
100-256

Cross Brace
100-244

100-371 Set of braces

100-214 Adjustable stiff arm.

100-430 Nine-foot insert

TOWERS

of any desired height are erected using nine-foot sections called inserts. The male end of the insert fits into a female sleeve. You can plug any insert into any base or any other insert. Basic towers, up to $29\frac{1}{2}$ feet high, erected by a fork lift. If necessary, additional inserts are added to reach full wall height. Inserts are added by men with a gin pole, or with a fork lift or crane.

A PAIR OF TOWERS

makes a free-standing unit. Making a pair requires a stringer brace at the bottom and a stringer and cross brace for every nine feet of height. These braces automatically space the tower $7\frac{1}{2}$ feet apart. Pairs of towers are connected by stringer braces to make continuous scaffolding for any desired length of wall.

Post and Beam Construction

In this chapter are presented some typical wood post and beam construction methods which are well adapted for modular residential construction where curtain wall (and roof planking) may be erected between the supporting posts and/or beams. There is nothing new about post and beam construction as may be attested to by the many existing architectural orders erected by the Greeks and the Romans.

11.1 POSTS AND BEAMS

Both posts and beams are designed to carry the superimposed loads, and although posts are usually of solid wood, the beams may be fabricated in several different ways to meet either cost, or appearance, or both. The types of beams most used in residential construction are as follows:

1. Solid single pieces of lumber which may be dressed, or faced with finished plywood, or cedar boards, and so on, see Fig. 11.1.

2. Laminated wood beams built up (on the job) with several pieces of lumber that are dry nailed together; such beams are often used as a main beam for underpinning the first floor joists, see Fig. 11.2.

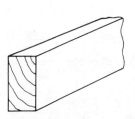

Fig. 11.1. Solid wood beam.

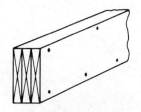

Fig. 11.2. Wood laminated beam.

3. Glue-laminated beams, built up from precision dressed pieces of lumber which are glued together under controlled conditions, and with the exposed faces sanded and stained to meet the aesthetic taste of the owner, see Figs. 11.3 and 11.4.

4. Built up solid plywood beams, see Fig. 11.5.

5. Box beams which are fabricated with select frame members separating the plywood skin finish

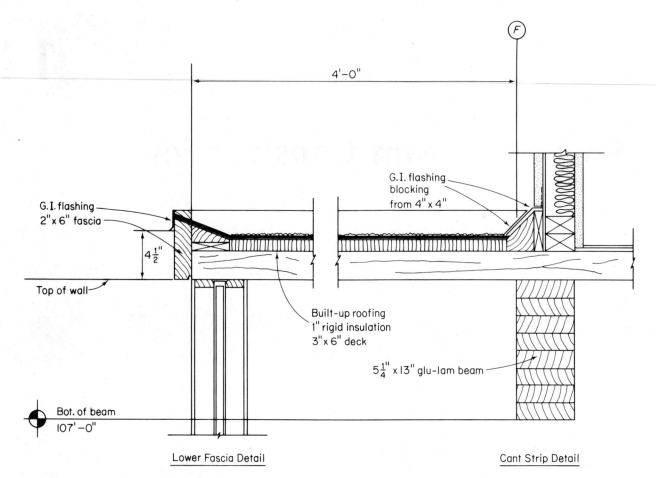

G.I. flashing
2" x 6" fascia

4½"

Top of wall

Bot. of beam
107'–0"

G.I. flashing
blocking
from 4" x 4"

Built-up roofing
1" rigid insulation
3" x 6" deck

5¼" x 13" glu-lam beam

4'–0"

F

Lower Fascia Detail

Cant Strip Detail

Fig. 11.3. Post and beam construction.

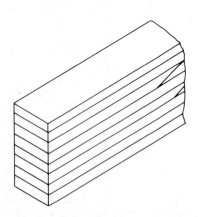

Fig. 11.4. Laminated on flat.

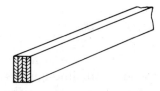

Fig. 11.5. Solid plywood beam.

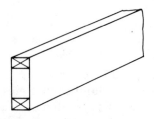

Fig. 11.6. Plywood box beam with wood frame members.

which may be nailed and/or glued together, see Fig. 11.6.

11.2 CURTAIN WALLS IN POST AND BEAM CONSTRUCTION

Where post and beam modular methods are used in residential construction, the posts are usually placed at 4'-0" OCs between which curtain walls are erected. Such curtain walls may be built from:

1. Standard wood wall framing.
2. Brick, or brick veneer.
3. Fiber, metallic, ceramic, or plastic standard 4'-0" x 8'-0" panels.
4. Stucco over other materials.
5. Stone or concrete block.
6. One modular section may lend itself for the accommodation of the chimney for a solid fuel fire grate.
7. Glass blocks.
8. Glass windows which are designed to complement the total modular design.

Brick and concrete block will lay-up into the exact width of the curtain walls. Modular brick will lay-up to 8" in length requiring six bricks between 4'-0" OC posts, and concrete block will lay-up three blocks of 16" units in the same length.

The construction of the perimeter walls and the roof and roofing will afford more tolerable working conditions for the workmen to erect free standing interior walls and partitions; additionally, this type of construction permits the future changing of the positions of interior walls and partitions to meet new needs, and without any temporary support for the beams during alteration work.

Article 11.3 and Figs. 11.7 and 11.8 are here reproduced through the courtesy of the Canadian Wood Council. The typical post and beam details, Fig. 11.3 and Figs. 11.9 through 11.30 (which are used in the construction of a residence) are here reproduced through the courtesy of Mr. Alton M. Bowers, architect, Calgary, Alberta.

11.3 POST AND BEAM: DESCRIPTION AND DESIGN FACTORS

General Description

The post and beam system is essentially a framework made up of decking, beams and posts sup-

ported on a foundation. The floor and roof decks transfer loads to the beams which, in turn, carry them to posts and on down to the foundation.

Post and beam construction dates back to some of the earliest buildings of Greece and shows up in traditional Japanese homes, half-timbered Tudor houses and early American and English Colonial homes. This construction was followed, in the middle of the 19th century in North America, by the development of conventional frame construction which used more and lighter construction members than post and beam construction; the construction members were concealed in the frame.

Posts and beams are larger and spaced farther apart than studs, joists and rafters of conventional framing (Figures 11.7 and 11.8). Horizontal spaces between beams are normally spanned by plank decking, but conventional joist construction is sometimes used. Spaces between posts can be filled with wall panels, glass or supplementary framing and sheathing.

In recent years, trends in architecture have separated the posts and beams from the conventional frame, and made them visible as a prominent part of the construction system. This concept is used in many types of buildings. Residential, commercial, industrial and recreational buildings are some examples where post and beam framing is used, either by itself or in combination with conventional framing. The architectural style of post and beam is both charming and impressive, but, like other systems, should be used with discrimination.

Post and beam, though similar in style, should not be confused with heavy timber construction; in building codes, 'heavy timber' is classed by itself, and calls for specific minimum sizes of beams, columns and other components.

Advantages of Post and Beam

SURFACE FINISHES

Exposed posts, beams and planks can be stained to preserve natural colour, texture and grain characteristics of the wood. Stain is long lasting and, unlike paint films, does not form a semi-permeable membrane.

Paint retards fluctuations of moisture content in the wood, thus minimizing checking. Colours can be chosen that blend with furniture and the surroundings. Unpigmented varnishes should not be used if exposed to the sunlight. Lack of pigments in the varnish allows ultra-violet rays to penetrate through the varnish to the wood surface. This causes the surface of the wood to deteriorate, leaving unsightly finishes.

FREEDOM IN PLANNING

Partitions in post and beam construction do not normally carry vertical loads; therefore, their location is

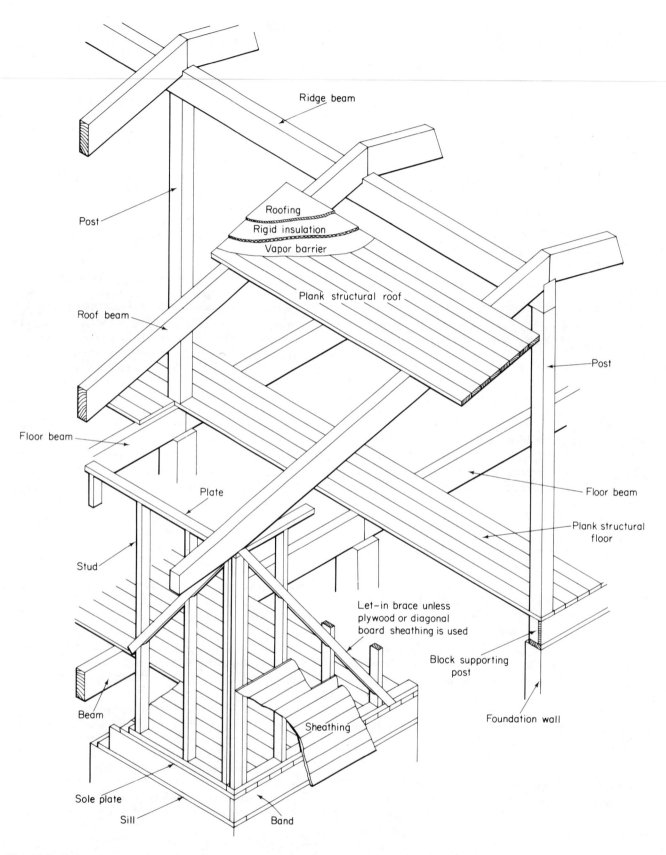

Fig. 11.7. Typical post and beam framing. *(Courtesy of Canadian Wood Council, Ottawa.)*

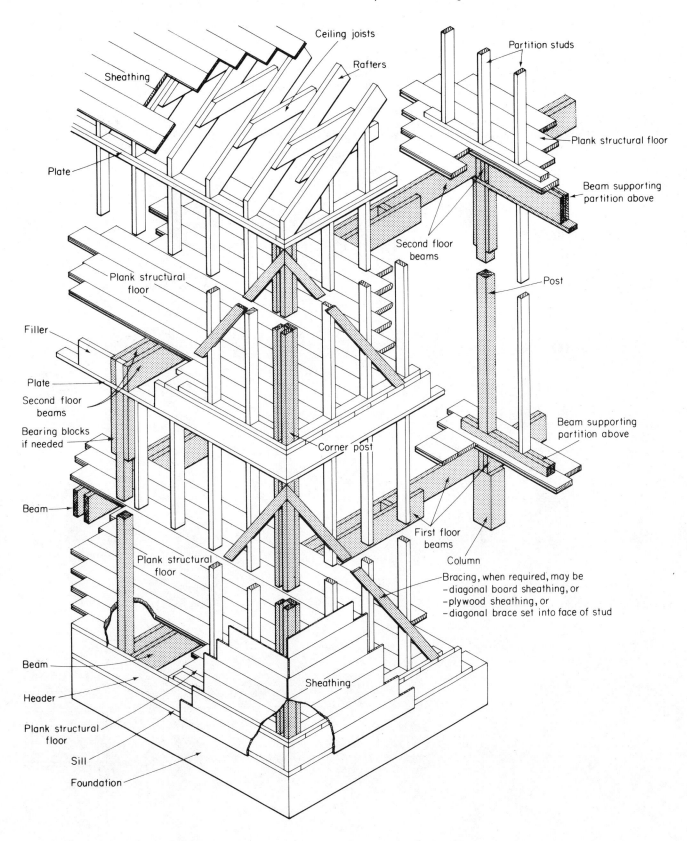

Fig. 11.8. Post and beam framing combined with conventional framing in two story house. *(Courtesy of Canadian Wood Council, Ottawa.)*

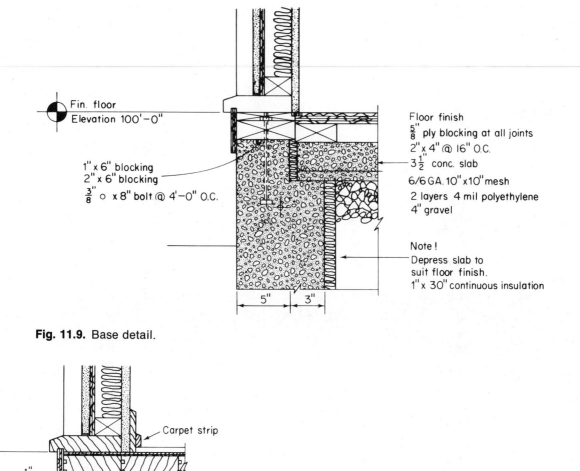

Fin. floor
Elevation 100'-0"

1" x 6" blocking
2" x 6" blocking
$\frac{3}{8}$" ⌀ x 8" bolt @ 4'-0" O.C.

Floor finish
$\frac{5}{8}$" ply blocking at all joints
2" x 4" @ 16" O.C.
$3\frac{1}{2}$" conc. slab
6/6 GA. 10" x 10" mesh
2 layers 4 mil polyethylene
4" gravel

Note!
Depress slab to
suit floor finish.
1" x 30" continuous insulation

5"   3"

Fig. 11.9. Base detail.

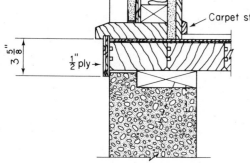

Carpet strip

$3\frac{5}{8}$"

$\frac{1}{2}$" ply

Fig. 11.10. Base detail.

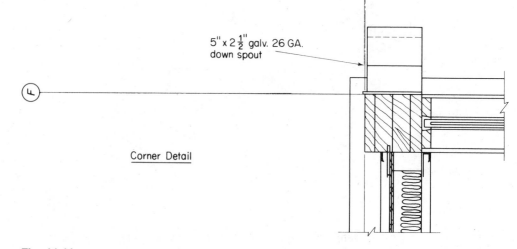

5" x 2$\frac{1}{2}$" galv. 26 GA.
down spout

Corner Detail

Fig. 11.11.

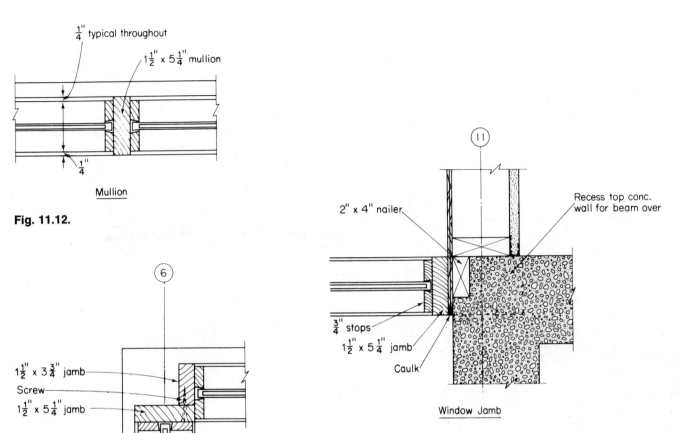

$\frac{1}{4}''$ typical throughout

$1\frac{1}{2}'' \times 5\frac{1}{4}''$ mullion

$\frac{1}{4}''$

Mullion

Fig. 11.12.

$1\frac{1}{2}'' \times 3\frac{3}{4}''$ jamb

Screw

$1\frac{1}{2}'' \times 5\frac{1}{4}''$ jamb

Corner Detail

Fig. 11.13.

2" x 4" nailer

Recess top conc. wall for beam over

$\frac{3}{4}''$ stops

$1\frac{1}{2}'' \times 5\frac{1}{4}''$ jamb

Caulk

Window Jamb

Fig. 11.14.

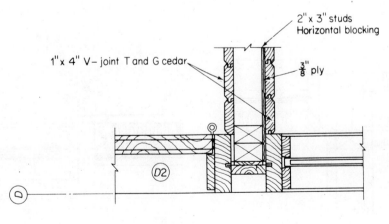

2" x 3" studs
Horizontal blocking

1" x 4" V-joint T and G cedar

$\frac{3}{8}''$ ply

Door-Window Jamb

Fig. 11.15.

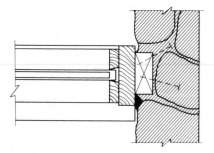

Window Jamb

Fig. 11.16.

Typical Wall Construction

$\frac{3}{4}$" stucco
Stucco wire
Building paper
$\frac{3}{8}$" sheathing
2" x 3" studs @ 16" O.C.
2" batt insulation
4 mil. polyethylene
Interior finish

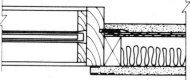

Fig. 11.17.

Typical Wall Construction

$\frac{3}{4}$" stucco
Stucco wire
Building paper
$\frac{3}{8}$" sheathing
2" x 3" studs @ 16" O.C.
2" batt insulation
4 mil. polyethylene
Interior finish

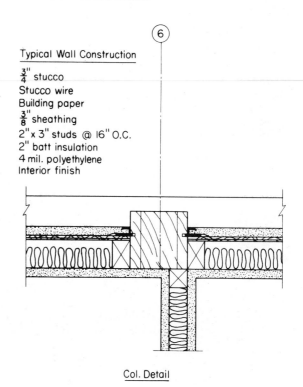

Col. Detail

Fig. 11.18.

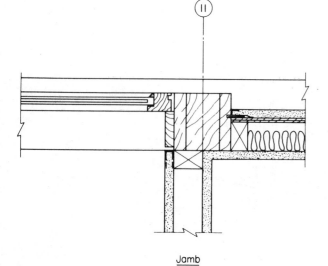

Jamb

Fig. 11.19.

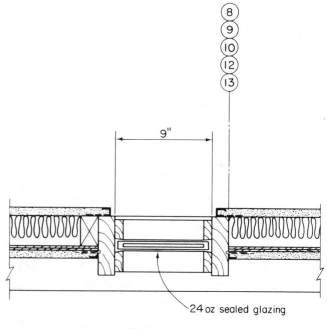

24 oz sealed glazing

Window Jambs

Fig. 11.20.

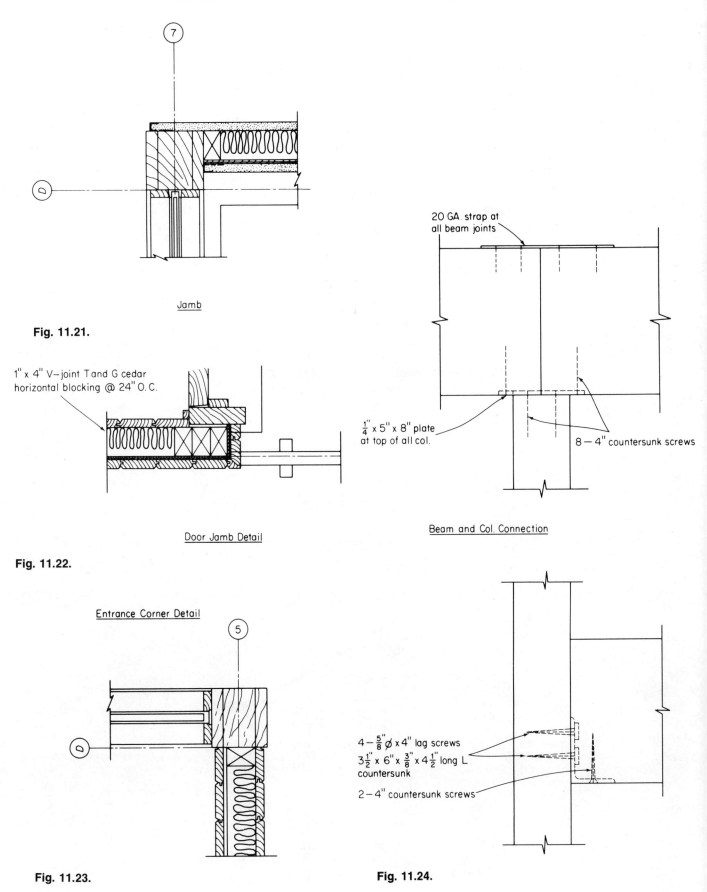

Jamb

Fig. 11.21.

1" x 4" V—joint T and G cedar
horizontal blocking @ 24" O.C.

Door Jamb Detail

Fig. 11.22.

20 GA. strap at
all beam joints

$\frac{1}{4}$" x 5" x 8" plate
at top of all col.

8 — 4" countersunk screws

Beam and Col. Connection

Entrance Corner Detail

Fig. 11.23.

4 — $\frac{5}{8}$" ⌀ x 4" lag screws
$3\frac{1}{2}$" x 6" x $\frac{3}{8}$" x $4\frac{1}{2}$" long L
countersunk

2 — 4" countersunk screws

Fig. 11.24.

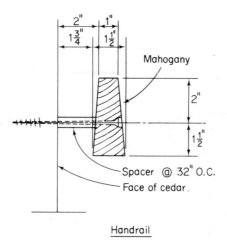

Mahogany

Spacer @ 32" O.C.

Face of cedar.

Handrail

Fig. 11.25.

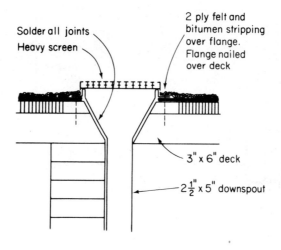

Solder all joints

Heavy screen

2 ply felt and bitumen stripping over flange. Flange nailed over deck

3" x 6" deck

2½" x 5" downspout

Roof Drain Detail

Fig. 11.26.

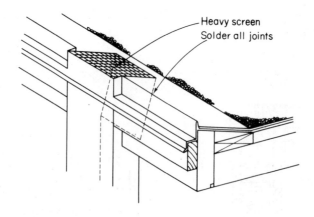

Heavy screen
Solder all joints

Garage Drain Detail

Fig. 11.27.

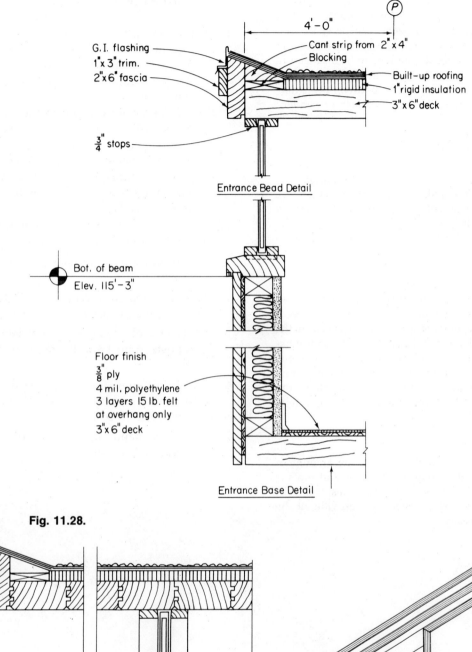

G. I. flashing
1"x 3" trim.
2"x 6" fascia

4'-0"

Cant strip from 2" x 4"
Blocking

Built-up roofing
1" rigid insulation
3"x 6" deck

$\frac{3}{4}$" stops

Entrance Bead Detail

Bot. of beam
Elev. 115'-3"

Floor finish
$\frac{3}{8}$" ply
4 mil. polyethylene
3 layers 15 lb. felt
at overhang only
3"x 6" deck

Entrance Base Detail

Fig. 11.28.

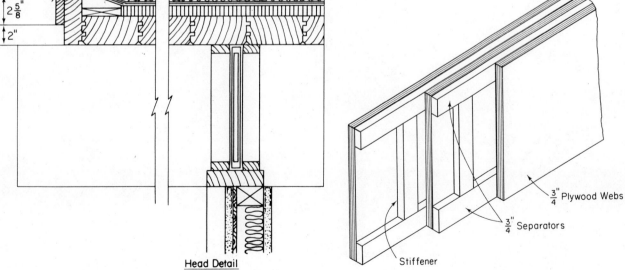

$\frac{7}{8}$"

$2\frac{5}{8}$"

2"

Head Detail

Fig. 11.29.

$\frac{3}{4}$" Plywood Webs

$\frac{3}{4}$" Separators

Stiffener

Fig. 11.30. Plywood web beam.

not controlled by structural considerations. This allows the designer freedom in layout of interior floor plans. Because posts and beams can be spaced far apart, the building interior is more spacious, allowing furniture and interior decorations to be arranged in many different ways. Posts, beams and planks may be exposed or closed in. Spaces between posts may be filled with glass, decorative panels or conventional framing.

SPEED OF ERECTION

Well-planned post and beam framing has fewer but larger-sized pieces than conventional framing. This results in simple details, fewer joints and therefore faster erection. The roof soon covers the floor, giving protection for all further work. The remainder of the house can be fabricated in the shop and inserted into the structure.

REDUCTION IN BUILDING HEIGHT

In post and beam framing, room height is measured from the floor to the underside of the plank, whereas in conventional construction, it is measured from the floor to the underside of the joist. The difference between thickness of the plank and depth of the joist provides a reduction in total height of the building. This makes possible a saving in sheathing, siding and length of studs, as well as down-spouting, stairs and other services.

THERMAL PERFORMANCE

Because wood is a good insulator, the amount of additional insulation required in walls and roofs is less than for other construction materials; if the climate is not too severe, wood may provide all the insulation necessary.

CHAPTER 11 REVIEW QUESTIONS

1. Of the many types of wood beams used in post and beam residential construction what type of beam would you suggest using for:
   (a) a beam supporting the first floor joists from a basement;
   (b) an open ceiling beam in the living room of a house finished in a rustic design;
   (c) an open ceiling beam for the living room of a house finished with decorative wood panels;
   (d) a beam spanning a wide span without intermediate supports between the outside supporting walls?

2. In less than 150 words, define modular construction and support its use in residential construction.

3. List six building materials adaptable to modular construction.

4. Make a sketch of a laminated post with a metal anchor, and a metal post cap. See pages 176 to 183.

5. What clear distance is usually left between upright posts in post and beam residential construction?

6. List six different types of curtain walls or other features that may be accommodated between the upright posts in residential construction.

7. Make a freehand sketch (without scale) of a section of an 8" x 18" glue-lam beam.

Masonry Units, Glass Blocks, and Gliding Doors

In this chapter is presented a selection of excerpts from manufacturers' literature on the subject of masonry units, their types and methods of construction. This is followed with information about glass block construction and gliding doors.

12.1 INTRODUCTION TO MASONRY UNITS

During the last few years great progress has been made in the manufacture of masonry units in varying sizes, shapes, weights, textures and colors. They may be:

(a) Manufactured for bearing walls, curtain or partition walls. (A curtain wall supports no compressive load other than its own weight. It is built between framing members of the building of which it forms part.)

(b) Light or heavy weight, with two or three cores, and in different sizes, shapes and colors to meet specific needs. See pages 208–215.

(c) Textured, natural concrete color, or glazed finish, also with a different color finish on opposite faces.

(d) Cut on the job for fitting around other fixtures.

(e) Filled with insulation.

(f) Rodent free.

(g) Easy for subtrades to chase for wiring such as electric, telephone, and television wiring. (To chase is to cut into the face of a structure to accommodate conduit and so on.)

(h) Fire resisting.

(i) Used for outside basement walls which may be water or dampproofed.

(j) Plastered or furred out to receive decorative wood or other panels. (Furring strips of wood may be applied to a surface in order to secure another material such as tiles to a ceiling or wood panels to a concrete block wall.)

(k) Used for wythes which may be anchored to each other. See pages 208–210.

(l) Used in inclement or hot weather to build the outside faces of curtain walls; the masons may then work under weather controlled conditions to build the inner wythes, or back-ups.

(m) Used for walls to be reinforced and bonded into any design, and be provided with control joints to prevent fracturing of walls due to settlement or small earth tremors. See page 212.

(n) Used ideally for modular construction.

(o) Used (when suitably reinforced) to form piers for framing. See Pilaster No. 9, page 210.

Where ground conditions are favorable, masonry units may be used to build basement walls, affording a saving in time and material for the construction, erection, removing, and reconditioning of concrete forms. Each concrete block when layed-up becomes an immediate integral part of the structure. A well designed basement may also be provided with party walls separating different areas, and some of the faces of such walls may be built with texture faced blocks with a different color on opposite faces. The outside walls of crawl spaces may be insulated against cold penetration from the outside and as a means of conserving heat within the walled enclosed area. See article 7.9.

12.2 ADJUSTABLE WALL TIES, REINFORCING, AND CONTROL JOINTS

The material for this section has been excerpted from bulletins published by Dur-O-Wall National Inc., and is here reproduced with permission. See pages 208 through 212.

DUR-O-WAL ADJUSTABLE WALL TIES, THEIR STRUCTURAL PROPERTIES AND RECOMMENDED USE*

DESCRIPTION

As shown in Fig. 1 the ties consist of two pieces, a double eye unit and a double pintle unit. They are made of 3/16 in. diameter cold drawn steel wire complying with ASTM A82. The wire is galvanized to retard corrosion of portions of the tie which are not embedded in mortar and thereby protected.

Adjustable ties are designed for use in double wythe masonry walls in which the facing and backing do not course-out at proper intervals as to permit tying with masonry headers or straight metal ties. Where this situation is met by bending straight ties to fit or by using flexible metal strips the results are not satisfactory from either a structural or economic standpoint.

The ties will accommodate variations in joint levels up to 1½ in. Practice in installation is to set the eye units in the wythe which is built up first, it being more convenient to engage the pintles in the eyes rather than vice versa. Generally the pintles are installed downward through the eyes although if necessary they can be installed in the reverse position. As indicated in Fig. 1 the looped portion or shank of each unit is embedded in the horizontal mortar joint.

STRENGTH OF TIES

The strength of the ties was investigated in a program of tests performed by the Armour Research Foundation of the Illinois Institute of Technology. Specimens consisting of single eyes and single pintles were loaded as shown schematically in Fig. 2. The average results of three or more tests are given in Table 1.

TABLE 1

Unit (3/16 in. diam. wire)	Load Position	y (Exposed Shank) in.	e (Eccentricity) in.	Yield Load, lb.
Single eye	—	—	—	550
Single pintle	1	1¾	0	490
do	2	1	1½	50
do	2	1¾	1½	40

The results indicate that the double pintle-eye tie will resist a compressive or tensile load of at least 80 lb. where applied at the maximum eccentricity (1½ in.) and greater loads up to about 980 lb. as the load eccentricity decreases to zero.

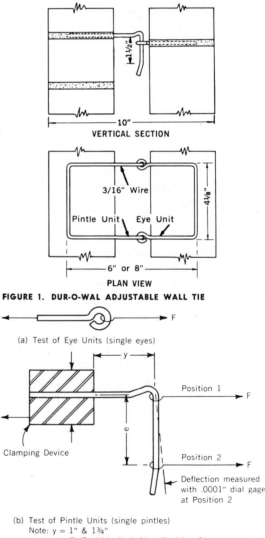

FIGURE 1. DUR-O-WAL ADJUSTABLE WALL TIE

(a) Test of Eye Units (single eyes)

(b) Test of Pintle Units (single pintles)
Note: y = 1" & 1¾"
e = 0 (Position 1) & 1½" (Position 2)

FIGURE 2. SCHEMATIC OF TESTS OF EYE AND PINTLE UNITS

TRUSS INSTALLATION AND PLACEMENT

PLACEMENT

Out to out spacing of side rods shall be approximately 2 inches less than the nominal thickness of the wall or wythe and shall be placed to insure a minimum of $\frac{5}{8}$" mortar cover on exterior face of walls and $\frac{1}{2}$" on the interior face.

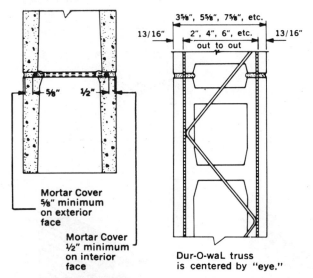

Mortar Cover $\frac{5}{8}$" minimum on exterior face

Mortar Cover $\frac{1}{2}$" minimum on interior face

Dur-O-waL truss is centered by "eye."

CORNERS AND TEES

Prefabricated or job fabricated corner and tee sections shall be used to form continuous reinforcement around corners, and for anchoring abutting walls and partitions. Material in corner and tee sections shall correspond to type and design of reinforcement used.

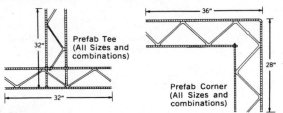

Prefab Tee (All Sizes and combinations)

Prefab Corner (All Sizes and combinations)

SPLICES

Side rods shall be lapped at least 6 inches at splices.

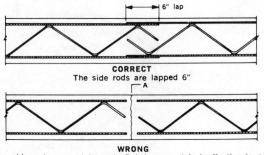

CORRECT
The side rods are lapped 6"

WRONG
The side rods are not lapped. Reinforcement is ineffective in preventing a crack from starting and opening up at A.

Proper lapping of rods at splices is essential to the continuity of the reinforcement so that tensile stress will be transmitted from one rod to the other across the splice.

FACED AND COMPOSITE WALLS

Dur-O-waL shall be centered over both wythes and the galvanized diagonal cross rods shall serve as ties. Dur-O-waL shall be spaced 16" o.c. vertically and the collar joint shall be filled solidly with mortar.

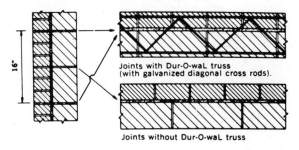

Joints with Dur-O-waL truss (with galvanized diagonal cross rods).

Joints without Dur-O-waL truss

WALLS NOT TIED WITH DUR-O-WAL TRUSS CROSS RODS

Place Dur-O-waL in (each wythe) (backing wythe) 16" o.c. of all faced, cavity or veneered wall not otherwise noted as being tied with the cross rods of Dur-O-waL.

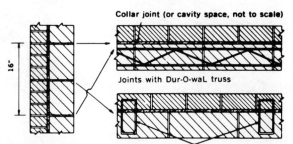

Collar joint (or cavity space, not to scale)

Joints with Dur-O-waL truss

Joints with either regular or Adjustable Rectangular or Z-Type ties.

EXTRA TIES

Provide extra ties at all openings in masonry walls by bending and hooking side rod or cross rod of Dur-O-waL or by adding either regular or adjustable wall ties in alternate courses with Dur-O-waL.

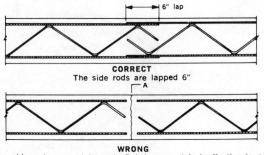

JOINT B
JOINT A

ELEVATION

Bend and hook side rod

ALTERNATE 1 — JOINT A

Bend and hook diagonal cross rod

ALTERNATE 2 — JOINT A

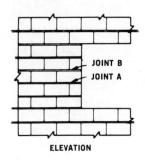

Either regular or Adjustable Rectangular or Z-Type ties

ALTERNATE 3 — JOINT B

TYPICAL TRUSS INSTALLATIONS

COMPOSITE WALLS

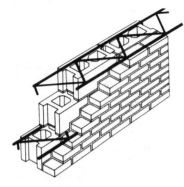

12″ Tied Wall No. 12 Dur-O-waL
16″ c. to c.

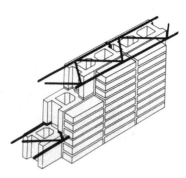

12″ Tied Wall with Stack Bond
Facing No. 12 Dur-O-waL 16″ c. to c.

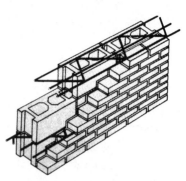

12″ Tied Wall Stack Bond Backup
No. 12 Dur-O-waL Trirod 16″ c. to c.

CAVITY WALLS

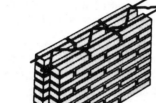

10″ Cavity Wall No. 10 Dur-O-waL
With Drip 16″ c. to c.

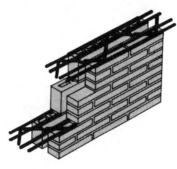

10″ Cavity Wall No. 10 Dur-O-waL
Double With Drip 16″ c. to c.

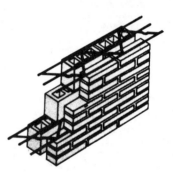

12″ Cavity Wall No. 12 Dur-O-waL
Trirod with Drip 16″ c. to c.

SINGLE WYTHE WALLS

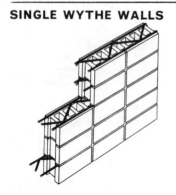

Stack Bond Load Bearing Walls.
Dur-O-waL—8″ c. to c. Top 3
Courses. 16″ c. to c. Remainder of
Wall. Non Load Bearing Walls.
Dur-O-waL—16″ c. to c.

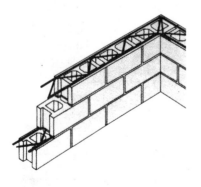

8″ Wall showing Corner No. 8
Dur-O-waL 16″ c. to c.

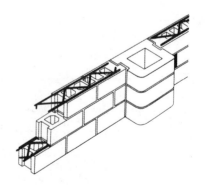

8″ Wall with Pilaster No. 8
Dur-O-waL 16″ c. to c.

WALL OPENINGS—Unless otherwise noted, Dur-O-waL shall be installed in the first and second bed joints, 8 inches apart immediately above lintels and below sills at openings and in bed joints at 16-inch vertical intervals elsewhere. Reinforcement in the second bed joint above or below openings shall extend two feet beyond the jambs. All other reinforcement shall be continuous except it shall not pass through vertical masonry control joints.

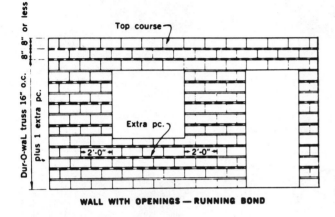

WALL WITH OPENINGS — RUNNING BOND

SINGLE WYTHE WALLS—Exterior and interior. Place Dur-O-waL 16″ o.c. and in bed joint of the top course.

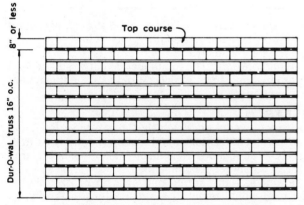

WALL WITH NO OPENINGS — RUNNING BOND

FOUNDATION WALLS—Place Dur-O-waL 8″ o.c. in upper half to two-thirds of wall.

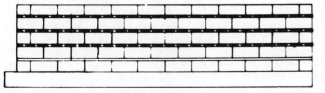

Foundation Wall

BASEMENT WALLS—Place Dur-O-waL in first joint below top of wall and 8″ o.c. in the top 5 bed joints below openings.

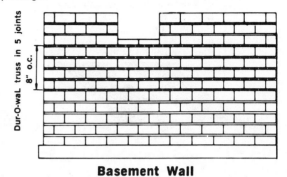

Basement Wall

STACK BOND—Dur-O-waL shall be placed 16″ o.c. vertically in walls laid in stack bond except it shall be placed 8″ o.c. for the top 3 courses in load bearing walls.

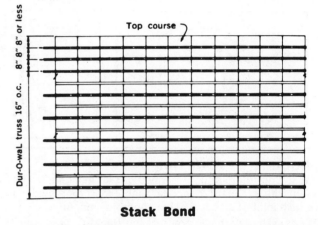

Stack Bond

CONTROL JOINTS—Unless as otherwise noted, all reinforcement shall be continuous except it shall not pass through vertical masonry control joints.

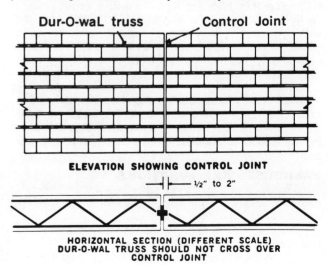

ELEVATION SHOWING CONTROL JOINT

½″ to 2″

HORIZONTAL SECTION (DIFFERENT SCALE)
DUR-O-WAL TRUSS SHOULD NOT CROSS OVER CONTROL JOINT

RAPID® CONTROL JOINT

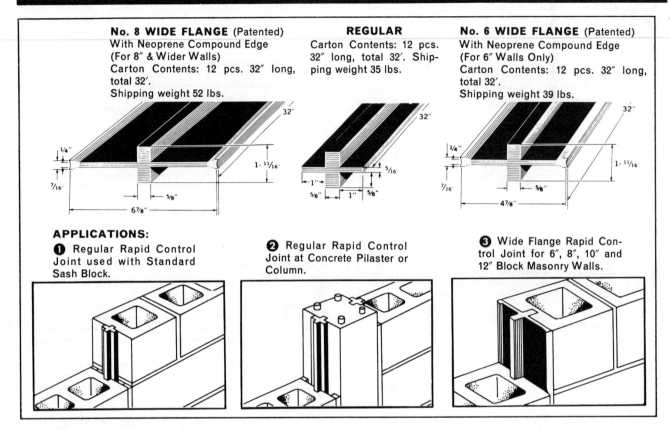

No. 8 WIDE FLANGE (Patented)
With Neoprene Compound Edge
(For 8″ & Wider Walls)
Carton Contents: 12 pcs. 32″ long,
total 32′.
Shipping weight 52 lbs.

REGULAR
Carton Contents: 12 pcs.
32″ long, total 32′. Shipping weight 35 lbs.

No. 6 WIDE FLANGE (Patented)
With Neoprene Compound Edge
(For 6″ Walls Only)
Carton Contents: 12 pcs. 32″ long,
total 32′.
Shipping weight 39 lbs.

APPLICATIONS:

❶ Regular Rapid Control Joint used with Standard Sash Block.

❷ Regular Rapid Control Joint at Concrete Pilaster or Column.

❸ Wide Flange Rapid Control Joint for 6″, 8″, 10″ and 12″ Block Masonry Walls.

PRODUCT DESCRIPTION

Rapid Control Joint is preformed gasket designed to be used with standard concrete sash block to provide a vertical joint of stress relief in concrete masonry walls.

Types—The two basic types of the Rapid Control Joint are the Regular and the Wide Flange. The Wide Flange utilizes a concave neoprene compound edge that can be compressed tightly in the joint.

Regular Rapid Control Joint is a factory-extruded solid section of rubber conforming to ASTM D-2000 2AA-805 with a durometer hardness of approximately 80 when tested in conformance with ASTM D-2240. The shear section is ⅝-inch in thickness. The 5/16-inch flange thickness is designed to provide a control-joint width of approximately ⅜-inch.

Wide Flange Rapid Control Joint is a factory molded product of rubber conforming to ASTM D-2000 2AA-805 with a compressible neoprene compound edge conforming to ASTM 2BC-310C12 with a durometer hardness of 30.

BASIC USES AND FUNCTIONS

To provide a vertical stress relieving joint in concrete masonry walls while providing adequate shear strength for lateral stability of the wall.

For control joints to perform adequately, the following three principles should be considered:
1. *Stress Relief:* The joint must cut the masonry wall completely from top to bottom, so as to form a truly stress relieving joint.

2. *Shear Strength:* The joint must be structurally sound in that sufficient strength is developed to provide for lateral stability.
3. *Weather Tight:* The joint must be either self-sealing or one that can be easily caulked to prevent moisture penetration.

RESEARCH

Shear tests were conducted on the two types of Rapid Control Joint at IIT Research Institute (Formerly Armour Research Foundation). Results in Table 1 indicate a large factor of safety for recommended control joint spacing. For example, with control joint spacing of 20′ a 20 psf wind load would result in a shear stress in the key of about 27 psi. This is significantly less than the shear strength of the material developed in the tests.

TABLE 14: SHEAR STRENGTH

Control Joint Type	Average Load per 8″ of Joint (lbs)	Shear Strength (psi)
Wide Flange	2350	470
Regular	2706	541

SPECIFICATION FOR RAPID CONTROL JOINT.

(Regular) (Wide Flange) Rapid Control Joint, a product of Dur-O-waL, or approved rubber material of equal shear strength shall be placed in masonry walls as noted on plans.

12.3 GLAZED MASONRY WALLS

The material for this section has been excerpted from literature supplied by The Burns & Russell Company, and is here reproduced with the permission of the copyright owner. *(See the following pages.)*

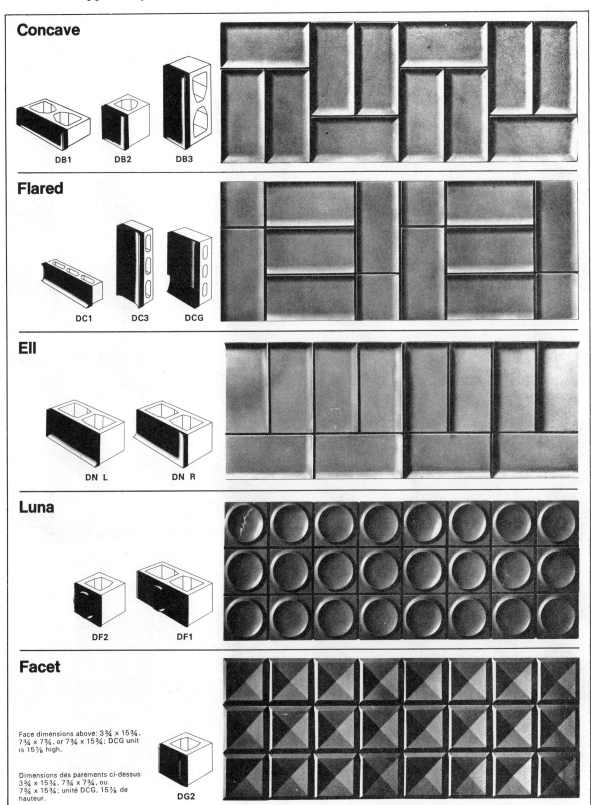

Concave

DB1   DB2   DB3

Flared

DC1   DC3   DCG

Ell

DN L   DN R

Luna

DF2   DF1

Facet

Face dimensions above: 3¾ x 15¾, 7¾ x 7¾, or 7¾ x 15¾; DCG unit is 15⅞ high.

Dimensions des parements ci-dessus: 3¾ x 15¾, 7¾ x 7¾, ou 7¾ x 15¾; unité DCG, 15⅞ de hauteur.

DG2

FUNCTIONAL SERIES   SERIES FONCTIONNELLES

CAP CHAPERON	FIELD CHAMP	COVE BASE BASE AVEC GORGE

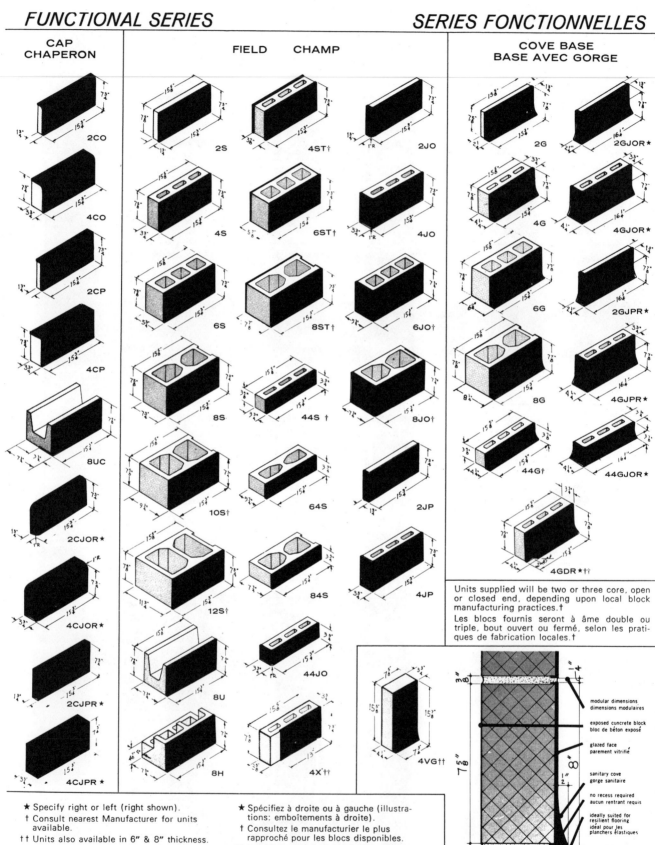

Units supplied will be two or three core, open or closed end, depending upon local block manufacturing practices.†

Les blocs fournis seront à âme double ou triple, bout ouvert ou fermé, selon les pratiques de fabrication locales.†

modular dimensions
dimensions modulaires

exposed concrete block
bloc de béton exposé

glazed face
parement vitrifié

sanitary cove
gorge sanitaire

no recess required
aucun rentrant requis

ideally suited for
resilient flooring
idéal pour les
planchers élastiques

★ Specify right or left (right shown).
† Consult nearest Manufacturer for units available.
†† Units also available in 6″ & 8″ thickness.

★ Spécifiez à droite ou à gauche (illustrations: emboîtements à droite).
† Consultez le manufacturier le plus rapproché pour les blocs disponibles.
†† Les blocs sont également disponibles en épaisseur de 6″ et 8″.

Build and Finish in one operation
Bâtissez et finissez en une seule opération

Spectra-Glaze
Glazed Masonry Units
Blocs De Béton Vitrifié Pour Maçonnerie

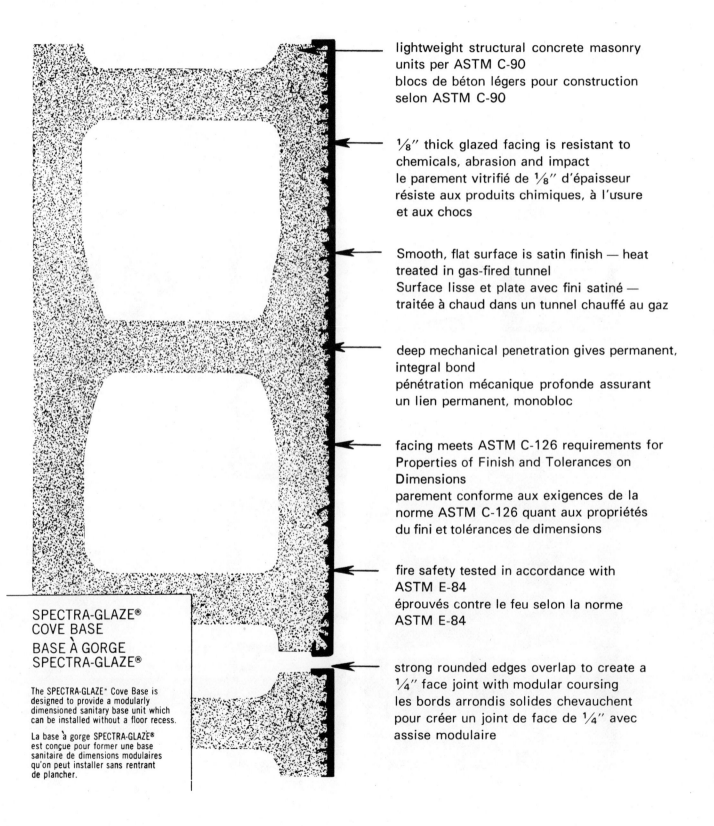

lightweight structural concrete masonry units per ASTM C-90
blocs de béton légers pour construction selon ASTM C-90

$\frac{1}{8}$″ thick glazed facing is resistant to chemicals, abrasion and impact
le parement vitrifié de $\frac{1}{8}$″ d'épaisseur résiste aux produits chimiques, à l'usure et aux chocs

Smooth, flat surface is satin finish — heat treated in gas-fired tunnel
Surface lisse et plate avec fini satiné — traitée à chaud dans un tunnel chauffé au gaz

deep mechanical penetration gives permanent, integral bond
pénétration mécanique profonde assurant un lien permanent, monobloc

facing meets ASTM C-126 requirements for Properties of Finish and Tolerances on Dimensions
parement conforme aux exigences de la norme ASTM C-126 quant aux propriétés du fini et tolérances de dimensions

fire safety tested in accordance with ASTM E-84
éprouvés contre le feu selon la norme ASTM E-84

strong rounded edges overlap to create a $\frac{1}{4}$″ face joint with modular coursing
les bords arrondis solides chevauchent pour créer un joint de face de $\frac{1}{4}$″ avec assise modulaire

SPECTRA-GLAZE®
COVE BASE
BASE À GORGE
SPECTRA-GLAZE®

The SPECTRA-GLAZE® Cove Base is designed to provide a modularly dimensioned sanitary base unit which can be installed without a floor recess.

La base à gorge SPECTRA-GLAZE® est conçue pour former une base sanitaire de dimensions modulaires qu'on peut installer sans rentrant de plancher.

12.4 GLASS BLOCKS, AND GLASS BLOCK PANEL CONSTRUCTION

The material for this section has been excerpted from booklets supplied by the Pittsburgh Corning Corporation, and is here reproduced with permission. *(See the following pages.)*

Glare and Heat Reducing Inserts

As indicated below, many of the PC glass units are available with fibrous inserts for medium control of brightness, solar heat gain and light transmission.
"LX" Insert—white fibrous insert for use where medium control is desired. SUNTROL® Insert—green fibrous insert for use where maximum control is desired.

Non-Light Directing Units

DECORA   **ARGUS**   **ARGUS PARALLEL FLUTES**   **VUE**

Non-Light Directing Units

DECORA®—The Decora unit gives high light transmission and can be installed in any position. Design is pressed into the inner faces, and the outer faces of the unit are smooth. This unit is almost transparent and is not recommended for sun exposures. Available in 6, 8, and 12-inch squares and with "LX" or Suntrol® insert.
ARGUS®—The outer faces of this unit are smooth. Rounded flutes on inner faces are at right angles to each other. Laid with flutes on one side either horizontal or vertical. 6, 8, and 12-inch squares.
ARGUS PARALLEL FLUTES—High light transmission is available with this unit. It is the same basic unit as the Argus, except flutes are parallel. This unit can be laid with flutes horizontal or vertical. Available in 6, 8, and 12-inch squares plain and 8 and 12-inch squares with the "LX" insert.
VUE®—Here is a glass unit that provides high light transmission and good visibility. It is frequently used in panels of other patterns to provide a vision area where desired. It can be laid without regard to which edge is side or top. Both the outer and the inner faces are smooth and clear. 8 and 12-inch squares.

Sculptured Glass Modules

These basic geometric shapes are pressed into the glass to a depth of approximately 1½ inches on both sides of the unit. All patterns are available in clear glass. Each unit is 12" square and weighs about 17 pounds. They are installed on the same proportionate module and in the same manner as all Pittsburgh Corning Glass Units. Modules like other Pittsburgh Corning glass units offer light, insulation value and a finished job in one operation.

4 x 12 Glass Blocks

This striking rectangular block is smooth on the outer faces, textured on the inner surfaces. In clear glass, it is offered in three degrees of light control. Plain (normal), "LX" (medium), or Suntrol (maximum). They can be integrated with standard blocks, Sculptured Glass Modules, or any number of other masonry building materials.

WEDGE   **PYRAMID**

HARLEQUIN   **LEAF**

Light-Diffusing Units

ESSEX®—Diffuses sunlight in all directions. For sun or non-sun exposures. Give maximum light transmission; moderate brightness and solar heat control. 8 and 12-inch squares. Available with "LX" or Suntrol inserts. 8 and 12-inch squares.

Light-Directing Units

PRISM B—Recommended on sun or non-sun exposures for maximum light transmission; moderate brightness and solar heat control. Available in 8 and 12-inch squares. Available with "LX" or Suntrol inserts in 8 and 12-inch squares. NOTE: Do not use light-directing blocks BELOW eye-level because they will direct light up into your eyes. EYE-LEVEL is considered to be 6 feet above the finish floor.

Glass Building Units

Details
(scale 1″ = 1′-0″)

These pages show elevations and sections of typical glass block panels. The large scale sections are typical head, jamb and sill details to show the architect principles of construction only.

Any structural members must be calculated for safe loading, and local building codes checked for any possible restrictions on panel sizes or detail. While single panels of glass block are limited to a maximum of 144 square feet, panel and curtain wall sections up to a maximum area of 250 square feet may be erected if properly braced to limit movement and settlement.

If chase construction as shown on page 9 cannot be used, substitute the panel anchor construction that is shown on page 10. Panel anchors are used to give lateral support for glass block panels.

Any glass block installation that is made in a frame construction shall have the wood adjacent to the mortar properly primed with Pittsburgh Corning Asphalt Emulsion.

For Underwriters' Listings refer to Underwriters Guide Cards R-2556A, R-2556B and R-2556C, Guide Number 120 IW7.

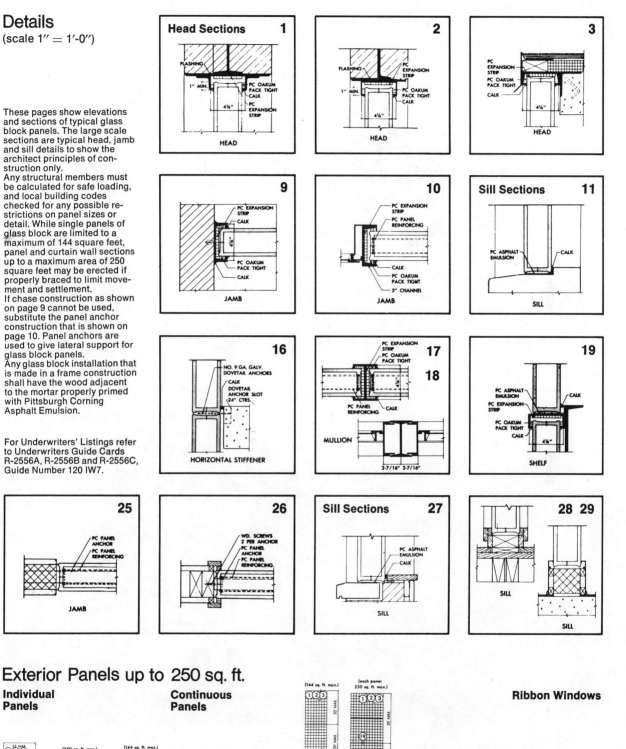

Exterior Panels up to 250 sq. ft.

Individual Panels

Continuous Panels

Ribbon Windows

12.5 MASONRY UNIT LINTELS

The material for this section has been excerpted from 'The Clay Masonry Manual', and is published with the permission of the 'Canadian Structural Clay Association.'

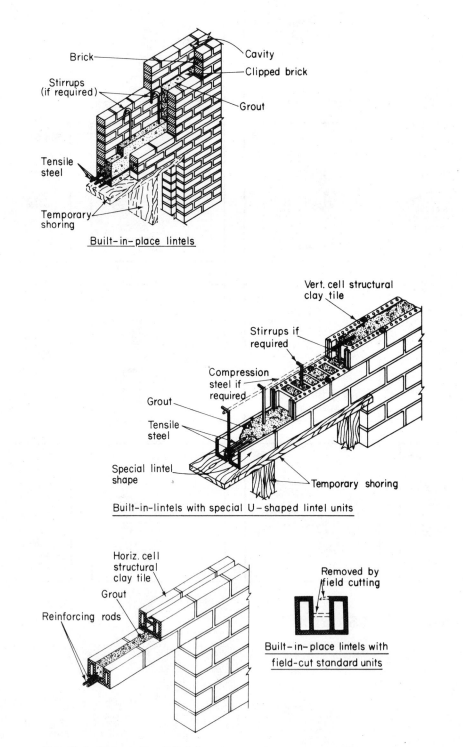

Fig. 12.1. Masonry unit lintels.

12.6 GLIDING DOORS

The material for this section has been excerpted from booklets supplied by the Andersen Corporation, and is here reproduced with the permission of the copyright holder. *(See the following pages.)*

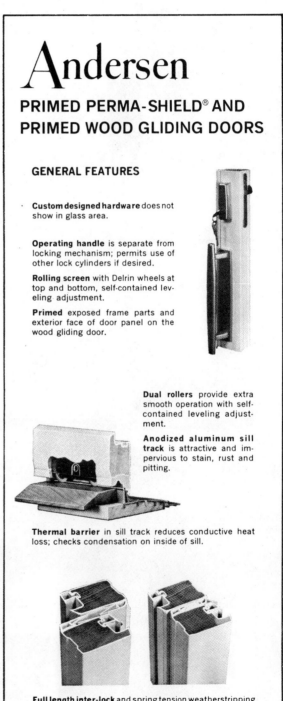

Andersen

PRIMED PERMA-SHIELD® AND PRIMED WOOD GLIDING DOORS

GENERAL FEATURES

Custom designed hardware does not show in glass area.

Operating handle is separate from locking mechanism; permits use of other lock cylinders if desired.

Rolling screen with Delrin wheels at top and bottom, self-contained leveling adjustment.

Primed exposed frame parts and exterior face of door panel on the wood gliding door.

Dual rollers provide extra smooth operation with self-contained leveling adjustment.

Anodized aluminum sill track is attractive and impervious to stain, rust and pitting.

Thermal barrier in sill track reduces conductive heat loss; checks condensation on inside of sill.

Full length inter-lock and spring tension weatherstripping pull meeting stiles together for a weathertight seal.

PRIMED WOOD
GLIDING DOOR

SHORT SPECIFICATIONS

All Gliding Doors shall be Andersen WOOD Gliding Doors as manufactured by the Andersen Corporation, Bayport, Minnesota.

DETAILED SPECIFICATIONS

a. Frame, exterior casings and door panels of selected wood and treated with a toxic water repellent preservative. Exposed exterior frame parts and exterior face of door panels primed. PVC screen head channel applied. Extruded anodized aluminum sill facing with PVC thermal barrier, stainless steel cap applied over operating door track.

b. Door panels, 1⅞" thick, stop glazed with tempered (safety) insulating glass.* Left or right hand operating (viewed from outside) as specified, triple door right hand only.

Check with Andersen distributor for types available and shipping schedule.

c. Weatherstripping, a combination of rigid vinyl (PVC), polypropylene woven pile and flexible vinyl.

d. Hardware, zinc die-cast exterior and interior handles, exterior key lock and interior locking lever with golden bronze tone finish, alternate: silver satin finish. Twin adjustable rollers, golden dichromate finish, tandem nylon wheels with ball bearings, applied to bottom of operating door. Stationary door brackets silver satin finish.

e. Screen, aluminum frame with white Perma-Clean® finish, Alclad aluminum cloth, gun-metal finish, 18 x 16 mesh. Delrin rollers, with spring mounting in top rail and an adjustable mounting in bottom rail, wool pile weatherstrip at meeting stile, all factory applied. Interior and exterior door pulls furnished.

Andersen Corporation reserves the right to change details or specifications without notice, in the interest of improving products or service.

WOOD DOOR SIZES

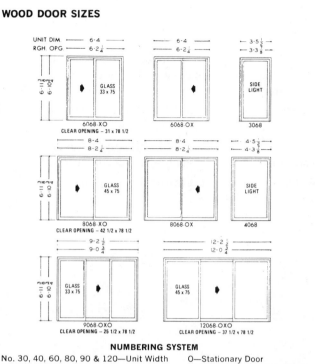

NUMBERING SYSTEM

No. 30, 40, 60, 80, 90 & 120—Unit Width
No. 68—Unit Height
X—Operating Door

O—Stationary Door
(Numbering figured as viewed from outside)

Primed WOOD gliding door unit

Includes frame, door panels, glazed tempered (safety), weatherstripped and treated with toxic water repellent preservative. Exterior frame parts and exterior face of door panels are factory primed. Bronze tone interior and exterior handles, keyed exterior lock and interior locking lever furnished. Oak threshold and flashing included.

Primed WOOD side light

A stationary unit with frame and panel milled to match operating door unit.

GLAZING ALTERNATES

*GLARE REDUCING safety insulating glass.

*HEAT ABSORBING safety insulating glass.

Check with Andersen distributor for types available and shipping schedule.

DOOR OPERATION may be specified either left or right hand (as viewed from outside). Triple door right hand only.

HARDWARE ALTERNATE

Silver satin may be specified instead of golden bronze tone.

OPTIONAL

SCREEN/Strong aluminum frame with white Perma-Clean™ finish, Alclad aluminum cloth, gun metal finish, 18 x 16 mesh. Interior and exterior door pulls furnished.

SILL SUPPORT/Treated and primed wood support strip for extended portion of metal sill facing.

CHAPTER 12 REVIEW QUESTIONS

1. Define a curtain wall.
2. List twelve characteristics of manufactured masonry units.
3. Why is it advantageous to build outside basement walls with concrete units where ground conditions permit their use?
4. Sketch an adjustable wall tie as used in building with manufactured masonry units.
5. List four distinct advantages of using adjustable wall ties in new concrete block wall construction.
6. Sketch and design a control joint as used in concrete unit built walls.
7. Draw a freehand plan of a concrete masonry wall built with a pilaster.
8. Sketch a built-in-place lintel with special U-shaped lintel units (reinforced), and arrow name the parts.
9. Sketch and define the following wood members of a window or gliding door: (a) sill; (b) jamb; (c) meeting rail; (d) support mullion.
10. Define: (a) a bearing wall; (b) a curtain wall; (c) dampproofing; (d) waterproofing; (e) a wythe; (f) bonding; (g) modular construction; (h) a pier; and (i) chasing.

13

Ceilings and Flat Roofs

In this chapter we shall discuss various types of ceiling framing, both for flat roofs which support a roof deck and ceiling and for pitch roofs; and methods and materials used in the building of roof decks. This is followed with information about softwood and heavy timber roof decks.

13.1 CEILING FRAMING

In most residential construction a ceiling of gypsum boards or plaster—but not a flat roof deck—can be attached to standard wood ceiling joists placed at 16" OCs. *It is imperative that these OCs be placed accurately to receive a modular sized plaster lath. Be familiar with your local building code!*

At Fig. 13.1a is shown two methods of securing the ends of ceiling joists over a load bearing partition. As an additional precaution, to prevent the ends of the joists from twisting and warping, a 1 x 4 may be nailed on top of the ends of the joists. Remember to place the crown sides of all ceiling rafters uppermost. At Fig. 13.1b is shown the outer ends of ceiling joists which may be secured to the rafters over the wallplates. This is excellent practice where the rafters and joists are placed with similar OC spacings. It lends stability to the ceiling joists, and gives a truss effect to the roof rafters.

At Figs. 13.2 and 13.3 are shown methods of framing ceiling joists to a flush beam; and at Fig. 13.4 is shown ceiling framing with the joists running in two directions.

At Figs. 13.5 and 13.6 are shown three methods of providing good nailing for gypsum and other boards at internal partitions.

Hanging Beam. Check your local building code for allowable spans (in different species of wood) for the various sizes of lumber, and permitted OC spacings for ceiling joists supporting a ceiling; also for joists supporting a roof deck, snow load, and ceiling. For some ceiling framing, stiffeners are mandatory such as bridging in floor joists (see Fig. 9.14 page 153) or by installing a hanging beam as follows:

Step 1: Read Fig. 13.7 with these steps.

Step 2: String a taut mason's line alongside the central ceiling joist (crown side up) and set it true to line.

Step 3: Place a 1 x 4 at right angles to the central joist and temporarily nail it.

Step 4: Starting from the central ceiling joist and working both ways, place and temporarily nail all the remaining joists *at their exact specified OCs.*

223

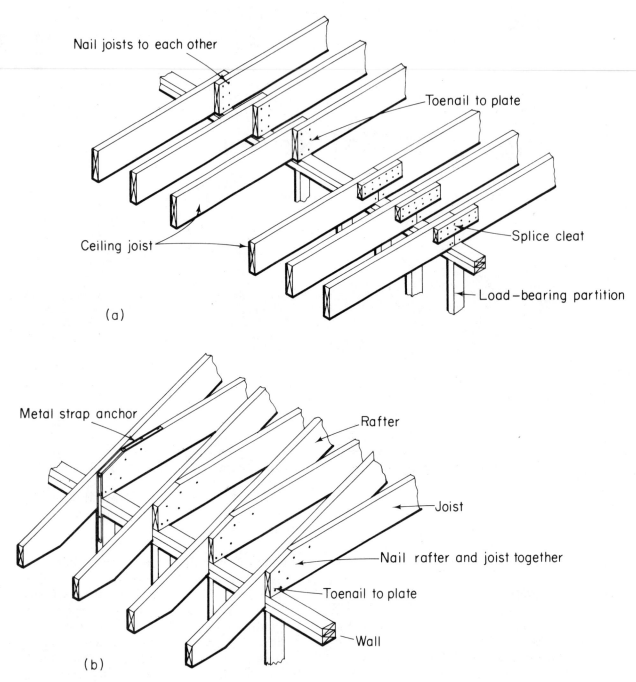

Fig. 13.1. Ceiling joist connections: (a) at center partition with joists lapped or butted, and (b) at outside wall. *(Courtesy of Forest Products Laboratory, U.S. Department of Agriculture.)*

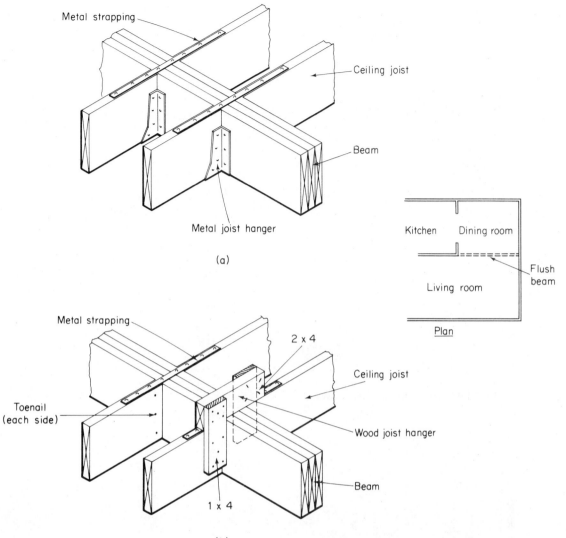

(a)

(b)

Fig. 13.2. Flush ceiling framing: (a) metal joist hanger, and (b) wood hanger. *(Courtesy of Forest Products Laboratory, U.S. Department of Agriculture.)*

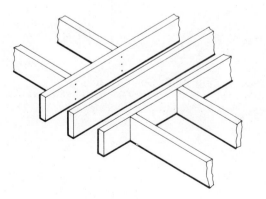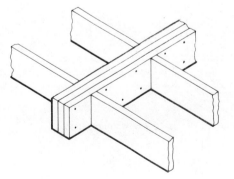

Fig. 13.3. Joists nailed to flush beam.

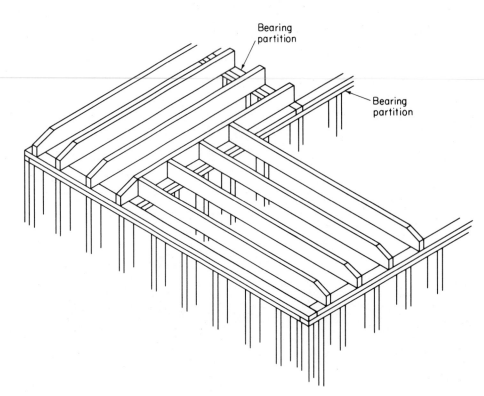

Fig. 13.4. Joists running two ways.

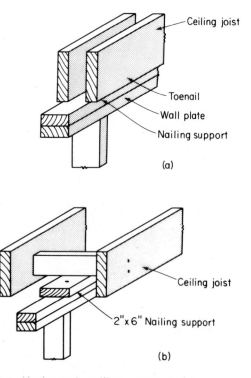

(a)

(b)

Fig. 13.5. Horizontal nailing support for interior finish. *(Courtesy of the Department of Forestry and Rural Development, Ottawa.)*

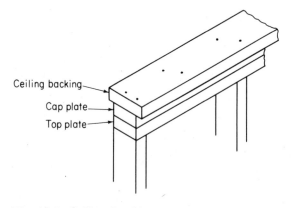

Fig. 13.6. Ceiling backing.

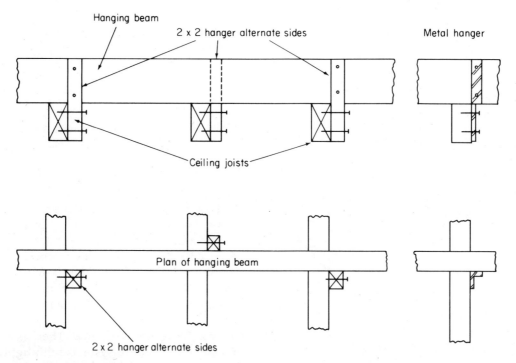

Fig. 13.7. A hanging beam.

Step 5: If a hanging beam is specified, select a sound piece of lumber of the correct dimensions, and with its crown side up, set it centrally at right angles over the ceiling joists as at Fig. 13.7. Toenail the ceiling joist to the hanging beam and secure it further by nailing 2 x 2 wooden hangers on alternate sides as shown on the drawing. Metal hangers may also be used for this purpose. This method will give rigidity to the whole assembly. The ends of the hanging beam are secured to opposite wall plates.

Step 6: Nail two or three boards closely to the top of the ceiling joists so that workmen may later walk or crawl to make periodic inspections.

Access Hatchway. All ceilings that have a space of 2'-0" or more between the top of the ceiling joists and the bottom of the roof rafters should be provided with an access hatchway. This is usually placed in an inconspicuous part of the residence, sometimes in the ceiling of a large clothes closet, or in a utility room. This is provided so that an inspection of the chimney, the insulation, ventilation, electric lines, and so on may be made for maintenance purposes. Read your local building code for the minimum size of an access hatchway to be provided; for apartment blocks it would be larger than for single residences. Such a hatchway must be provided with a trap door and the door must be insulated. There is also a mandatory minimum area to be made for ventilation in all roofs.

At Fig. 13.8 is shown two flat roof designs where the roof rafters may serve also as ceiling joists; and at Fig. 13.9 is shown a framing method for a flat roof overhang. Note carefully the centrally drilled ventilation holes for ceiling ventilation. Any holes drilled in any wooden construction member for the accommodation of service lines or ventilation must be drilled centrally in order to cause a minimum of weakening in the member.

13.2 TILED CEILINGS

Furring strips of wood, usually 1 x 2, may be nailed to the underside of wooden ceiling joists to support ceiling tiles. Assuming that the tiles are 12" x 12", to line up the furring strips, proceed as follows:

Step 1: Measure the length of each ceiling joist **which is nearest to the opposite walls of the room,** and mark the center.

Step 2: Assume that the width of the furring strip is exactly 2"; on each wall ceiling joist center line, lay off 1" and set a nail; but be sure that these are both spotted on the same (say north) side.

Step 3: From nail to nail, strike a chalk line with the mason's line; to this, align and nail the first full length furring strip.

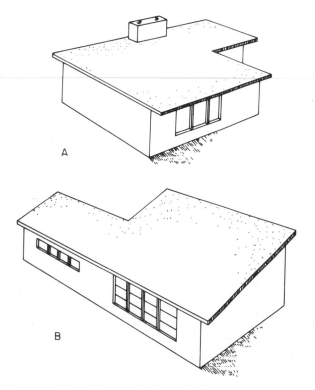

Fig. 13.8. Two flat-roof designs. Rafters may also serve as ceiling joists. *(Courtesy of the Department of Forestry and Rural Development and the Central Mortgage and Housing Corporation, Ottawa.)*

Step 4: Using a 6'-0" straightedge, shim down with a shingle (between the top of the furring strip and the bottom of the ceiling joist) the high spots until the first furring strip is true to line and level on its underside plane.

Step 5: Nail all the other 1 x 2 strips to their exact 12" OCs, and set them true to line and shimmed down to bring the total area into one even plane.

Step 6: Find the exact center of the room, and from this center lay up all the tiles. Note carefully that by working from the exact center of the room, the perimeter or marginal tiles will be of the same width at opposing walls. Supposing that the room measures 16'-9" x 10'-8", then the tiles on the 16'-9" wall sides would be 4" in width and the 10'-8" wall sides would have $4\frac{1}{2}$" wide tiles, and the ceiling would appear to be symmetrical. *This is important!*

Ceiling Suspension Systems. There is a wide choice of suspended ceiling systems with lighting systems installed above, below, or as inserts in the panels forming the ceiling. This type of installation is used extensively in the vestibules and hallways of apartment buildings.

The suspended "T's" may be finished with baked enamel in a variety of colors, and the panel textures and colors may be selected to match any intended decor. The removable panels lend themselves for

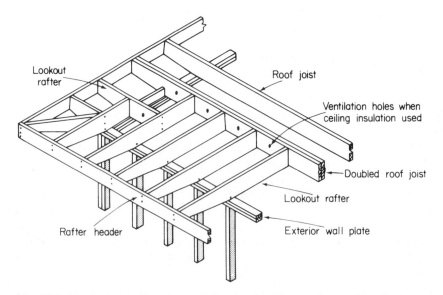

Fig. 13.9. Typical construction of flat roof with overhang. Ventilation holes are drilled in framing members to provide continuous ventilation over ceiling insulation. *(Courtesy of the Department of Forestry and Rural Development and Central Mortgage and Housing Corporation, Ottawa.)*

easy access to many service lines such as telephone, cable TV, electric, and plumbing. The installation of suspended ceiling systems is a subtrade.

13.3 TONGUE AND GROOVE SOFTWOOD PLYWOOD ROOF DECKS

The material for this article has been excerpted from one of the publications of the Plywood Manufacturers of British Columbia and is here reproduced with permission. Similar construction for roof decks is used in all States of America, and in all Provinces of Canada.

Tongue and Groove Western Softwood Plywood for Roof Decks and Subfloors

Tongue and Groove (T&G) western softwood plywood is a Sheathing or Select Sheathing grade plywood panel with a factory-machined tongue along one of the long edges and a groove along the other. T&G plywood is usually manufactured from a standard 48"x96" panel blank in thicknesses of ½", ⅝", and ¾", although net 48" wide panels and other grades, sizes, and thicknesses are available on special order.

Lightweight T&G plywood panels can be easily handled and quickly applied over joists or rafters. In addition to eliminating costly blocking at panel edges, T&G plywood panels save on nails and nailing time. Time studies have shown that a typical floor or roof area can be sheathed with T&G panels in approximately two-thirds the time usually required for square-edge panels and edge blocking — and fewer nails are required.

T&G plywood panels interlock to ensure the effective transmission of loads across panel joints, eliminating differential deflection between panel edges and providing a strong, uniform deck suitable for any finish flooring or roofing material.

For complete installation instructions for T&G plywood see the PMBC publication "Tongue and Groove Plywood."

FLOORS

Floors may be constructed of glued laminated or solid sawn plank. When laid flat, the floor planks should be tongued and grooved, or splined. When laid on edge, they should be spiked together (Figure 4). Planks are splined together by inserting strips of thinner wood (called splines) into grooves cut into opposing edges of abutting planks to form a continuous joint.

Planks should be laid so that mid-span end joints are staggered; a continuous line of end joints is permissible only over points of support. Floor planks should be covered by one-inch nominal tongued and grooved flooring laid crosswise or diagonally, or by half-inch tongued and grooved phenolic-bonded plywood. A half-inch clearance to end walls should be allowed for expansion, and the gap between flooring and end wall should be fire stopped at top or bottom.

flexibility for the designer and builder. Recent fire tests in the United States indicate that 1⅛ inch plywood performs as well under fire exposure as traditional heavy timber roof deck materials. Noncombustible decking may also be substituted for wood material, as provided for in previous editions of the Code.

Tongued and grooved or splined plank roofs must be at least two inches thick. Laminated decks of planks set on edge, with joints staggered as required for floors, must be at least three inches thick; and if tongued and grooved plywood is used for roof decking, it must be at least 1⅛ inches thick.

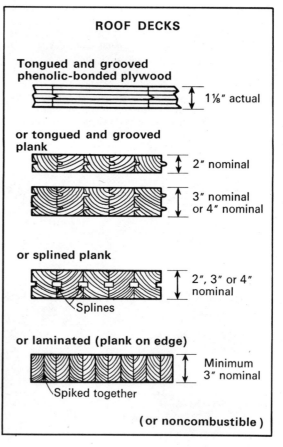

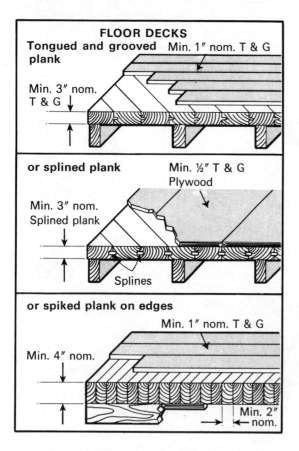

ROOFS

Heavy timber roof decks may be built of solid sawn or glued laminated material, or tongued and grooved phenolic-bonded plywood (Figure 5). Use of plywood as a heavy timber roof deck material is new to the 1970 National Building Code, providing greater

13.4 HEAVY TIMBER ROOF DECKS

The material for this article is here reproduced through the courtesy of the Canadian Wood Council.

TECTUM ROOF DECK tile

Design flexibility is the keynote to Tectum Roof Tile. The custom length feature of the tile product allows the architect to design a roof deck with no end joints exposed. Tile has rabbeted edges along its length, designed for use with bulb-tees, trussed tees or pre-cast joists. The rabbeted edges rest on the flanges of tees or on the top of concrete joists. The spaces between tile and tees or tile and anchors in concrete joists are filled with Tectum Grout which provides excellent anchorage and uplift resistance. The top surface of the tile has a factory applied membrane that serves as an ideal surface for a mopped on roof.

PREMIXED TECTUM GROUT Tectum Grout is a specifically designed grout for use on roof deck systems. Tectum Grout is designed for special mixing to provide a pourable, fast setting, structurally strong material to complement Tectum Tile roof deck systems. Tectum Grout completely fills the void around bulb tees without bleeding through the face of Tectum. Tectum Grout sets in approximately 30 minutes thus minimizing the possibility of workmen displacing tile or unset grout. A minimum compressive strength of 500 p.s.i. gives Tectum Grout superior uplift resistance.

Grout should be applied as the tile is laid to provide a uniform stable deck ready to receive roofing.

ERECTION Bulb-tees should be accurately positioned, by templates, to the required spacing. They should have a minimum bearing of 1″ on the main framing members. The bulb-tee flanges should be welded to the main members with a ¾″ fillet weld at each crossing, alternating from one side of the bulb-tee to the other, and on both sides at the ends. Allow expansion joints as directed by the structural engineer or provide a minimum of ⅛″ end clearance to allow for expansion.

Tectum Tile is laid between the bulb-tees with the rabbeted edges resting on the flanges. The tile should have a minimum bearing of ½″ on the bulb-tee flanges, except sizes 112 and 158 where the minimum bearing should be 7 16″.

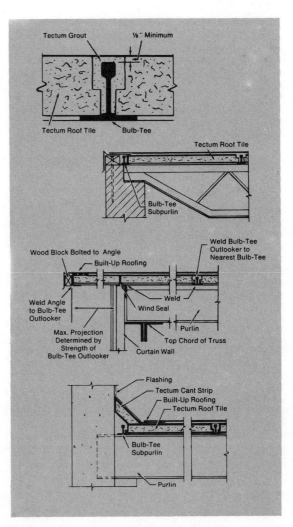

TECTUM TILE DESIGN DATA

Product Classification		T-150	T-200	T-250	T-300	T-350	T-400
Thickness (In.)		1½	2	2½	3	3½	4
Maximum C to C Bulb Tee Spacing (In.)		24	33	36	42	54	60
Ultimate Load (P.S.F.)		200	200	200	200	200	200
Approximate Weight (P.S.F.)		2.8	3.7	4.7	5.5	6.0	6.9
Total Roof Deck Weight Incl. 6 P.S.F. for Built-Up Roof		9	10	11	12	12	13
"U" Value. Incl. Built-Up Roof and Surfaces	Heat Flow Up	.27	.22	.18	.16	.14	.12
	Heat Flow Down	.24	.20	.17	.15	.13	.12
"R" Value		2.63	3.50	4.38	5.25	6.13	7.00

LONG SPAN TECTUM ROOF DECK

Long Span Tectum Roof Deck Plank is an adaption of the standard Tectum Roof Deck System which permits greater areas of the exposed Tectum surface to remain unbroken by subpurlins.

Long Span Tectum Plank is available only in the 3″ thickness. The tongue and groove edge of Long Span Tectum is designed to accept 1½″ hot-dipped galvanized 16 gauge steel channels. These structural members permit a significant increase in the allowable clear span of Tectum.

As with other Tectum Roof Deck products, Long Span Tectum Roof Plank is provided with a factory applied asphalt felt membrane. A silicone treatment to repel water is applied to the wood fibers of the Tectum product. The felt top surface facilitates the application of built-up roofs.

The addition of the 1½″ 16 gauge channel permits spans of up to 6′ for the 3″ thickness. The inherent strength of the Long Span Tectum Plank is further developed by the use of Teks Screws. The S 25 Teks Screw — a self-drilling, self-tapping screw is driven through the Tectum and into the supporting steel. The use of Teks Screws provides the Long Span Tectum Roof Deck Plank System with an excellent resistance to up-lift resulting from high velocity winds. The elimination of bulb trees in the Long Span Tectum System makes it a more economical roof deck. Normal operations such as grouting and painting are thus avoided.

Despite the substantial increase in clear span, Gold Bond Long Span Tectum retains the same structural characteristics as standard Tectum. Long Span Tectum is engineered to carry an ultimate design load of 140.

installation instructions

Reinforced Long Span Tectum may be secured to steel, wood or concrete framing on spans up to 72″ o.c. It may be anchored either by nailing or by use of the Tectum S-25 screw and washer assembly. The Tectum S-25 screw is a self-drilling, self-tapping screw and is driven through the Tectum and into the supporting steel. (Steel thickness not to exceed 5/16″ with a hardness not to exceed 80 on the Rockwell B scale). Galvanized cork nails are driven through the plank and into the framing where steel nailer joists or wood framing are used. When concrete beams are used, steel or wood nailer inserts are cast in the concrete. For the required spacing of nails or screws see detail on this page. Reinforced Long Span Tectum is applicable to either flat or pitched roof construction. Built-up roofing is applied over Tectum decks in accordance with the roofing manufacturer's instructions.

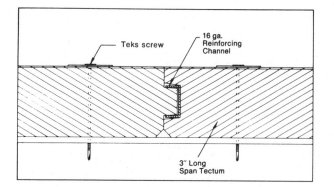

Components of Long Span Tectum Roof Deck

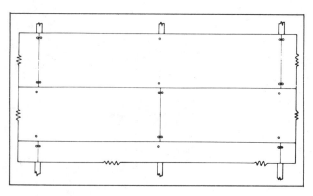

Screw pattern for Long Span Tectum Roof Deck

13.5 ROOF DECK TILES

The following material has been excerpted from one of the publications of the National Gypsum Company and is here reproduced with permission.

Tectum II

Tectum having a ¾" thick foam surface grouted in place on No. 200 steel bulb-tee subpurlins or Keystone 000-5-14-2 Truss Tees spaced 33"o.c. Polyurethane foam strips ¾" thick were installed over the grouted joints to maintain a continuous insulative layer. The roof covering consisted of a 4 ply built-up asphalt roof.

Available in Tile or Plank

Tectum II is available from Newark, Ohio plant:

Product	Thickness Tectum*	Actual Width	Length
PLANK	2"		
	2½"	30¾", 46½"	Up to 12'
	3"		
TILE	2"	31¼"	
	2½"	31¼", 35¼"	Up to 12'
	3"	31¼", 35¼" 41¼"	
LONG SPAN	3"	30¾"	Up to 12'

* Add foam thickness (minus ⅛" for penetration) to obtain total product thickness.

Tectum II Tile is specified with a grout system and a filler strip. The polyurethane foam filler strip will give a uniform insulation factor.

A typical Tectum II Plank installation. Simple tongue and groove plank allows an uncomplicated roof

Application of felt to a Tectum II Plank roof.

A Tectum II Tile roof deck. showing grout treatment for joints over bulb tees.

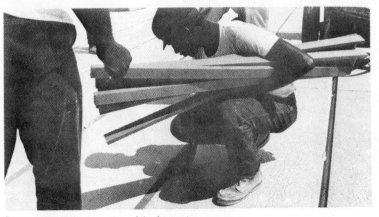

Shown here is the insertion of the foam strip over the unset grout at the bulb tee.

Fiberglass Reinforced Plastic Forms for Beams

The following material has been excerpted from literature published by "Symons Corporation" and is here reproduced with the permission of the copyright holder.

Symons/Bergan fiberglass reinforced plastic BEAMFORMS are reusable and especially designed for forming concrete joists and beams. In addition to providing excellent structural strength, these new fiberglass forms produce a smooth, fin-free soffit — a much superior finish than that achieved by ordinary pan joist construction methods.

BEAMFORMS are available in lengths up to 15 feet (or even longer on special order). Individual joist forms can be designed wide and deep, minimizing the need for bridging ribs. When placed on wide centers of five to nine feet, the number of joists required can be conveniently reduced. In long spans, BEAMFORMS will decrease dead weight and increase load bearing capacity of the joist slab — the amount of concrete required for cover to give an acceptable fire rating is put to work structurally.

BEAMFORMS and Symons heavy-duty frames are excellent in combination as a dependable support system for horizontal forming, permitting fast, easy erection and quick, precise adjustments wherever needed.

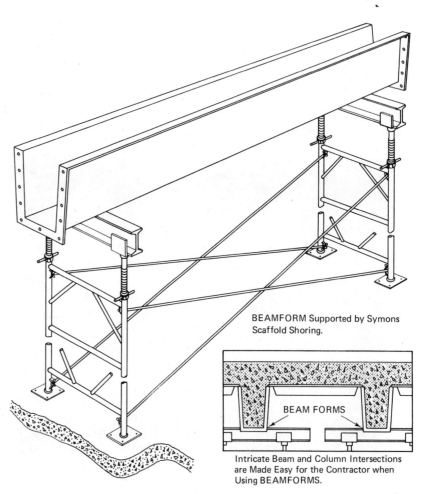

BEAMFORM Supported by Symons Scaffold Shoring.

BEAM FORMS

Intricate Beam and Column Intersections are Made Easy for the Contractor when Using BEAMFORMS.

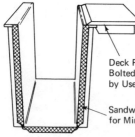

Deck Plate with Bolted Joist Stringers by User

Sandwich Construction for Minimal Deflection

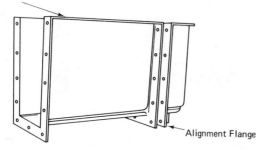

Flange for Mounting Deck Plate

Alignment Flange

Extra strength sandwich construction provides considerable resistance to deflection, and permits clear spans up to 12 feet. Basically, this construction consists of three layers of fiberglass, one inch end grain wood blocks, then another three layers of fiberglass.

Deck plates should be bevel cut, and the edges sealed for longer life and easy stripping. Joists on deck plates reduce requirements for knee bracings.

Space is allowed between flanges to accommodate a wood (or metal) stripping key. Minor length adjustments are made at this point by varying the width of the wooden block bolted between flanges.

CHAPTER 13 REVIEW QUESTIONS

1. At Fig. 13.1A is shown two methods of connecting ceiling joists over a load bearing partition. State which method you prefer and support your view in not more than 100 words.

2. Sketch the type of metal fasteners you would select in order to secure the ends of the ceiling joists in question 1 above.

3. Under what circumstances is it advantageous to connect a supporting ceiling beam with flush ceiling joists?

4. What trades will be advantaged by flushing ceiling joists to a supporting ceiling beam? What trades may be disadvantaged by this method?

5. Examine Fig. 13.4 page 226 and give the reason why ceiling joists are sometimes placed to run both ways.

6. Define ceiling backing as portrayed in Fig. 13.6 page 226.

7. Make a freehand sketch of a plan, front and side elevation of a hanging beam and define it.

8. Make a freehand sketch of a section of a splined plank as used in roofing.

9. What is the minimum sized access hatchway to ceilings in houses, and also in apartment blocks in your area?

10. Sketch the plan of the wood framing for a ceiling access hatchway and arrow-name the parts.

Pitch Roofs, Chimney Flashings, and Roofing Materials

In this chapter we shall discuss in depth how to cut (on the ground) wood framing members such as common rafters, hip rafters, and jack rafters. This is followed by factory (or on the job) made wood roof trusses. Roofing materials such as shingles, hand split shakes, cedar roof decking, permalite concrete over a corrugated deck are dealt with, and the chapter concludes with wall and roof flashing.

14.1 TYPES OF ROOFS

a) Flat Roof. This is a roof in which heavy ceiling joists are used as rafters. It has a minimum slope for drainage. It must be well supported at the walls. It must be heavily waterproofed.

Fig. 14.1. Flat roof.

b) Shed or Lean-to Roof. This is a roof which has one sloped surface only. The slope is across the width of the building. The horizontal distance over which the slope passes is the RUN. The lean-to roof is so named because it leans against another building or wall. It saves the expense of building one wall. It is the simplest roof that a carpenter may have to build.

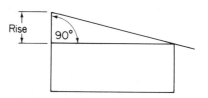

Fig. 14.2. Lean-to or shed roof.

c) Gable Roof. This is a roof which has two sloped surfaces meeting at the RIDGE. The horizontal distance from the foot of one rafter to the foot of the opposite rafter is called the SPAN. Half the span over which each rafter passes is the RUN. This roof is like two lean-to roofs placed together.

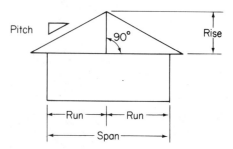

Fig. 14.3. Gable roof.

d) Hip Roof. This is a roof which has a sloped surface from each wall towards the RIDGE. The rafters that fit against the ridge are called COMMON RAFTERS (afterwards in the text they are named CRs). The main part of the roof whose members are CRs is like two lean-to roofs placed together.

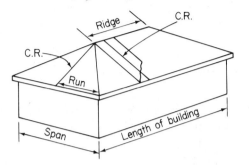

Fig. 14.4. Hip roof.

e) Gambrel or Barn Roof. This is a gable-type roof which has more than one slope on one face. It is used extensively by farmers. The upper portion is like two lean-to roofs placed together. The lower slopes are like lean-to roofs.

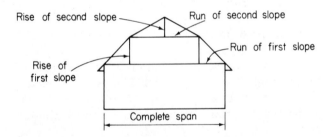

Fig. 14.5. Gambrel or barn roof.

f) Saw-tooth Roof. This is a series of roofs which, when viewed from the ends, resembles the angles of saw teeth. It is a factory type of roof. It allows for the placing of glass on one slope to light the floor area below. Each pair of slopes resembles two lean-to roofs of different runs and slopes, but of the same height.

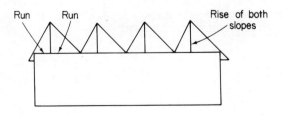

Fig. 14.6. Saw-tooth roof.

A plumb line is a vertical line. A run line is a horizontal line. The angle between a plumb line and a run line is 90°.

Roof framing principles are based on right angle triangulation.

14.2 DEFINITIONS OF ROOF TERMS

Starting at the top of Fig. 14.7, the definitions are as follows:

(a) *Ridge board* — The uppermost member of a roof against which the CRs (common rafters) fit. The rafters are shortened half the thickness to allow for it.

(b) *Back of the rafter* — The top side of a rafter to which the roof sheathing is nailed. *The crown side of all placed uppermost.*

(c) *Center line ℄* — The hypotenuse of a right triangle having the run of a rafter for its base; and the rise of the roof for its height.

(d) *Rise* — The perpendicular distance from the top of the wall plate to its intersection with the center line.

(e) *Seat of birdsmouth* — The horizontal and plumb cuts of a rafter at its nailing point to the cap plate.

(f) *Wall stud* — The perpendicular framing stud supporting the plates and rafters.

(g) *Run* — The run is half the span.

(h) *Span* — The shortest horizontal distance between opposite outside walls.

(i) *Wall plate* — The first plate of the wall framing over which is nailed the cap plate.

(j) *Cap plate* — The top plate which is nailed to the framing plate.

(k) *Rafter tail* — That part of the rafter that extends beyond the wall plates.

(l) *Plumb and level cuts* — The design of the rafter tail to receive a fascia board and a soffit.

(m) *Pitch* — The slope of a roof. It is the relationship of the rise of a roof to the span. A quarter pitch roof has a rise equal to one-quarter of its span, and a one-third pitch roof has a rise equal to one-third of its span, and so on. It is usually shown on the drawing of a roof by a right angle triangle with the rise shown in inches and the run of the rafter shown in inches. These figures are used by the carpenter to lay out rafters with a framing square. See Fig. 14.7.

The line length of a common rafter is measured on the center line, and extends from No. 1 plumb cut line to its intersection with lines 2 and 3 at the birds-mouth as shown at Fig. 14.13, page 241.

14.3 THE CARPENTER'S FRAMING SQUARE

(a) The body of the square is two inches wide and twenty-four inches long.

(b) The tongue is one and a half inches wide and sixteen inches long.

(c) The tool has two rectangular sides that meet at a right angle. It usually has the name of the manufacturer and the rafter tables on the face side.

(d) The square has eight edges with different fractions of inches shown on each.

At Fig. 14.8 is shown two squares back to back with a quarter and a one-third pitch outlined. To find the slope length of a CR per foot run, of a one-third pitch roof (8" rise to 12" run) as shown at Fig. 14.8, measure diagonally across the square from the 12" mark to the 8" mark. It measures, say, $14\frac{3}{8}$", or 14.42". Don't accept this. Try it! Now try it for a one-quarter pitch roof.

Rafter Tables

Figure 14.9 shows in broken form a square for the easy reading of the rafter tables during the study of the text.

First line — Length of a common rafter per foot run.

Second line — Length of hip or valley per foot run.

Third line — Difference in length of jack rafter 16" centers.

Fourth line — Differences in length of jacks at 24" centers.

Fifth line — Side cut of jacks use.

Sixth line. — Side cut of hip or valley use.

Under the 8" mark of the rafter tables at Fig. 14.9 read 14.42; this is the ready reckoned hypotenuse, (the length of the slope of the rafter per foot run) of a right angle with a base of 12", a rise of 8", and a hypotenuse of 14.42. See Fig. 14.7 where a right angle with a one-third pitch is shown above the left hand CR.

Now measure from the 12" mark to the 6" mark on the framing square and compare your reading of the tape with the reading under the 6" mark of the square.

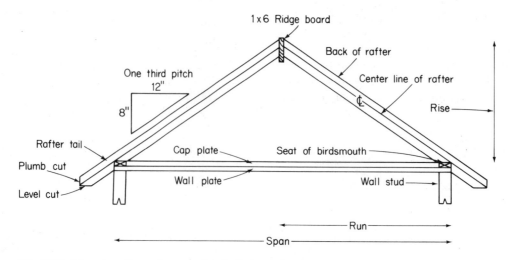

Fig. 14.7. The elevation of a one-third pitch roof using 2 x 4's for rafters.

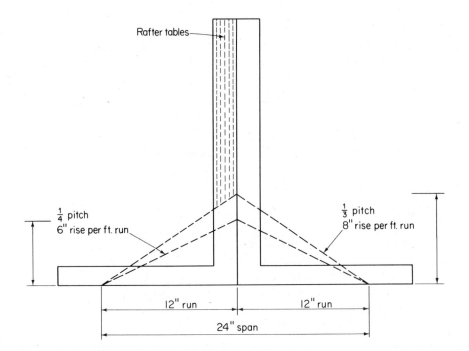

Fig. 14.8. Two framing squares back-to-back.

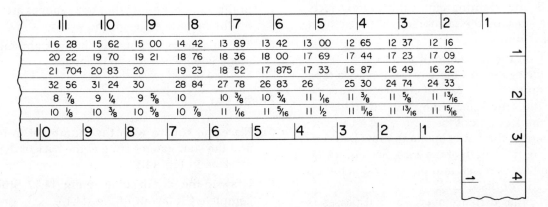

	23	22	21	20	19	18	17	16	15	14	13	12
LENGTH COMMON RAFTERS PER FOOT RUN						21 63	20 81	20 00	19 21	18 44	17 69	16 97
LENGTH HIP OR VALLEY PER FOOT RUN						24 74	24 02	23 32	22 65	22 00	21 38	20 78
DIFF. IN LENGTH OF JACKS 16 INCHES CENTERS						28 84	27 74	26 66	25 61	24 585	23 588	22 625
DIFF. IN LENGTH OF JACKS 2 FEET CENTERS						43 27	41 62	40	38 42	36 08	35 38	33 94
SIDE CUT OF JACKS USE						6 $\frac{11}{16}$	6 $\frac{15}{16}$	7 $\frac{3}{16}$	7 $\frac{1}{2}$	7 $\frac{13}{16}$	8 $\frac{1}{8}$	8 $\frac{1}{2}$
SIDE CUT HIP OR VALLEY USE						8 $\frac{1}{4}$	8 $\frac{1}{2}$	8 $\frac{3}{4}$	9 $\frac{1}{16}$	9 $\frac{3}{8}$	9 $\frac{5}{8}$	9 $\frac{7}{8}$
	22	21	20	19	18	17	16	15	14	13	12	11	10

	11	10	9	8	7	6	5	4	3	2	1
	16 28	15 62	15 00	14 42	13 89	13 42	13 00	12 65	12 37	12 16	
	20 22	19 70	19 21	18 76	18 36	18 00	17 69	17 44	17 23	17 09	
	21 704	20 83	20	19 23	18 52	17 875	17 33	16 87	16 49	16 22	
	32 56	31 24	30	28 84	27 78	26 83	26	25 30	24 74	24 33	
	8 $\frac{7}{8}$	9 $\frac{1}{4}$	9 $\frac{5}{8}$	10	10 $\frac{3}{8}$	10 $\frac{3}{4}$	11 $\frac{1}{16}$	11 $\frac{3}{8}$	11 $\frac{5}{8}$	11 $\frac{13}{16}$	
	10 $\frac{1}{8}$	10 $\frac{3}{8}$	10 $\frac{5}{8}$	10 $\frac{7}{8}$	11 $\frac{1}{16}$	11 $\frac{5}{16}$	11 $\frac{1}{2}$	11 $\frac{11}{16}$	11 $\frac{13}{16}$	11 $\frac{15}{16}$	
	10	9	8	7	6	5	4	3	2	1	

Fig. 14.9.

Once the rise of a roof per foot run has been established (e.g., 8"), all the framing square information for that particular roof will be found under that same figure, in this case 8". Under 8" read length of common rafter per foot run 14.42; also under 8" read length of hip rafter per foot run 18.76 and so on.

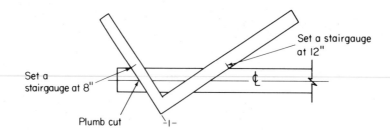

Fig. 14.10. Stair gauges set for one-third pitch.

14.4 LAYOUT OF A COMMON RAFTER FOR A ONE-THIRD PITCH ROOF

Fig. 14.10 shows a framing square with the stair gauges set for the layout of the CR for a $\frac{1}{3}$ pitch roof. The square is shown in position for the marking of the plumb cut. Stair gauges may be purchased in the tool section of most hardware stores, and they are almost indispensable in rafter framing with the square.

It is important to see that all the (numbered) layout lines of the small figures in this section correspond with those shown on the complete rafter layout in Fig. 14.13.

Specifications

(a) Assume that a $\frac{1}{3}$ pitch gable roof with a 1 x 6 ridge is to be erected on a building with a span of 36'-0". The overhang is 16 inches.

(b) Select a straight clean piece of 2 x 4 of suitable length. Examine the stock for the crown side by holding it with the wide face flat in the palm of the hand and sighting along the narrow face. Place the stock on the sawhorses with the crown side away from you.

(c) Draw a ₡ on the wide face of the 2 x 4.
To find the center of a piece of 2 x 4, place a tape square across the $3\frac{1}{2}"$ face of the stock. Traverse the tape until the 4" mark may be read on the top edge of the stock. Spot the center. Set an adjustable square to the center spot and draw in a center line the full length of the 2 x 4. See Fig. 14.11. When the CR is raised into place, the ₡ is the LL (line length) of the CR.
The horizontal distance over which it passes is the run.
The height that it is raised over the plate is the rise. See Fig. 14.7, page 238.

(d) Set the stair gauges for a $\frac{1}{3}$ pitch roof by attaching one gauge to the 8" mark on the tongue of the square and the other on the 12" mark on the body

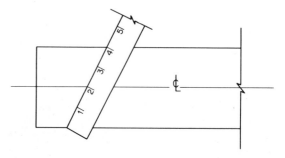

Fig. 14.11.

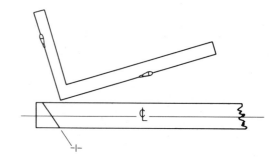

Fig. 14.12.

of the square. Be sure that both gauges are placed on the outside edges of the square.

(e) The imaginary hypotenuse of the right angle triangle is between the stair gauges.

Plumb Cut

(a) Place the square with the plumb cut to the left and the stair gauges snug to the back of the rafter stock as in Fig. 14.10.

(b) Draw in the plumb cut as in Fig. 14.12. Study the complete CR layout in Fig. 14.13.

Line Length

(a) A little to the right of the plumb cut and actually on the CR, work out mathematically the LL of the CR. Multiply 14.42 (as read on the rafter tables)

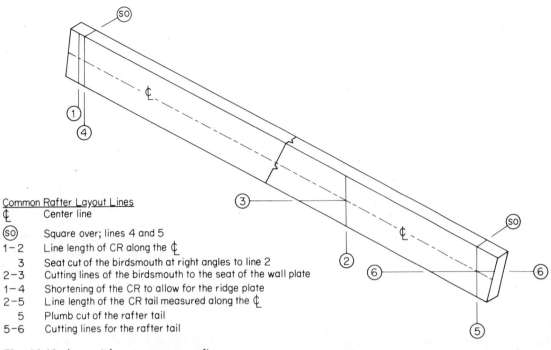

Common Rafter Layout Lines
¢   Center line
(SO)   Square over; lines 4 and 5
1–2   Line length of CR along the ¢
3   Seat cut of the birdsmouth at right angles to line 2
2–3   Cutting lines of the birdsmouth to the seat of the wall plate
1–4   Shortening of the CR to allow for the ridge plate
2–5   Line length of the CR tail measured along the ¢
5   Plumb cut of the rafter tail
5–6   Cutting lines for the rafter tail

Fig. 14.13. Layout for a common rafter.

by the actual number of feet of run in the CR; that is $14.42 \times 18 = 21'\text{-}7\frac{5}{8}''$. The run is 18'-0" and the total span is 36'-0".

(b) With a steel tape, measure on the center line of the rafter stock $21'\text{-}7\frac{5}{8}''$ and draw the plumb cut line No. 2 as shown at Fig. 14.13.

Checking the LL of the CR by the Step Off Method

(a) An excellent additional checking method, and one that should always be used by carpenters is the step off method. Place the stair gauges for the pitch of the roof. Set the framing square at Line 1 as in Fig. 14.10.

Mark a short line on the run of the square to intersect with the top edge of the rafter stock.

Slip the square along the rafter stock until the 8" plumb marking line intersects with the run line already drawn.

Repeat this operation until the number of steps taken is the number of feet of run in the CR.

The last application at the run end of the framing square should reach to within $\frac{3}{8}''$ of the LL as determined from the rafter tables.

When the step off method checks to such a tolerance, accept the LL as reckoned from the rafter tables on the framing square and as measured by the tape.

The carpenter should use the step off method for checking only—not for the original layout.

(b) When using the framing square, the run is always 12". The number of steps to be taken would therefore always be the number of times that 12" (or one foot) is contained in the run of the CR—in this case 18 times for a run of 18 feet.

(c) With the plumb cut of the framing square to the left, draw in another plumb cut at the $21'\text{-}7\frac{5}{8}''$ mark as in Fig. 14.14. Also see Fig. 14.13, line 2.

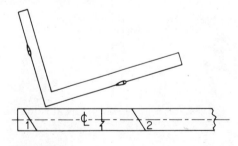

Fig. 14.14.

The Birdsmouth

(a) Slip the framing square to the left until the level cut may be drawn to intersect with No. 2 plumb line at the ¢ as in Fig. 14.15.

(b) The meeting of Lines 2 and 3 at the ¢ lays out the birdsmouth. Lines 2 and 3 should be squared under the bottom edge of the rafter. After the birdsmouth has been removed, the level and plumb cuts fit to the wall plate.

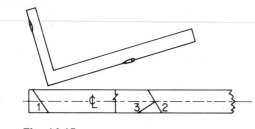

Fig. 14.15.

Shortening of the CR

(a) Referring to the specifications, you will see that the ridge board sawn size is 1 x 6. The actual size of this member would be $\frac{3}{4}''$ x $5\frac{1}{2}''$. See Fig. 14.16.

(b) It is possible to erect the roof without using a ridge board, but a ridge board simplifies the actual erection of a roof. An economy in labor time is effected, apart from any other reason.

(c) It will be seen at Fig. 14.16 that the introduction of a ridge board necessitates the shortening of the CRs. The amount of shortening necessary on each CR is half the thickness of the ridge board, *measured at right angles to the plumb cut.*

(d) The shortening is measured at right angles (that is on the run) to the plumb Line 1.

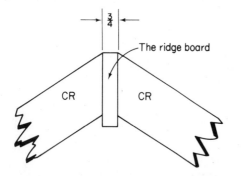

Fig. 14.16. Shortening of CR at ridge board.

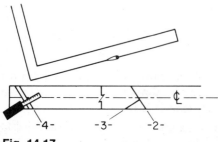

Fig. 14.17.

(e) Place the framing square with the plumb marking line to the left on Line 1. Slip the square so that it is $\frac{3}{8}''$ back at right angles to Line 1 as in Fig. 14.17. The tape may be seen placed in the position of checking the shortening. Draw in Line 4 as at Fig. 14.17. See also Fig. 14.13, Line 4.

(f) Square over Line 4 to the back of the rafter. This is the cutting line.

Tail LL of a CR

The tail length is reckoned in a manner similar to the LL of the CR.

(a) Referring to the specifications on page 240 you will recall that the overhang is 16" and that the pitch is $\frac{1}{3}$.

(b) On the CR, just to the right of Line 2, make the calculations for the line length of the tail for the overhang.

(c) Reading on the framing square for a $\frac{1}{3}$ pitch roof, we see that the rise per foot of run is 8". On the first line of the rafter tables read 14.42 under 8".

(d) To find the LL of the tail of the CR, multiply the length of the rafter per foot run, 14.42, by the actual number of feet of run in the tail of the CR, which is 16", or one and one-third feet of run.

(e) 14.42" $\times$ $1\frac{1}{3}''$ = $19\frac{1}{4}''$, the LL of the tail. The tape may be seen in the position for checking the LL of the tail as at Fig. 14.18. Draw in Line 5.

Checking the LL of the Tail by the Step Off Method

With the plumb marking side of the framing square to the left and placed on Line 2, measure 1'-0" on the run. Slip the square to the right and measure a further 4" as in Fig. 14.19. The 12" and the 4" together equal 16", the given overhang *measured on the run.*

It will be seen that the procedure for the layout of the LL of the tail of a CR from Line 2 is exactly the same as for the layout of the LL for the CR.

Plumb Cut of the Tail of CR

(a) When the rafter stock is of just sufficient length for the CR, it is necessary to turn the square over to draw Line 5. The square should be turned over so that the heel is away from the carpenter. The plumb side of the square should be to the right and the stair gauges should fit snug against the bottom of the rafter stock.

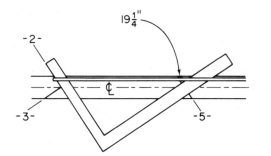

Fig. 14.18.

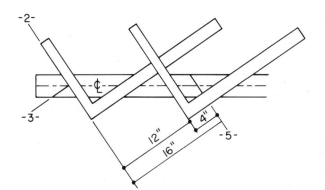

Fig. 14.19.

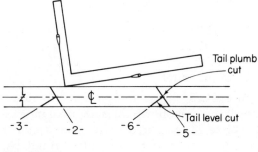

Fig. 14.20.

(b) Figure 14.20 shows the measured LL of the tail of the rafter as calculated from the rafter tables, and as checked by stepping off with the framing square from Line 2 to Line 5.

Level Cut of the Tail of the CR

(a) Where a plumb and level cut is to be made at the end of a CR, place the square with the plumb side to the left. Adjust the position of the square until it will intersect with Line 5 at the ℄ .

(b) The complete CR rafter layout with the Line numbering as in this chapter is shown at Fig. 14.13, page 241.

Actual Sawing Lines of the CR

The actual cutting lines are on:
  Line 4 for the shortened plumb cut.
  Line 3 and lower part of Line 2 for the birdsmouth.
  Lines 5 and 6 for the plumb and level tailcut.

14.5 A REGULAR HIP ROOF

Assume that a regular hip roof with a 1 x 8 ridgeboard has a 24'-0" span, a length of 36'-0", and an overhang of 16" at the eaves, and a one-third pitch. Some simple facts can be deduced:

(a) All slopes of the roof have the same pitch.

(b) The rise of the CRs is the same as the rise of the HRs (hip rafters).

(c) The runs of the six CRs shown at Fig. 14.21 are all equal. Two CRs are shown on each long wall, and one CR is shown on each short wall. The shaded portions are dealt with in paragraph (i).

(d) The run of the HR is equal to the diagonal of a square whose sides are equal to the run of a CR. Reread this!

(e) Assume that a skeleton plan for a doll house is drawn with a span of 2'-0". The run of any CR would be equal in length to the side of a 12" square, and the run of the HR would be the *diagonal* of a square with 12" sides. Measure across one side of a framing square from the 12" marks on the outside edges (or you could lay it out across the desk at which you are sitting). It measures 17". Try it!

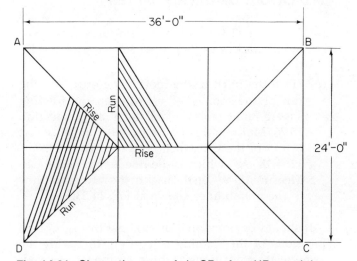

Fig. 14.21. Shows the runs of six CRs, four HRs, and the slopes of the CR and HR projected to a plane surface.

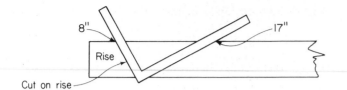

Fig. 14.22. Shows the square set for the layout of a HR having a one-third pitch.

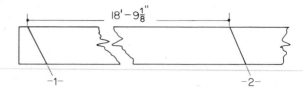

Fig. 14.23. Shows the line length of the CR.

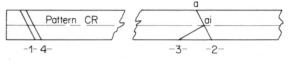

Fig. 14.24.

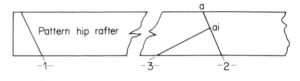

Fig. 14.25.

(f) For the layout of a CR with a one-third pitch we set the stair gauges at 8″ and 12″; see Fig. 14.10, page 240. For a HR for the same roof pitch we set the gauges at 8″ and 17″. See Fig. 14.22.

(g) To find the length of a CR per foot run for this roof we read on the first line of the rafter tables under the 8″ mark, 14.42 as at Fig. 14.9, page 239.

(h) To find the length of the hip or valley per foot run of the CR, we also read under the 8″ mark **on the second line of the rafter tables, 18.76.** Thus, for this roof the right angle for the CR has a rise of 8″, a run of 12″, and a framing square rafter table reading of 14.42. In a similar manner the HR has a rise of 8″, and a run of 17″, and a framing square rafter table reading of 18.76.

(i) Figure 14.21 shows the slope of a CR and HR projected to a flat plane. The rise in each case is equal to one-third of the span of the roof (one-third pitch). For a bit of fun, make a tracing of this drawing, cut on the slopes and rise lines of the shaded portion, and fold the triangles up on their base lines (runs). Examine your work!

14.6 LAYOUT OF THE HIP RAFTER

(a) Figure 14.22 shows a framing square on a piece of HR stock with its gauges set at 8″ rise and 17″ run.

(b) Since the pitch of the roof is one-third and the run of the roof is 12′-0″, read on the framing square rafter tables under the 8″ mark and find 18.76. Multiply this figure by the number of feet of run of the CR which is twelve. Thus $18.76 \times 12 = 18'\text{-}9\frac{1}{8}''$, which is the line length of the HR. Measure $18'\text{-}9\frac{1}{8}''$ on the back of the rafter and draw in the plumb lines shown at Fig. 14.23.

It should be noted that throughout this section all positions of the framing square are in exactly the same relative positions as those for the larger drawings of the hip rafter layout shown on page 249, Fig. 14.43.

14.7 HIP RAFTER ORIGINAL LEVEL LINE AT THE BIRDSMOUTH

(a) Figure 14.24 shows the pattern common rafter. On Line 2 at *a-ai* is shown the perpendicular height of the CR from the seat cut, Line 3, to the back of the rafter.

(b) Figure 14.25 shows the pattern hip rafter. On Line 2 at *a-ai* is shown the amount of rafter above the plate as measured from the common rafter and reproduced on the hip rafter. It will be seen that, since a 2 x 6 is used for the hip rafter, the latter will require a larger birdsmouth to be removed.

(c) Figure 14.26 shows the framing square in position for drawing the level line. Draw the level Line 3.

14.8 SHORTENING OF THE HIP RAFTER

(a) From an inspection of Fig. 14.27, it will be seen that the run of the shortening of the hip rafter is equal to half the diagonal thickness of the common rafter as at Lines 1–4.

(b) It should be constantly kept in mind that the hip rafter is shortened against the roof member against which it fits. It fits against the common rafters.

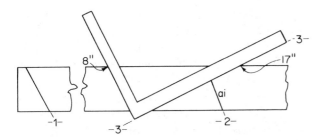

Fig. 14.26.

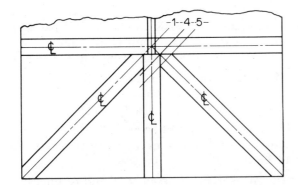

Fig. 14.27. Shortening of the hip rafter.

Fig. 14.28.

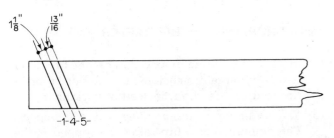

Fig. 14.29.

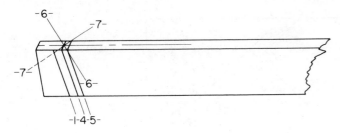

Fig. 14.30.

(c) Measure back at right angles to Line 1, half the diagonal thickness of the CR. See Fig. 14.28. In this case, as a 2 x 4 is specified for the CR, half the diagonal of this member is $1\frac{1}{8}''$. Draw in Line 4 measured at right angles to Line 1.

14.9 SIDE CUTS (CHEEK CUTS) OF THE HIP RAFTER

(a) Upon close inspection of related plates and figures, it will be seen that the run of the side cuts of the hip rafter is equal to half the thickness of the hip rafter stock. The run of the side cut is measured on plan between Lines 4 and 5 as in Fig. 14.27.

(b) Measure back at right angles from Line 4 half the thickness of the hip rafter stock. In this case, since a 2 x 6 is specified for the hip, half the thickness of the hip is $\frac{13}{16}''$. Draw in Line 5 as in Fig. 14.29. Lines 6 and 7, Fig. 14.30 are the hip rafter cheek-cutting lines.

14.10 BACKING (BEVELING) OF THE HIP RAFTER

(a) The sheathing of a roof is the material which is secured to the backs of the rafters. Usually, roofs are sheathed with common boards, shiplap, or plywood.

(b) The sheathing of adjacent sides of the roof meets at a point above the center of the back of the hip rafter. *The steeper the pitch of the roof, the more acute will be the angle at the meeting place of the adjacent pieces of sheathing.*

(c) Theoretically, and sometimes in actual practice for special projects, the back of the hip rafter must have the edges beveled to receive the sheathing. Unless some provision is made, the roof sheathing will not lie flat where it comes in contact with the edges of the hip rafter.

(d) Closely examine Fig. 14.33 at Line 2. A perpendicular line through the center of the back of the hip on Line 2 at *a* would intersect with the inside intersection of the wall plates.

(e) This perpendicular height corresponds to the perpendicular height of the common rafter from the seat cut Line 2 of the CR. See Figures 14.24 and 14.25.

(f) The length of the run between Lines 2 and 8 on the plan is equal to half the thickness of the hip rafter stock. The length of the run is also equal to the run between Lines 4 and 5. The length of the run is shown between Lines 2 and 8, Fig. 14.31.

(g) On the pattern hip rafter draw Line 8 forward from Line 2, a distance of $\frac{13}{16}''$ as at Line 8 Fig. 14.32. *This is very important.*

(h) On the plan at Fig. 14.33, the top of the outside edges of the HR are vertical over the plates as at b and c on Line 8; but since the HR on plan is at an angle of 90° to the CR, when it is raised into

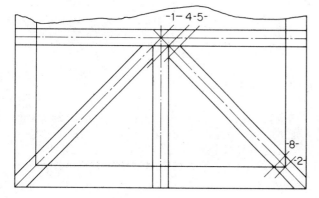

Fig. 14.31. Backing of the hip rafter.

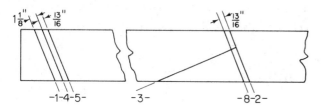

Fig. 14.32.

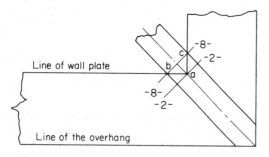

Fig. 14.33.

position, its outside edges become higher than the plane of the CRs. *The steeper the pitch, the greater the difference.* See Fig. 14.34 where a HR with a steep pitch has been presented to emphasize this point.

(i) The height of the (center) of the hip rafter above the wall plate was located as at *a-ai* on Line 2 of the pattern hip rafter layout. This is the same height as is measured from the back of the CR on Line 2 of the common rafter layout. See Figs. 14.24, 14.25 and 14.43.

(j) The measurement from the wall plates to the outside edges of the HR where the HR passes directly over the wall plate is *a-ai* and must be measured from the back of the rafter on Line 8.

(k) Measure down from the back of the hip rafter on Line 8 the same distance as on Line 2 at *a-ai*. Spot *bi* on Line 8 as in Fig. 14.34, and 14.43.

(l) Place the framing square with the run to intersect Line 8 at *bi*. Draw a level line as at Line 9, Fig. 14.35.

(m) The measurement between *bi* and *b2* (measured vertically) on Line 8 is the amount of bevel re-required on the hip rafter. From the back of the rafter on Line 8, gauge the amount of backing (beveling) required. See Fig. 14.36.

(n) With the carpenter's combination square, gauge from the back of the rafter stock an amount equal to *bi-b2* measured as at *c-c* on Line 8. Draw a ₵ down the back of the rafter. The amount to be backed is shown in Fig. 14.37. Note *bi-b2* equals *b-c*.

14.11 DROPPING THE HIP RAFTER

(a) In general building practice, unless the hip rafter is of very large dimensions, it is the usual practice to drop the rafter, instead of backing it.

(b) *Rule: A hip rafter must be either backed or dropped. The amount that a hip rafter is dropped is the amount that otherwise it would be backed.*

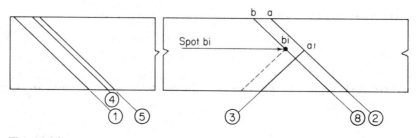

Fig. 14.34.

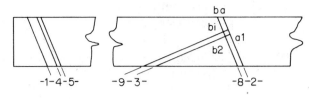

Fig. 14.35.

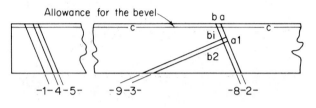

Fig. 14.36.

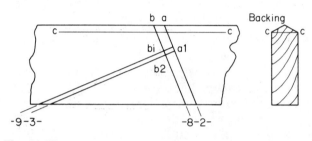

Fig. 14.37.

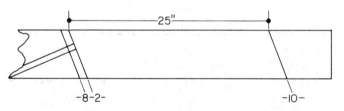

Fig. 14.38.

(c) Since the amount to be dropped is equal to the amount that otherwise it would be backed, it follows that if the rafter were to be dropped, the seat cut would be on Line 9. See Fig. 14.25.

(d) If the rafter is to be backed, the seat cut should be on Line 3.

(e) A hip rafter is either backed or dropped so that its edges will line up with the main plane of the roof.

14.12 REVIEW OF THE LL OF THE TAIL OF THE COMMON RAFTER

(a) Upon reviewing the operation for finding the LL of the tail of the common rafter, it will be recalled that reading under 8″ on the first line of the rafter tables we find 14.42. To find the LL of the

CR tail, multiply 14.42 by the number of feet and portions of feet of run in the CR overhang.

(b) The number of feet and portions of feet of run in the CR overhang is 1′-4″, or $1\frac{1}{3}$ feet, or 1.33 feet. 14.42″ × $1\frac{1}{3}$ is $9\frac{1}{4}$″ which is the LL of the tail of the common rafter to the nearest $\frac{1}{8}$″.

14.13 THE LL OF THE TAIL OF THE HIP RAFTER

(a) The LL of the tail of the hip rafter is laid off from the plumb Line 2 of the original birdsmouth.

(b) Make the calculations on the face of the rafter a little to the right of Line 2.

(c) Read on the rafter tables of the framing square under 8″ on the second line:

LENGTH OF HIP OR VALLEY PER FOOT RUN ... 18.76

(It will be recalled that this roof is specified to have a span of 24′-0″, the pitch $\frac{1}{3}$ and the overhang 16″.)

(d) Multiply 18.76″ by the number of feet of run in the *common rafter tail*. The LL of the tail is therefore 18.76″ × $1\frac{1}{3}$′ which equals 25″ to the nearest $\frac{1}{8}$″.

(e) Measure 25″ along the back of the hip rafter from Line 2, and draw the plumb Line 10. This is the LL of the tail from Line 2. See Fig. 14.38.

14.14 CHECK THE LL OF THE HIP RAFTER BY THE STEP OFF METHOD

(a) The run of the overhang of the tail of the hip rafter is the **diagonal** of the run of the overhang of the common rafter.

(b) For every 1′-0″ of run of the common rafter tail, the run of the hip rafter tail will be 17″. (17″ is the diagonal of 12″ and 12″.) Set the stair gauges to 8″ and 17″ for a $\frac{1}{3}$ pitch hip rafter layout.

(c) For every portion of 1′-0″ of run of the common rafter tail, the hip rafter tail will require the diagonal of that portion of the common rafter tail. See Fig. 14.39.

(d) Apply the framing square to the rafter stock with the plumb marking side of the square to the left, and placed on Line 2. Measure 17″ on the run. This is the first step. Slip the square to the right so that the 8″ setting of the tongue is at the 17″ mark on the rafter. Draw a fine plum line. See Figs. 14.40 and 14.41.

(e) Slip the framing square further to the right from the marked plumb line and measure $5\frac{5}{8}$″ on the run as in Fig. 14.41. The measuring edge of the framing square should now be within $\frac{1}{8}$" of Line 10. See Figs. 14.38 and 14.42.

(f) The LL of the tail of the hip rafter (25") is measured on the back of the HR stock from Line 2 to Line 10. The run of the overhang is measured on the blade of the square and is $22\frac{5}{8}$".

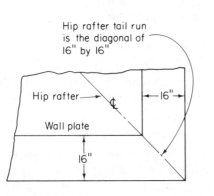

Fig. 14.39.

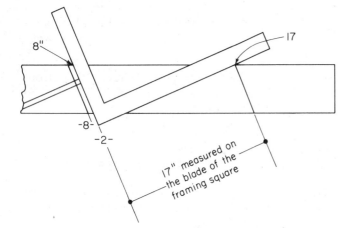

Fig. 14.40.

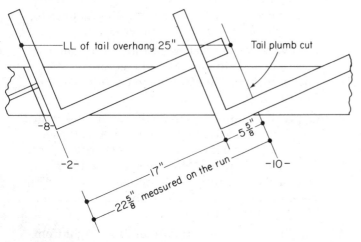

Fig. 14.41.

14.15 THE TAIL END SIDE CUTS OF THE HIP RAFTER

(a) Where the eaves are to be boxed in, the tail end side cuts of the hip rafter will be the reverse of the side cuts at the ridge end of the hip rafter.

(b) From Line 10 on the plan, lay off half the thickness of the rafter stock as at Line 11, Fig. 14.42.

(c) On the hip rafter, measure back, at right angles from Line 10 towards the birdsmouth, half the thickness of the rafter stock. See Fig. 14.43. In this case, as a 2 x 6 is specified for the hip rafter, half the thickness of this member is $\frac{13}{16}$". Draw Line 11.

(d) Draw a ₵ on the back of the rafter as at x-x. From Line 11 draw Lines 14 and 15 to intersect at the ₵ on Line 10. See Fig. 14.43, page 249.

(e) Draw from Line 10 two level lines in the same relative positions as are Lines 3 and 9 from the back of the rafter at the birdsmouth. The final cutting level line at the end of the rafter will correspond to the cutting line at the birdsmouth. If the rafter is to be dropped, cut on Line 9 at the birdsmouth and on Line 13 at the tail end. If the rafter is to be backed, cut on Line 3 and on Line 12.

(f) Study Fig. 14.43b.

(g) It is common practice in building operations to allow the hip rafter tail end to run out past the tail ends of the CRs and then to mark the actual length with a straight edge lined up with the tail ends of the CRs, from both sides of the roof.

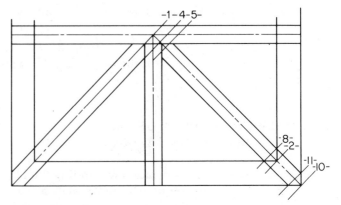

Fig. 14.42.

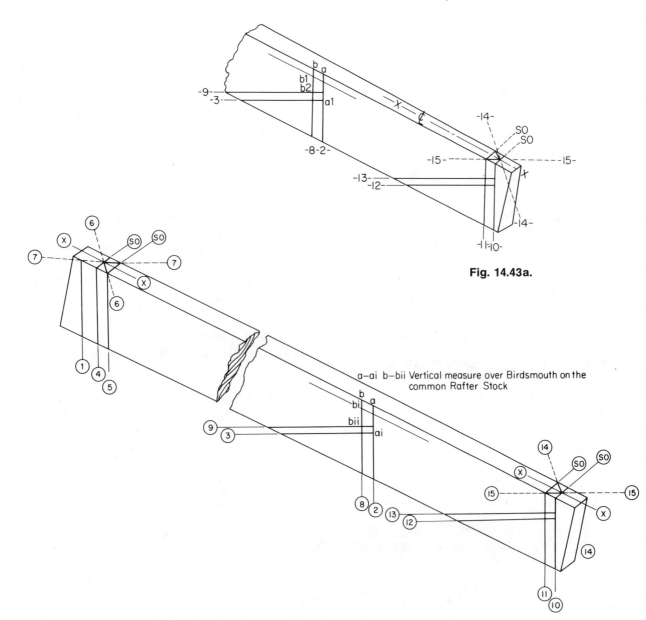

Fig. 14.43a.

a–ai b–bii Vertical measure over Birdsmouth on the common Rafter Stock

Layout of the Hip Rafter

Lines
1—2 Line length of Hip Rafter
2—3 Original Birdsmouth
1—4 Shortening half the diagonal of the CR (measured at right angles to the Plumb Line
4—5 Half the Hip Rafter Stock (measured at right angles to the Plumb Line
X—X Center Line on the Hip
SO Square over
6—6 Side Cut (Cheek Cut) of Hip
7—7 Side Cut (Cheek Cut) of Hip
8—2 Half the thickness of the Hip Rafter Stock (as in 4—5)
a—ai Vertical measure over Birdsmouth to back of rafter as on the CR Birdsmouth layout
b—bii Vertical measure over Birdsmouth to back of rafter as on the CR Birdsmouth layout

3—ai Original Seat Cut of the Birdsmouth
9—bii Developed Seat Cut for determining the Dropping
9—3 Vertical measure of Dropping
b—bi Vertical measure equal to 9—3 used if the rafter is to be backed
2—10 LL of the Hip Rafter Tail
2—10 Run of the overhang when measured at right angles to the plumb line
10—11 Half the Hip Rafter Stock (measured at right angles to the plumb line)
10—12 The Layout of the Original Birdsmouth for Level Cut, used when the Hip Rafter is to be Backed
10—13 The Layout of the Level Cut used when the Hip Rafter is to be Dropped
X—X Center line on the back of Hip
14—14 Side Cut (Cheek Cut) for the Tail of the Hip Rafter
15—15 Side Cut (Cheek Cut) for the Tail of the Hip Rafter

Fig. 14.43b.

14.16 JACK RAFTERS FOR A REGULAR HIP ROOF

Assume that a regular hip roof has a one-third pitch and a 24'-0" span. Some simple facts are as follows:

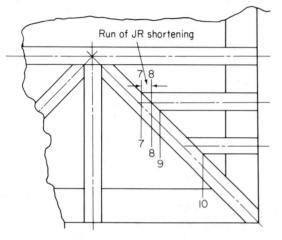

Fig. 14.44. Run of any JR equals its distance from outside corner of wall plates.

(a) The run of any JR (jack rafter) is equal to its length from the outside edge of the wall plate to its intersection at the center line of the HR. See Fig. 14.44.

(b) *With the exception of the cheek cuts for a JR, all other cuts are identical with those for the CR of the same roof.*

(c) The common differences in the length of the JRs with 16" and 24" OCs (on centers) are given on the rafter tables of the framing square.

(d) The example given in this section is based on JRs with 16" OCs. The common differences in lengths of adjoining JRs placed at 16" OCs is 19.23 as shown on the rafter tables, say, 19¼". Look under the 8" mark for this roof at Fig. 14.9, page 239.

(e) The layout of the JRs is made on the pattern CR, see Figs. 14.45 and 14.47.

(f) All JRs are shortened half the diagonal thickness of the HR against which they fit; if the HR stock is 1⅝", the diagonal thickness *(on plan)* is 1⅛".

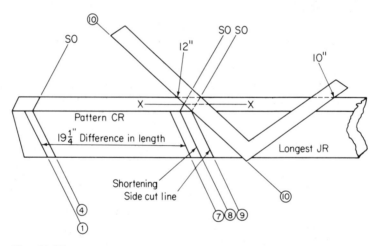

Fig. 14.45.

The longest JR is 19¼" shorter than the longest CR

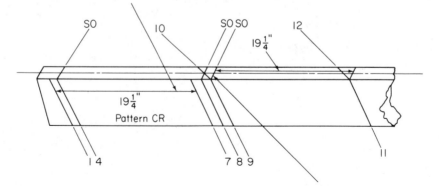

Fig. 14.46.

14.17 LAYOUT OF JACK RAFTERS

(a) For a one-third pitch the *longest* JR is $19\frac{1}{4}''$ shorter than the CR, and each succeeding JR is $19\frac{1}{4}''$ shorter still.

(b) Since the JR fits against the HR at 45° *on plan*, it must be shortened half the diagonal thickness of the HR stock **on plan.** For an HR that is $1\frac{5}{8}''$ thick, half its diagonal **on plan** is $1\frac{1}{8}''$. See Fig. 14.44 for the run (plan) of the shortening ($1\frac{1}{8}''$) of the JR between Lines 7 and 8; then see how it is measured *at right angles* from Line 7 to 8 at Figs. 14.45 and 14.47.

(c) To lay out the side (cheek) cut of the JR, measure at right angles from Line 8 half the thickness of the JR stock (say $\frac{3}{4}''$ for a 2 x 4). Draw Line 9 and square it over (SO). This is shown on Figs. 14.45 through 14.47.

(d) Draw a center line on the back of the JR which will intersect with the square over (SO) Line 8, as shown at X—X at Fig. 14.47.

(e) Across the back of the JR draw Line 10, which extends from Line 9 and intersects Line 8 at its intersection with the center line X—X as shown at Fig. 14.47.

(f) To obtain the same line with the framing square, see Fig. 14.9, read on the fifth line of the rafter tables under the 8" mark for, "Side cut of jacks use . . .", 10. Set the stair gauges at 12" on the blade of the square and 10" on the tongue; with the 12" gauge on the left, apply the square across the back of the JR as shown at Fig. 14.45. It is the same line as in (e) above. Use either method. In effect, the cheek cut is a line which is pivoted on the back of the JR at the central spot on Line 8 to Line 9, producing the cutting line 10.

LAYOUT OF THE JACK RAFTER

NOTE   Jack Rafters are laid out on the Pattern Common Rafter

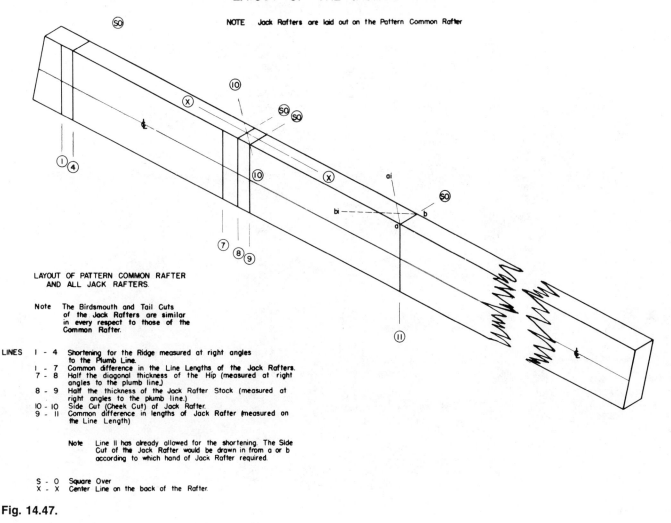

LAYOUT OF PATTERN COMMON RAFTER
AND ALL JACK RAFTERS.

Note   The Birdsmouth and Tail Cuts
of the Jack Rafters are similar
in every respect to those of the
Common Rafter.

LINES   1 - 4   Shortening for the Ridge measured at right angles
to the Plumb Line.
1 - 7   Common difference in the Line Lengths of the Jack Rafters.
7 - 8   Half the diagonal thickness of the Hip (measured at right
angles to the plumb line.)
8 - 9   Half the thickness of the Jack Rafter Stock (measured at
right angles to the plumb line.)
10 - 10   Side Cut (Cheek Cut) of Jack Rafter.
9 - 11   Common difference in lengths of Jack Rafter (measured on
the Line Length)

Note   Line 11 has already allowed for the shortening. The Side
Cut of the Jack Rafter would be drawn in from a or b
according to which hand of Jack Rafter required.

S - O   Square Over
X - X   Center Line on the back of the Rafter.

Fig. 14.47.

Application Recommendations

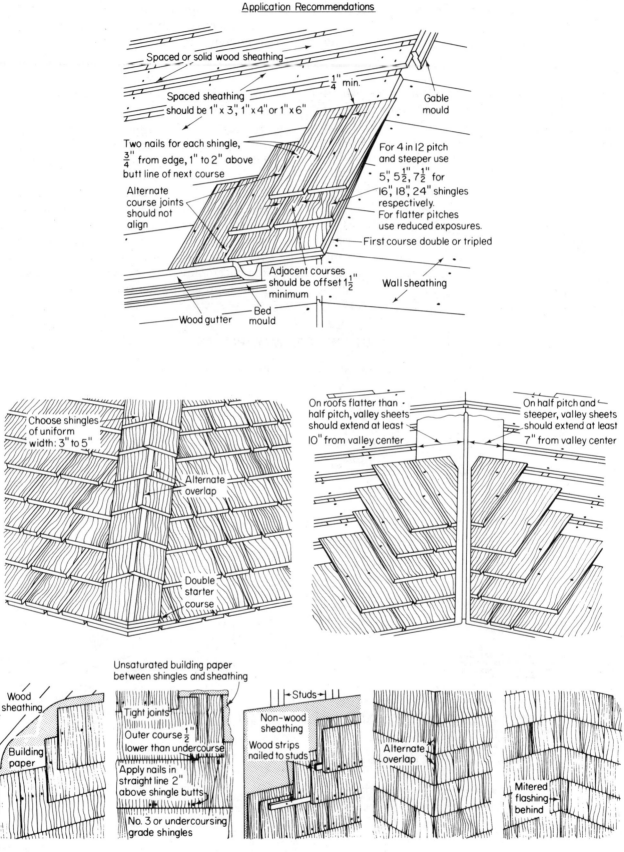

Fig. 14.48.

With an understanding of hip roof layout, a builder is able to apply his knowledge in hundreds of other ways to develop the length and fittings of other construction framing members.

14.18 WOOD ROOF TRUSSES

Strong, durable, and inexpensive wood roofs may be rapidly constructed with the use of (off-site) shop prefabricated quality controlled roof trusses. As a further advantage, where extremes of climate are encountered, the perimeter walls of housing units may be erected free standing; the roof trusses placed,

and the roof covered, giving early tolerable working conditions for workmen to complete the erection of internal walls and all services.

The following material has been excerpted from one of the booklets published by Timber Engineering Company and is here reproduced with the permission of the copyright holder on pages 254–258.

Then follows on pages 259 through 261 a broad outline of shingles and handsplit shakes, this is followed on pages 262–263 with information on Cedar Roof Decking; then there is a presentation of some roof systems suitable for apartments and condominiums. The last article shows some typical wall sections with flashings both for weather and termites. Study them and see also Chapter 4.

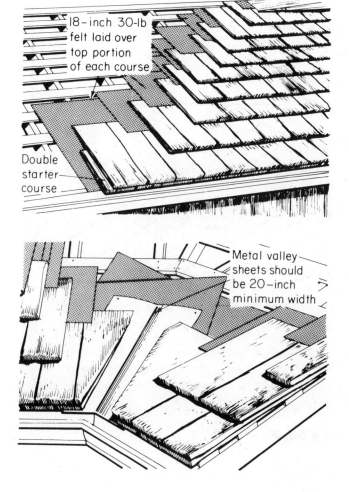

Fig. 14.49.

TECO Nail-On Truss Plates

WHERE USED:

Where single plane assembly of residential, commercial, and farm trusses is desired, TECO truss plates provide economical clear span construction without the requirement for costly fabricating equipment. For residential construction, the TECO plate system is covered by HUD (FHA) Bulletin SE-297. This bulletin covers designs in accordance with HUD (FHA) construction standards for spans 16' to 32' and roof slopes 3:12 to 7:12. Design information is also available for other uses.

DESCRIPTION:

TECO Nail-On truss plates are manufactured from 20 gauge zinc coated sheet steel having the same physical properties as grade "A" structural steel (ASTM A-446). They are manufactured in flat and flanged styles. The plates have pre-punched 0.128" diameter holes.

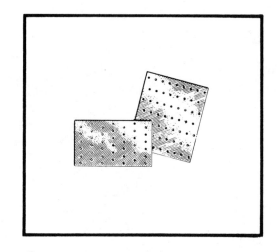

SUGGESTED SPECIFICATION:

Trusses shall be fabricated using TECO prepunched Nail-On truss plates as manufactured by Timber Engineering Company, Washington, D.C. Plate sizes, plate locations, and nail quantities shall be as shown on truss fabrication drawings. The nails are to be of uncoated steel wire 8d diameter (0.131") by 1½" long. Truss members shall be tight fitting at all joints.

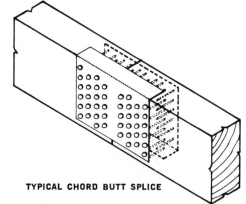

TYPICAL CHORD BUTT SPLICE

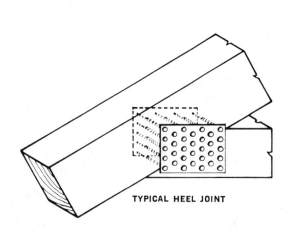

TYPICAL HEEL JOINT

Copyright Timber Engineering Company 1970

WOOD ROOF TRUSSES

The versatility and engineering scope of wood, as expanded through the TECO connector system, have made possible a steady and substantial growth in engineered timber construction. Today, wood roof trusses span from 20 to 250 feet and are built in a wide variety of types and configurations to provide "clear span" construction. This economical form of construction enables the architect and engineer to design with greater freedom; the builder to assemble his component parts with increased speed, efficiency and economy; and the building owner (or occupant) to utilize more effectively the living or working area available to him.

THE TECO SYSTEM

The TECO system of construction is based on the use of split ring and shear plate connectors that spread loads more evenly throughout timber joints. By making it possible to utilize up to 100% of the strength of the pieces joined, TECO connectors remove the limitation of joint weakness in fastening pieces together. One split ring or shear plate used in conjunction with a single bolt is capable of doing the job of a multitude of bolts or nails . . . *and*, there is far less wood removed for the connection. Higher joint efficiency permits the use of smaller members, thus providing considerable saving in material.

TECO split ring connectors are steel rings with a tongue and groove or "split" in the metal. The special metal cross-section is tapered from the center to the edges so that a "wedge-fit" is obtained when the connectors are inserted in pre-cut grooves. The grooves in turn are tapered to conform to the ring section. When installed, one-half of the ring is in each of two lapping members. With such an assembly method, automatic alignment of members is provided and no special jigs or presses are required. Manufactured of mild steel, the rings are available in 2½" and 4" diameters.

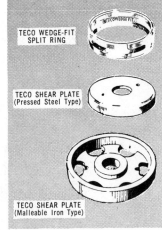

TECO WEDGE-FIT SPLIT RING

TECO SHEAR PLATE (Pressed Steel Type)

TECO SHEAR PLATE (Malleable Iron Type)

TECO shear plates are circular metal plates with a flange on one face around the perimeter and are available in two sizes. The 2⅝" diameter plate is pressed steel. The 4" diameter plate is malleable iron. The connectors fit into pre-cut daps so that they are flush with the wood surface. Shear plates are used in wood-to-wood connections where demountability is desired, or in wood-to-steel connections such as foundation anchor straps, gusset plates or strap and pin connections.

WIDE CHOICE OF TRUSS TYPES

The versatility of timber and the excellent timber design data available make it possible to build almost any type of roof truss with wood. Good engineering practices and a recognition of the characteristics of wood, however, dictate the choice of one type in preference to another under certain conditions.

On the following pages are illustrated the major types of wood roof trusses built with TECO connectors: trussed rafters, pitched trusses, bowstring trusses, flat trusses, Lank-Teco (combination pitched and flat) trusses, scissor trusses, and utility frames. Listed on page 7 are typical designs available without charge to architects and engineers.

TRUSSED RAFTERS 20 to 50 ft.

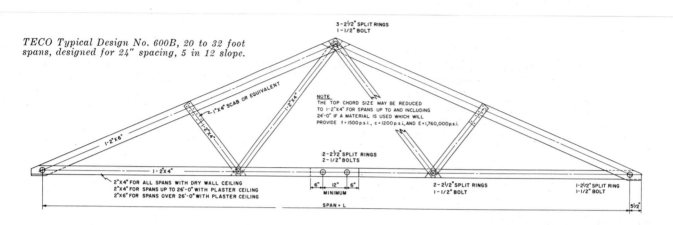

TECO Typical Design No. 600B, 20 to 32 foot spans, designed for 24" spacing, 5 in 12 slope.

There is a TECO trussed rafter to suit most any small building project. The simple Fink is used in residential construction, for machine shops, warehouses, sheds, community buildings and motels. Inverted, the Fink becomes an economical single slope trussed rafter for sawtooth factory buildings or one-story school classrooms.

The bowstring trussed rafter provides for standardization of the glued upper chord. Manufactured by lumber fabricators, it is used in stores, small theaters, garages and farm buildings. The flat trussed rafter provides minimum depth of roof structure. It, too, is used for stores, recreation buildings, and for community buildings.

Two single slope trussed rafters may be joined over a central corridor for economical one-story school classroom buildings.

Scissor and raised chord trussed rafters are most popular for church construction but they are also economical for community buildings.

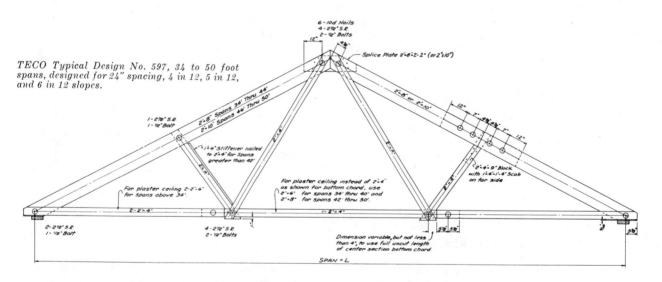

TECO Typical Design No. 597, 34 to 50 foot spans, designed for 24" spacing, 4 in 12, 5 in 12, and 6 in 12 slopes.

Actually the "trussed rafter" is not a rafter but a roof truss. The term is, however, a commonly accepted description of a roof truss to which sheathing and usually ceiling finish may be applied directly without additional framing members. Most trussed rafters are spaced 24 inches on centers, since limitations on ceiling materials and roof sheathing make this spacing most efficient. However, designs are available for greater spacing.

TECO trussed rafters using the ring connector system provide economical clear span construction for spans up to 50'. They are easily and efficiently shipped to the job site either fully assembled, folded, or knocked down. TECO rings placed in grooves made in adjoining members automatically align truss members so that there is no question that the truss is assembled properly.

PITCHED TRUSSES 30 to 80 ft.

Pitched roof trusses are popular for industrial and commercial buildings and warehouses for spans up to 80 feet with spacing 15 to 20 feet. They may be built for longer spans but they are most economical compared with other types of trusses in spans to about 60 feet.

For normal slopes or depth of truss of ¼ to ⅕ the span, the commonly used web member systems are the Belgian, Pratt and Howe. There is little difference in economy between the three types. The Pratt and Howe types are preferred for roofs with monitors since the vertical members may be extended to form an integral part of this framing. The Fink truss is used for very steep slopes while Belgian framing permits simple attachment of purlins or joists at upper chord panel points.

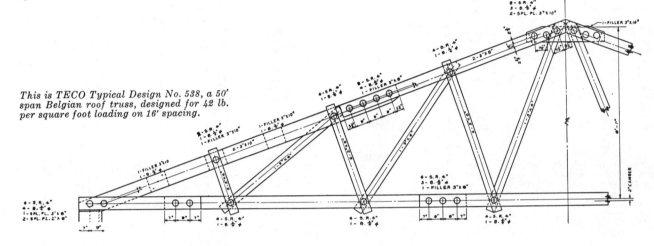

This is TECO Typical Design No. 538, a 50' span Belgian roof truss, designed for 42 lb. per square foot loading on 16' spacing.

BOWSTRING TRUSSES

Bowstring trusses are the most versatile of all timber truss types. They have a wide economic range from 30 to 40 feet to as much as 250 feet.

The spacing of bowstring trusses is limited generally only by the length of roof joists available. When trusses are used with columns, knee braces should be provided.

Lumber fabricators have specialized in the design and construction of bowstring trusses since they have proven to be the most efficient of all truss types. Some frabricators can furnish a complete service including design, fabrication, assembly, and erection.

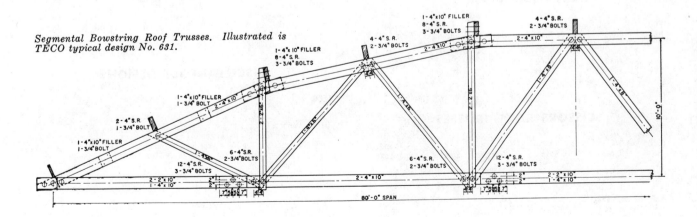

Segmental Bowstring Roof Trusses. Illustrated is TECO typical design No. 631.

TECO TYPICAL DESIGNS

The following are typical designs which are available on request without charge. (For special conditions not covered by these designs, please feel free to consult with the TECO Engineering Department.)

TRUSSED RAFTERS

TECO No.	Type	Span	Depth	Spacing	Live and Dead Load
600A	Pitched	20' to 32'	4 in 12 slope	2' - 0''	45
600B	Pitched	20' to 32'	5 in 12 slope	2' - 0''	45
600C	Pitched	20' to 32'	6 in 12 slope	2' - 0''	45
600D	Pitched	20' to 32'	7 in 12 slope	2' - 0''	45
612	Raised Chord	20' to 40'	8 in 12 slope	2' - 0''	40
654	Flat	20' to 40'	0.11 x span	2' - 0''	45
602	Scissors	30' - 0''	15' - 0''	2' - 0''	50
597	Pitched	34' to 50'	6 in 12 slope	2' - 0''	45
636	Segmental Bowstring	34' to 50'	0.134 x span	4' - 0''	45
606	Hip and L details	20' to 32'	4, 5, 6, 7 in 12		
607	Eave details				
608	Sawtooth	20' to 32'	4 in 12 slope	2' - 0''	45
609	Cantilevered	30' to 50'	4 in 12 slope	2' - 0''	45
616	Pitched	32' to 40'	5 in 12 slope	4' - 0''	45
617	Pitched	20' to 30'	5 in 12 slope	4' - 0''	45
620	1½ Story Frame	26' - 0''	10 in 12 slope	2' - 0''	Special
623	Arch Frame	20' to 30'	Varies	2' - 0''	30
625	Scissors	20' to 36'	8 in 12 slope	2' - 0''	50
681A	Pitched	20' to 32'	8 in 12 slope	2' - 0''	45
681B	Pitched	20' to 32'	9 in 12 slope	2' - 0''	45
681C	Pitched	20' to 32'	10 in 12 slope	2' - 0''	45
681D	Pitched	20' to 32'	11 in 12 slope	2' - 0''	45
682	Cantilevered	16' to 24'	4 in 12 slope	2' - 0''	45
683	Arch Frame	20' to 32'	5 in 12 slope	4' - 0''	30

TRIANGULAR ROOF TRUSSES

TECO No.	Type	Span	Depth	Spacing	Live and Dead Load
622	Triangular	25' & 30'	4 in 12 slope	8' - 0''	50
653	Belgian	30' to 50'	5⅜ in 12 slope	13' - 0''	35
569	Belgian	30' - 0''	7' - 6''	16' - 0''	40
624	Pratt	30' - 0''	8' - 7''	16' - 0''	40
605	Sawtooth	30' - 0''	8' - 0''	8' - 0''	40
637	Sawtooth	30' - 0''	4 in 12 slope	16' - 0''	45
517	Belgian	35' - 0''	6' - 7''	12' - 0''	40
564	Belgian	40' - 0''	8' - 0''	12' - 0''	40
562	Belgian	40' - 0''	7' - 0''	16' - 0''	45
345	Fink	40' - 0''	8' - 0''	16' - 0''	40
638	Sawtooth	40' - 0''	4 in 12 slope	16' - 0''	45
357	Belgian	44' - 0''	8' - 0''	16' - 0''	40
611	Cantilever	44' - 0''	4' - 6''	15' - 0''	40
570	Belgian	45' - 0''	10' - 6''	16' - 0''	45
621	Pratt	50' - 0''	10' - 0''	16' - 0''	40
538	Belgian	50' - 0''	8' - 9''	16' - 0''	42
559	Fink	50' - 0''	10' - 0''	15' - 0''	45
644	Howe	50' - 0''	12' - 0''	16' - 0''	45
519	Belgian	57' - 0''	12' - 0''	15' - 0''	45
33	Fink	60' - 0''	12' - 0''	15' - 0''	40
640	Belgian	60' - 0''	10' - 0''	8' - 0''	45
595	Belgian	60' - 0''	12' - 0''	14' - 0''	40
529	Belgian	64' - 0''	14' - 0''	16' - 0''	45
537	Belgian	70' - 0''	16' - 0''	16' - 0''	45
646	Belgian	80' - 0''	12' - 0''	12' - 0''	40
533	Belgian	80' - 0''	15' - 0''	16' - 0''	47

BOWSTRING ROOF TRUSSES

TECO No	Type	Span	Depth	Spacing	Live and Dead Load
626	Segmental Bowstring	30' - 0''	3' - 9''	16' - 0''	45
627	Segmental Bowstring	40' - 0''	5' - 0''	16' - 0''	45
628	Segmental Bowstring	50' - 0''	6' - 3''	16' - 0''	45
629	Segmental Bowstring	60' - 0''	7' - 6''	16' - 0''	45
630	Segmental Bowstring	70' - 0''	8' - 9''	16' - 0''	45
631	Segmental Bowstring	80' - 0''	10' - 0''	16' - 0''	45
632	Segmental Bowstring	90' - 0''	11' - 3''	16' - 0''	45
633	Segmental Bowstring (built up flat top)	100' - 0''	12' - 6''	16' - 0''	45

FLAT ROOF TRUSSES

TECO No.	Type	Span	Depth	Spacing	Live and Dead Load
565	Warren	30' - 0''	3' - 6''	8' - 0''	40
639	Pratt	30' - 0''	4' - 0''	12' - 0''	40
641	Warren (single slope)	30' - 0''	3' - 6''	8' - 0''	45
635	Howe	40' - 0''	4' - 3''	16' - 0''	45
642	Warren	40' - 0''	4' - 6''	16' - 0''	45
567	Howe	45' - 0''	5' - 0''	15' - 0''	46
520	Howe	48' - 0''	7' - 0''	14' - 0''	40
598	Howe	50' - 0''	5' - 6''	16' - 0''	45
221	Pratt	50' - 0''	6' - 0''	15' - 0''	40
634	Pratt	60' - 0''	6' - 6''	8' - 0''	45
645	Howe	60' - 0''	6' - 6''	16' - 0''	45
384	Pratt	64' - 2''	7' - 0''	16' - 0''	50
649	Howe	70' - 0''	7' - 6''	16' - 0''	45
650	Pratt	80' - 0''	8' - 6''	16' - 0''	45

LANK-TECO TRUSSES

TECO No.	Type	Span	Depth	Spacing	Live and Dead Load
582	Lank-TECO	40' - 0''	4' - 0''	16' - 0''	40
583	Lank-TECO	48' - 0''	5' - 3''	16' - 0''	40
584	Lank-TECO	60' - 0''	7' - 0''	16' - 0''	45

SCISSORS ROOF TRUSSES

TECO No	Type	Span	Depth	Spacing	Live and Dead Load
575	Scissors	30' - 0''	15' - 0''	15' - 0''	40
589	Scissors	35' - 0''	11' - 8''	15' - 0''	40
573	Scissors	40' - 0''	20' - 0''	15' - 0''	40
560	Scissors	42' - 0''	21' - 0''	15' - 0''	50
577	Scissors	50' - 0''	25' - 0''	15' - 0''	40

MISCELLANEOUS DESIGNS

TECO No.	Description
613	General Cross Sections for School Designs
614	General Bracing Requirements for Roof Trusses
665	Lumber Storage Rack—For Bundle Lengths 8' - 24'
647	3-Hinged Arch, 40' - 0'' Span, 8' - 0'' Spacing, 20' - 0'' Depth
648	3-Hinged Arch, 40' - 0'' Span, 8' - 0'' Spacing, 10' - 0'' Depth
678	Portable Loading Ramp—48'' high, 35' - 9'' long
679	Pole Frame Structure. 30' - 0'' to 40' - 0'' Span, 11' - 0'' Spacing. 25 lbs. Live and Dead Load
680	Umbrella Shed. 28' - 0'' Span, 9' - 0'' Spacing. 25 lbs. Live and Dead Load

14.19 SHINGLES AND HANDSPLIT SHAKES

The following material has been excerpted from one of the booklets published by the Red Cedar Shingle & Handsplit Shake Bureau, Seattle, Washington; and Vancouver, Canada, and is here reproduced by permission.

SUMMARY OF GRADES, SIZES AND SHIPPING WEIGHTS

Certigrade Shingles

Grade	Length	Thickness (At Butt)	No. Courses Per Bdl.	Bdls./Cartons Per Sq.	Shipping Weight		Description
No. 1 BLUE LABEL	16" Fivex 18" Perfections 24" Royals	.40" .45" .50"	20/20 18/18 13/14	4 bdls. 4 bdls. 4 bdls.	144 lbs. 158 lbs. 192 lbs.		The premium grade of shingles for roofs and sidewalls. These top-grade shingles are 100% heartwood, 100% clear and 100% edge-grain.
No. 2 RED LABEL	16" Fivex 18" Perfections 24" Royals	.40" .45" .50"	20/20 18/18 13/14	4 bdls. 4 bdls. 4 bdls.	144 lbs. 158 lbs. 192 lbs.		A proper grade for many applications. Not less than 10" clear on 16" shingles, 11" clear on 18" shingles, and 16" clear on 24" shingles. Flat grain and limited sapwood permitted. Reduced weather exposures recommended.
No. 3 BLACK LABEL	16" Fivex 18" Perfections 24" Royals	.40" .45" .50"	20/20 18/18 13/14	4 bdls. 4 bdls. 4 bdls.	144 lbs. 158 lbs. 192 lbs.	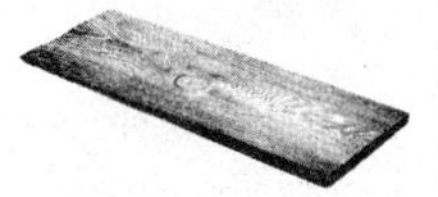	A utility grade for economy applications and secondary buildings. Not less than 6" clear on 16" and 18" shingles, 10" clear on 24" shingles. Reduced weather exposures recommended.
No. 4 UNDER-COURSING	16" Fivex 18" Perfections	.40" .45"	14/14 or 20/20 14/14 or 18/18	2 bdls. 2 bdls. 2 bdls. 2 bdls.	60 lbs. 72 lbs. 60 lbs. 79 lbs.		A utility grade for undercoursing on double-coursed sidewall applications or for interior accent walls.
No. 1 or No. 2 REBUTTED-REJOINTED	16" Fivex 18" Perfections 24" Royals	.40" .45" .50"	33/33 28/28 13/14	1 ctn. 1 ctn. 4 bdls.	60 lbs. 60 lbs. 192 lbs.		The same specifications as above but machine trimmed for exactly parallel edges with butts sawn at precise right angles. Used for sidewall application where tightly fitting joints between shingles are desired. Also available with smooth sanded face.

Approximate coverage of one square of shingles based on following weather exposures:

Length and Thickness	3½"	4"	4½"	5"	5½"	6"	6½"	7"	7½"	8"	8½"	9"	9½"	10"	11"	11½"	12"	13"	14"	15"	16"
16"x5/2"	70	80	90	100*	110	120	130	140	150†	160	170	180	190	200	220	230	240‡	–	–	–	–
18"x5/2¼"	–	72½	81½	90½	100*	109	118	127	136	145½	154½†	163½	172½	181½	200	209	218	236	254½‡	–	–
24"x4/2"	–	–	–	–	–	80	86½	93	100*	106½	113	120	126½	133	146½	153†	160	173	186½	200	213‡

NOTES: *Maximum exposure recommended for roofs.
          †Maximum exposure recommended for single-coursing on sidewalls.
          ‡Maximum exposure recommended for double-coursing on sidewalls.

Certigroove Shakes

Grade	Length	Thickness (At Butt)	No. Courses Per Ctn.	Cartons Per Sq. **	Shipping Weight		Description
No. 1 BLUE LABEL	16" 18" 24"	.40" .45" .50"	16/17 14/14 12/12	2 ctns. 2 ctns. 2 ctns.	60 lbs. 60 lbs. 85 lbs.		Machine-grooved shakes are manufactured from shingles and have striated faces and parallel edges. Used exclusively double-coursed on sidewalls.

Approximate coverage of one square of machine-grooved shakes based on following weather exposures:

Length	10"	11"	12"	13"	14"	15"	16"	16½"
16"	83	92	100*	–	–	–	–	–
18"	–	78	86	93	100*	–	–	–
24"	–	–	–	–	85	91	97	100*

NOTES: *Maximum exposure recommended for double-coursing on sidewalls.
          **Also marketed in one-carton squares.

Certi-Split Shakes

Grade	Length and Thickness	20" Pack		18" Pack**		Shipping Weight	Description
		Courses Per Bdl.	Bdls. Per Sq.	Courses Per Bdl.	Bdls. Per Sq.		
	15" Starter-Finish	8/8	5	9/9	5	225 lbs.	These shakes have split faces and sawn backs. Cedar logs are first cut into desired lengths. Blanks or boards of proper thickness are split and then run diagonally through a bandsaw to produce two tapered shakes from each blank.
No. 1 HANDSPLIT & RESAWN		10/10	4				
	18" x ½" to ¾"	10/10	4	9/9	5	220 lbs.	
	18" x ¾" to 1¼"	8/8	5	9/9	5	250 lbs.	
	24" x 3/8"	10/10	4	9/9	5	225 lbs.	
	24" x ½" to ¾"	10/10	4	9/9	5	280 lbs.	
	24" x ¾" to 1¼"	8/8	5	9/9	5	350 lbs.	
No. 1 TAPERSPLIT	24" x ½" to 5/8"	10/10	4	9/9	5	260 lbs.	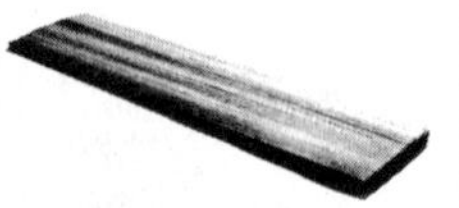 Produced largely by hand, using a sharp-bladed steel froe and a wooden mallet. The natural shingle-like taper is achieved by reversing the block, end-for-end, with each split.
No. 1 STRAIGHT-SPLIT	18" x 3/8" True-Edge*	14 Str.	4			120 lbs.	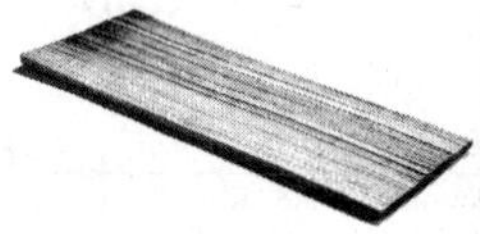 Produced in the same manner as tapersplit shakes except that by splitting from the same end of the block, the shakes acquire the same thickness throughout.
	18" x 3/8"	19 Str.	5			200 lbs.	
	24" x 3/8"	16 Str.	5			260 lbs.	

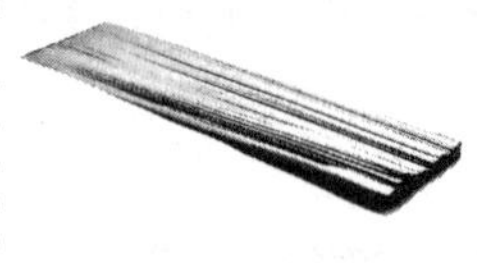

NOTES: *Primarily sidewall product, with parallel edges.
          **Pack used for majority of shakes.

Approximate coverage of one square of handsplit shakes based on following weather exposures:

	5½"	6½"	7"	7½"	8"	8½"	10"	11½"	13"	14"	15"	16"
18" x ½" to ¾" Handsplit-and-Resawn	55*	65	70	75**	80	85†	–	–	–	140‡	–	–
18" x ¾" to 1¼" Handsplit-and-Resawn	55*	65	70	75**	80	85†	–	–	–	140‡	–	–
24" x 3/8" Handsplit	–	65	70	75***	80	85	100††	115†	–	–	–	–
24" x ½" to ¾" Handsplit-and-Resawn	–	65	70	75*	80	85	100**	115†	–	–	–	–
24" x ¾" to 1¼" Handsplit-and-Resawn	–	65	70	75*	80	85	100**	115†	–	–	–	–
24" x ½" to 5/8" Tapersplit	–	65	70	75*	80	85	100**	115†	–	–	–	–
18" x 3/8" True-Edge Straight-Split	–	–	–	–	–	–	–	–	–	100	106	112‡
18" x 3/8" Straight-Split	65*	75	80	90	95	100+	–	–	–	–	–	–
24" x 3/8" Straight-Split	–	65	70	75*	80	85	100	115†	–	–	–	–
15" Starter-Finish Course				Use as undercourse at eaves, finish course at ridge.								

Notes: *Recommended maximum weather exposure for 3-ply roof construction.
      **Recommended maximum weather exposure for 2-ply roof construction.
      ***Recommended maximum weather exposure for roof pitches of 4/12 to 8/12.

†Recommended maximum weather exposure for single coursed wall construction.
††Recommended maximum weather exposure for roof pitches of 8/12 or steeper.
‡Recommended maximum weather exposure for double-coursed wall construction.

Quality Control

More than 250 mills meet rigid standards of in-plant and field inspection for the privilege of applying these labels under the bandsticks and to the cartons of their red cedar shingle and shake products. The specifications and inspections are established, carried out and enforced by the Red Cedar Shingle & Handsplit Shake Bureau. These labels are a *guarantee of quality by grade.* Specify these brands! CERTIGRADE – CERTI-SPLIT – CERTIGROOVE.

APPLICATION RECOMMENDATIONS

Shingles

ROOF APPLICATION—Shingles normally are applied in straight, single courses. But application may be varied for the sake of achieving certain effects (thatch, serrated, weave and ocean wave applications are common styles). The following applies regardless: Shingles must be doubled at all eaves, and butts of first-course shingles should project 1½" beyond the first sheathing board. Spacing between adjacent shingles (joints) should be ¼". Joints in any one course should be separated not less than 1½" from joints in adjacent courses, and joints in alternate courses should not be in direct alignment.

ROOF SHEATHING—Red cedar shingles may be applied over open or solid sheathing. Breather-type building paper—such as deadening felt—may be applied over either type sheathing, although paper is not used in most applications.

ROOF PITCH AND EXPOSURES—Proper weather exposure is important, and depends largely on roof pitch. On roof slopes of 4/12 and steeper, the standard exposures for No. 1 grade shingles are: 5" for 16" shingles, 5½" for 18" shingles and 7½" for 24" shingles. On roof slopes less than 4/12, to a minimum of 3/12, recommended exposures are 3¾", 4¼" and 5¾" respectively. Reduced exposures also are recommended on all roof pitches for No. 2 and No. 3 grade shingles.

HIPS AND RIDGES—All hips and ridges should be of the "Boston type," with protected nailing. Factory-assembled hip and ridge units are available. Be sure to use longer nails of sufficient length to penetrate the underlying sheathing, the 6d (2") length normally adequate.

VALLEYS—For roofs with one-half pitch or steeper, valley flashing should extend not less than 7" on each side of the valley center. On roofs of less than one-half pitch, flashing should extend at least 10" on each side. Shingles extending into the valley should be sawed to proper miter. Do not break joints into valley, or lay shingles with grain parallel with center line of valley. Use center-crimped and painted galvanized or aluminum valleys, if available.

NAILING—Apply each shingle with two (only) rust-resistant nails (hot-dipped zinc or aluminum). Each nail should be placed not more than ¾" from the side edge of the shingle and not more than 1" above the exposure line. Use 3d (1¼") nails for 16" and 18" shingles and 4d (1½") for 24" shingles. Drive the nails flush, but not so the head crushes the wood.

SIDEWALL APPLICATION—There are two basic methods of shingle sidewall application—single-course and double-course. In single-coursing, shingles are applied much as in roof construction, but greater weather exposures are permitted. Shingle walls have two layers of shingles at every point, whereas shingle roofs have 3-ply construction. Double-coursing allows for the application of shingles at extended weather exposures over undercoursing-grade shingles. Double-coursing gives deep, bold shadow lines. When double-coursed, a shingle wall should be tripled at the foundation line (by using a double underlay). When the wall is single-coursed, the shingles should be doubled at the foundation line.

NAILING—For double-coursing, each outer course shingle should be secured with two 5d (1¾") small-head, rust-resistant nails driven about two inches above the butts, ¾" in from each side, plus additional nails about four inches apart across the face of the shingle. *Single coursing* involves the same number of nails, but they can be shorter (3d/1¼") and should be driven not more than 1" above the butt line *of the next course*. Never drive the nail so hard that its head crushes the wood.

CORNERS—Outside corners should be constructed with an alternate overlap of shingles between successive courses. Inside corners should be mitered over a metal flashing, or they may be made by nailing an S4S 1½" or 2"-square strip in the corner, after which the shingles of each course are jointed to the strip.

REBUTTED-AND-REJOINTED SHINGLES—These are shingles whose edges have been machine-trimmed so as to be exactly parallel, and butts re-trimmed at precisely right angles. *They are used on sidewalls with tight-fitting joints to give a strong, horizontal line.* They are available with the natural "sawed" face, or with one face sanded smooth. These shingles may be applied either single or double-coursed.

SPECIALTY SHINGLES—Although red cedar shingles normally are produced in random widths, ranging generally from 4" to 14", so-called "dimension" shingles also are available. These dimension shingles are of uniform width, either 5" or 6". They are used mainly for walls where special patterns are desired. A modification of dimension shingles is achieved by trimming their butts in various textures or patterns—round, diamond, half cove, octagon, acorn, fish-scale, etc. Known in the trade as "fancy-butt" shingles, they are finding increased use by architects who seek to soften the harsh lines of some types of modern architecture.

Handsplit Shakes

ROOF APPLICATION—A 36-inch wide strip of 30-lb. roofing felt should be laid over sheathing boards at the eave line. The beginning or starter course of shakes should be doubled; for extra texture it can be tripled. The bottom course or courses can be of 15" or 18" shakes—the former being made expressly for this purpose. After applying each course of shakes, an 18" wide strip of 30-lb. roofing felt should be laid over the top portion of the shakes, extending onto the sheathing. Position the bottom edge of the felt above the butt at a distance equal to twice the weather exposure. For example, 24" shakes laid with 10" exposure would have felt applied 20" above the shake butts; thus the felt will cover the top four inches of the shakes, and will extend out 14" onto the sheathing.

When straight-split shakes are used, the "froe-end" (the end from which the shakes have been split and which is smoother) should be laid uppermost, i.e. toward the ridge. Roofing felt interlay is not necessary when straight-split or tapersplit shakes are applied in snow-free areas at weather exposures less than one-third the total shake length (3-ply roof).

ROOF SHEATHING—Red cedar handsplit shakes may be applied over open or solid sheathing. When spaced sheathing is used, 1 x 4s (or wider) are spaced on centers equal to the weather exposure at which the shakes are to be laid—but never more than 10 inches. In areas where wind-driven snow conditions prevail, a solid roof deck is recommended.

ROOF PITCH AND EXPOSURES—Proper weather exposure is important. As a general rule, a 7½" exposure is recommended for 18" shakes, 10" exposure for 24" shakes. (See shake coverage chart.) The minimum recommended standard roof pitch for handsplit shakes is 4/12, but there have been numerous satisfactory installations on lesser slopes, climatic conditions and care and mode of application being mitigating factors.

HIPS AND RIDGES—Either site-applied or factory-assembled hip and ridge units may be used. Weather exposures should be the same as roof shakes. Be sure to use longer nails, sufficient to penetrate the underlying sheathing.

VALLEYS—Valley and flashing metals that have proved reliable in a particular region should be selected. It is important that valley metals be used whose longevity will match that of cedar. Metal valley sheets should be center-crimped, of 20-inch minimum width, and for maximum life should be either underlaid with a strip of 30-lb. roofing felt applied over the sheathing or painted with a good grade metal paint.

NAILING—Secure each shake with two (only) rust-resistant nails (hot-dipped zinc or aluminum) driven at least one inch from each edge, and one or two inches above the butt line of the course to follow. Adequate nail penetration into sheathing boards is important. The two-inch length (6d) normally is adequate but longer nails should be used if shake thickness or weather exposure dictates. Do not drive nailheads into shakes.

SIDEWALL APPLICATION—Maximum recommended weather exposure with single-course wall construction is 8½" for 18" shakes and 11½" for 24" shakes. The nailing normally is concealed in single-course applications—that is, nailing points slightly above the butt line of the course to follow. Double-course application requires an underlay of shakes or regular cedar shingles. Weather exposures up to 14" are permissible with 18" resawn shakes, and 20" with 24" resawn or tapersplit shakes. If straight-split shakes are used, the double-course exposure may be 16" for 18" shakes and 22" for 24" shakes. Butt nailing of shakes is required with double-course application. Do not drive nailheads into the shake surface.

General

FINISHES AND WEATHERING QUALITIES—Red Cedar shingles and handsplit shakes are well equipped by nature to endure without any protective finish. However, under some circumstances (high humidity, lack of attic ventilation, low roof pitch, overhanging trees, etc.) the use of a wood preservative such as pentachlorophenol or phenyl-mercury-oleate is justified. When untreated, the wood will eventually weather to a silver or dark gray. The speed of change and final shade depend mainly on atmosphere and climate conditions. Bleaching agents may be applied, in which case the wood will turn an antique silver gray. So-called natural finishes, which are lightly pigmented and maintain the original appearance of the wood, are available commercially. Stains, whether solid color or semi-transparent, are most compatible with cedar, and on exterior walls paints also are suitable. Quality finishes are strongly recommended, and will prove most economical on a long-term basis.

RED CEDAR SHINGLE & HANDSPLIT SHAKE BUREAU

For additional information contact: Red Cedar Shingle & Handsplit Shake Bureau, 5510 White Bldg., Seattle, Wash. 98101, Phone 206-623-4881.
(In Canada: 1055 West Hastings St., Vancouver 1, B.C., Phone 604-684-0211)

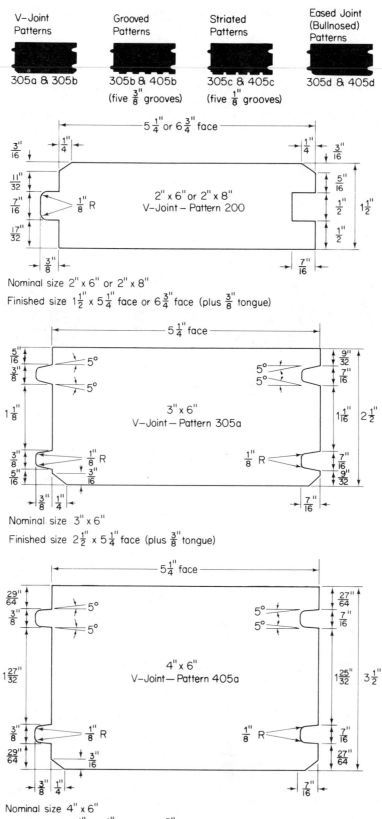

Fig. 14.50. Standard western red cedar backing. *(Courtesy of the Council of Forest Industry of British Columbia.)*

14.20 CEDAR ROOF DECKING

The following material has been excerpted from one of the booklets of The Council of the Forest Industry of British Columbia and is here reproduced by permission.

Western Red Cedar Decking

Western red cedar decking provides a solid, permanent roof deck and a handsome, ready to finish interior ceiling. It serves as an excellent base for any roofing material. It is available in grades, patterns and sizes suitable for both residential and commercial construction. Select grade decking is ideally suited for exposed ceiling areas in homes, schools, churches, motels and restaurants, wherever attractive surface appearance is important. Commercial grade is suitable for use in warehouses, service stations and other structures where appearance is not a prime factor. Roof decking may also be used in solid wall construction and is especially popular for summer cabins.

Standard Patterns

Western red cedar decking is supplied with a double tongue and groove in nominal widths of 6″ (milled to $5\frac{1}{4}″$ face) and in nominal thicknesses of 3″ and 4″ (milled to $2\frac{1}{2}″$ and $3\frac{1}{2}″$ respectively). It is available in the pattern illustrated below. Single tongue and groove decking in nominal 2″ x 6″ and 2″ x 8″ sizes is supplied only in V-joint pattern. All patterns illustrated are as specified in NLGA rules (see Fig. 14.50).

Application

Western red cedar roof decking in 3″ and 4″ thicknesses is supplied pre-drilled on 30″ centers for concealed lateral nailing. These starter holes are $\frac{1}{4}″$ in diameter and are drilled in the center of the piece from the tongue side, approximately $\frac{3}{4}$ of the way through the piece. They are designed to speed lay-up time, guide proper nailing procedures and prevent any splitting of the piece. To enhance the finished appearance, each piece is carefully trimmed at an approximate 2 degree angle, using a special saw, to provide a bevelled cut from fact to back. This ensures a tight face butt joint when laid in a random length pattern.

Two inch decking is single tongue and grooved and, because of its relative thickness, has no provision for side nailing.

On sloping roofs install decking with tongues up; on flat roofs install with tongues away from the applicator. Bright common nails may be used but dip galvanized common nails have better holding power and reduce hazard of rust streaks.

If there are three or more supports for the decking a "Controlled Random" laying pattern may be used. This is an economical pattern because it makes use of random plank lengths, but the following rules must be observed:

1. Stagger end joints in adjacent planks as widely as possible and not less than 2 feet.

2. Separate joints in the same general line by at least two intervening courses.

3. Minimize joints in the middle third of all spans.

4. Make each plank bear on at least one support.

5. Minimize joints in the end spans.

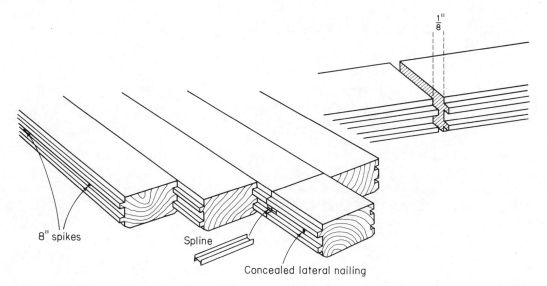

Fig. 14.51. Standard western red cedar backing with metal splines. *(Courtesy of the Council of the Forest Industry of British Columbia.)*

For better alignment, appearance and strength where 3″ and 4″ material is used, metal splines are recommended.

Fasten decking at each support with common nails twice as long as nominal plank thickness. Use one toe-nail and one face-nail for widths 6″ and less and one toe-nail and two face-nails for widths over 6″. Side nail 3″ and 4″ decking through the pre-drilled holes with 8″ spikes. At end joints not over a support a side nail should be within 10″ of each plank end. (See Fig. 14.51 on page 263.)

14.21 PERMALITE CONCRETE OVER CORRUGATED STEEL DECK

The following material has been excerpted from one of the publications of the Pennsylvania Perlite Corporation, and is published with permission.

Permalite®

Concrete Aggregate

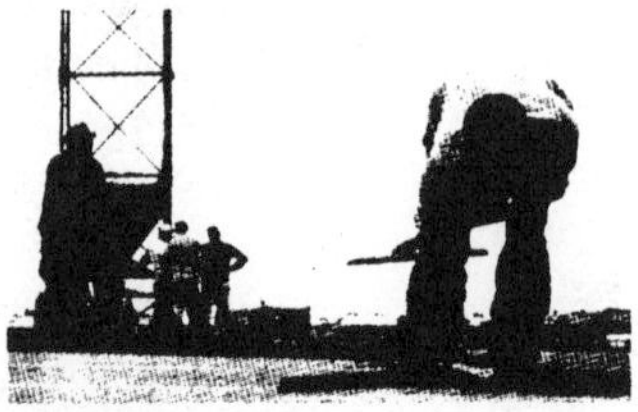

Insulating: Permalite expanded perlite is one of the most efficient insulating materials known. When mixed with portland cement it produces concrete that offers up to 20 times more thermal insulation than ordinary concrete.

Light weight: Permalite concrete can be mixed to weigh from 20 to 40 lbs. per cu. ft., thus significantly reducing the weight of roof decks and floor fills.

TWO WAYS

Permalite concrete is generally used in two ways:

1. as a monolithic drainage and insulating fill
2. as a light structural insulating slab cast in place over permanent forms

Roof Deck Systems

Fill Over CORRUGATED STEEL DECK

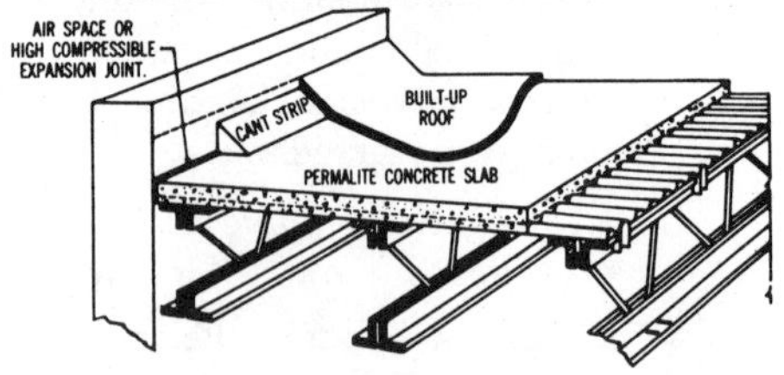

27 POUND DENSITY PERMALITE CONCRETE OVER CORRUGATED STEEL DECK

Thickness over Corrugations	INSULATING "U" VALUE*			DEAD LOAD—Lbs./Sq. Ft.**		
	Depth of Corrugation			Depth of Corrugation		
	7/8″ or 9/16″	7/8″ or 15/16″	1 5/16″ or 1 3/8″	7/8″ or 9/16″	7/8″ or 15/16″	1 5/16″ or 1 3/8″***
2″	.200	.192	.177	6.00	6.50	7.45
2½″	.171	.165	.154	7.10	7.75	8.60
3″	.149	.144	.136	8.25	8.85	9.70
3½″	.132	.128	.122	9.35	10.00	10.80
4″	.118	.115	.110	10.50	11.10	11.95

*Includes built-up roof, Permalite concrete, steel deck, and inside & outside air films.

**Includes dry Permalite concrete and steel deck. Add 6.00 psf for built-up roof and 0.12 psf for reinforcing mesh if specified.

***Add .37 psf for 22 gauge; add .70 for 20 gauge; add 1.42 for 18 gauge.

Fill Over STRUCTURAL CONCRETE

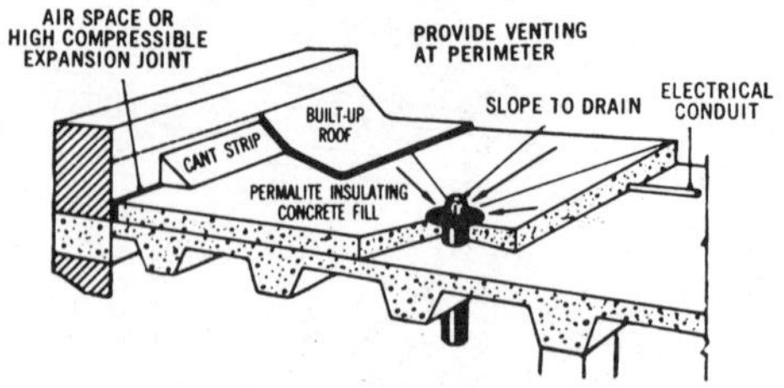

27 POUND DENSITY PERMALITE CONCRETE OVER STRUCTURAL CONCRETE

Permalite Thickness	Wt. of Fill lbs./sq. ft.	INSULATING "U" VALUE**			
		2″ Structural Concrete	4″ Structural Concrete	2″ Precast Plank*	Channel Slab 1″ Thick Web*
2″	4.50	.209	.200	.187	.202
2½″	5.62	.177	.171	.161	.172
3″	6.75	.154	.149	.141	.150
3½″	7.87	.136	.132	.126	.133
4″	9.00	.122	.118	.114	.119

*80 pcf lightweight concrete, "k" of 2.50 (Btu/hr./sq. ft./°F)

**Includes built-up roof, Permalite concrete, structural concrete and inside & outside air films.

TYPICAL DESIGN DATA AND SPECIFICATION PROPERTIES

TYPICAL PROPERTIES

* Oven Dry Density lbs. per cu. ft.	* Compressive Strength P.S.I. at 28 Days	**† Thermal Conductivity "k"	Coefficient Thermal Expansion Per Unit Per °F
36	440	0.77	0.0000061
30	250	0.64	0.0000055
27	180	0.58	0.0000048
24††	130	0.54	0.0000045
22	95	0.51	0.0000043

SPECIFICATION PROPERTIES†

Wet Density Range When Placed	Dry Density Range 28 Days	Compressive Strength Range 28 Days Min. Max.
50.5 ± 2.0	36 ± 2.0	350 to 520
45.5 ± 2.0	30.5 ± 2.0	230 to 340
40.5 ± 2.0	27 ± 2.0	125 to 200
38.0 ± 2.0	24 ± 2.0	90 to 165
36.5 ± 2.0	22 ± 2.0	70 to 125

MATERIALS PER CU. YARD†

Cement Sacks	Permalite Cu. Ft.	Water Gal.	*** A.E.A. Pts.
6.75	27	61	7
5.40	27	59½	7
4.50	27	54	7
3.85	27	54	7
3.38	27	54	7

*Based on impartial laboratory test data of Robert W. Hunt Co. Engrs. under sponsorship of the Perlite Institute. Average density of the aggregates used was 8.0 lb. per cu. ft. Strength data based on ASTM Type 1 Portland Cement.

**Perlite Institute data from report dated April 8, 1953, of Armour Research Foundation of Illinois Institute of Technology.

***As supplied or recommended by the Permalite processor.

†Quantities are approximate, subject to variation in mixers and job conditions.

††All data for this density are interpolated.

†(Btu/hr./sq. ft./°F)

Grefco Building Products Division produces other fine building products; principally Permalite Sealskin Roof Insulation, Permalite Pk composite urethane and Permalite board, Permalite urethane board, and Metalastic Mark II Expansion and Seismic Joint Cover.

See Sweet's files 7.10/Gr and 7.15/Gr.

FM

PRODUCERS COUNCIL

PERMALITE is a registered brand name of premium quality expanded perlite produced by licensed franchisees of GREFCO, Inc., a subsidiary of General Refractories Company.

PERLITE is a type of volcanic rock. When heated above 1600°F the crude perlite particles expand and turn white—much like popcorn—as trapped moisture in the ore vaporizes to form microscopic cells or voids in the glass-like material. The resulting cellular structure accounts for the light weight and excellent thermal insulation of expanded perlite.

Photomicrograph Permalite perlite particle

Treated Masonry Fill Insulation

The glass-like particles of PERMALITE perlite insulation are treated during furnacing to prevent moisture absorption. Tested by Structural Clay Products Research Institute and leading university engineering laboratories, walls insulated with perlite have proven to be virtually unaffected by wind-driven rain and condensation. Masonry walls with Permalite fill can have more than double the insulation value of uninsulated walls, offering

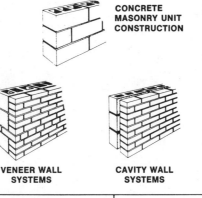

CONCRETE MASONRY UNIT CONSTRUCTION

VENEER WALL SYSTEMS

CAVITY WALL SYSTEMS

substantial fuel and air conditioning cost savings.

Specifications: Construction and design of insulated masonry walls shall be the same as for uninsulated walls. Weep holes shall be screened with copper, galvanized steel, or fibreglass screen. Flashings, joist wall plates and wall ties shall be as for uninsulated walls.

Masonry fill insulation shall be PERMALITE treated perlite manufactured by licensed PERMALITE franchisees.

For additional information send for Catalog No. MF3.

	SQUARE FEET OF WALL AREA	NUMBER OF BAGS				
		1" CAVITY	2" CAVITY	8" BLOCK	10" BLOCK	12" BLOCK
APPROXIMATE COVERAGE	100 square feet	2 bags	4 bags	7 bags	10 bags	13 bags
PER 4 CU. FT. BAG	200 square feet	4 bags	8 bags	14 bags	20 bags	25 bags
PERMALITE MASONRY	500 square feet	10 bags	20 bags	33 bags	48 bags	60 bags
FILL INSULATION	1,000 square feet	21 bags	41 bags	65 bags	96 bags	120 bags

CONCRETE MASONRY UNIT CONSTRUCTION

	"U" UNINSULATED	"U" INSULATED
6" Lightweight Block	0.36	0.20
8" Lightweight Block	0.33	0.15
12" Lightweight Block	0.29	0.10
8" Sand & Gravel Block	0.53	0.34
12" Sand & Gravel Block	0.48	0.22

VENEER WALL SYSTEMS

VENEER WALL SYSTEMS		"U" UNINSULATED	"U" INSULATED
INTERIOR BACK-UP	EXTERIOR VENEER		
6" Lightweight Concrete Block	Common Brick	0.30	0.20
	Face Brick	0.33	0.22
	4" Stone	0.35	0.23
	2½" Terra Cotta	0.29	0.20
8" Lightweight Concrete Block	Common Brick	0.27	0.13
	Face Brick	0.30	0.14
	4" Stone	0.32	0.14
	2½" Terra Cotta	0.27	0.13
8" Sand & Gravel Concrete Block	Common Brick	0.36	0.26
	Face Brick	0.42	0.29
	4" Stone	0.44	0.30
	2½" Terra Cotta	0.36	0.26

CAVITY WALL SYSTEMS

	INTERIOR WYTHE	"U" UNINSULATED	"U" INSULATED		
			2½"	3"	4"
COMMON BRICK	4" Common Brick	0.29	0.11	0.10	0.08
	4" Tile	0.27	0.11	0.09	0.08
	4" Concrete Block	0.30	0.11	0.10	0.08
	4" Lightweight Block	0.24	0.10	0.09	0.07
	6" Lightweight Block	0.23	0.10	0.09	0.07
FACE BRICK	4" Common Brick	0.33	0.12	0.10	0.08
	4" Face Brick	0.37	0.12	0.10	0.08
	4" Tile	0.30	0.11	0.10	0.08
	4" Concrete Block	0.34	0.12	0.10	0.08
	4" Lightweight Block	0.27	0.11	0.09	0.07
	6" Lightweight Block	0.25	0.10	0.09	0.07
CONCRETE BLOCK	4" Concrete Block	0.31	0.11	0.10	0.08
	4" Lightweight Block	0.25	0.10	0.09	0.07
	6" Lightweight Block	0.23	0.10	0.09	0.07
STONE	4" Common Brick	0.34	0.12	0.10	0.08
	4" Tile	0.31	0.11	0.10	0.08
	4" Concrete Block	0.35	0.12	0.10	0.08
	4" Lightweight Block	0.27	0.11	0.09	0.07
	6" Lightweight Block	0.25	0.10	0.09	0.07

(4" EXTERIOR WYTHE)

NOTE:
"U" values expressed in Btu/hr/sq ft/deg. F.
Masonry dimensions are nominal; cavity dimensions actual.

"U" values determined by Pennsylvania State University or calculated using maximum thermal conductivity "k" factor of 0.38 accepted for publication in the current A.S.H.R.A.E. Guide.

14.22 FLASHING

The following material has been excerpted from one of the publications of Wasco Products, Inc. of Sanford, Maine, and is here reproduced by permission. Many apartment buildings have similar typical wall sections as the following.

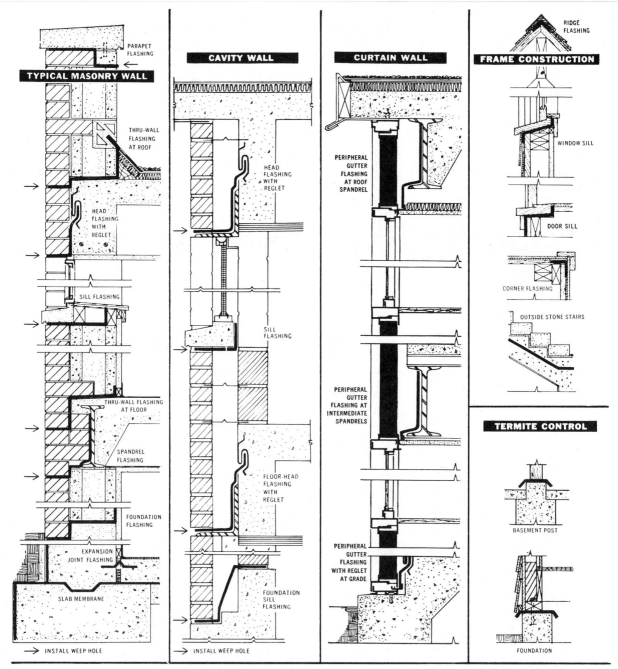

All standard Wascoflex Flashing products are guaranteed against defects in material and workmanship in the production and fabrication processes for a period of 5 years from the date of shipment of such WASCO Flashing, provided that such WASCO Flashing is installed in accordance with WASCO installation recommendations. Liability of Wasco under such guarantee shall not exceed the original purchase price of the defective unit or item exclusive of shipping expense and installation costs.

The following are registered U.S. Trademarks:
WASCO, WASCOSEAL, WASCOFLEX, COPPERSEAL, COP-R-TEX, FABRICOTE

WASCOFLEX® FLASHING

WASCOFLEX® ELASTIC ROOF FLASHING

SPECIFICATION DESCRIPTION

WASCOFLEX® Elastic Roof Flashing; a non-reinforced waterproofed impermeable sheet, scientifically compounded, based on a polyvinyl chloride resin and stabilized to give resistance to sunlight, heat and cold.

FEATURES: Self-extinguishing, resistant to fungus, mildew and bacteria. Can be stretched and shaped to form a tight-fitting one-piece flanged sleeve around vents, columns, and difficult roof contours. May be used as general roof flashing, base flashing, flashing for wall expansion joints, roof drains.

SIZES: Rolls 50′ long in widths of 12″, 18″, 24″ and 36″. Standard thickness—62 mil. Other thicknesses available.

PHYSICAL PROPERTIES

1. Specific Gravity: 1.32
2. Tensile Strength: Original 2200 PSI Min. After 500 hours weatherometer ASTM E 42-57

   (Type E and Federal Test Standard 601, Method 4111)

   Exposed 0% Elongation 2000 PSI Min.

   Exposed 50% Elongation 2000 PSI Min.
3. Hardness: Shore "A" 83 + 5% @ 76°F.
4. Flammability: Self-extinguishing
5. Recommended End Use Temperature: 180°F. to — 40°F.
6. Moisture Vapor Transmission (Perms): 0.104
7. Volatility, 24 hrs. @ 212° F   2.0 max.
8. Affect of asphalt & pitch   None
9. Resistant to fungus, mildew & bacteria.

WASCOFLEX® ADHESIVE

SPECIFICATION:

Adhesive for bonding WASCOFLEX® Elastic Roof Flashing to itself or other building materials shall be WASCOFLEX® Adhesive as manufactured by WASCO, Sanford, Maine.

DESCRIPTION:

WASCOFLEX® Adhesive has been compounded and designed to be used with WASCOFLEX® Elastic Roof Flashing. The tough, permanent bonds it will produce have superior resistance to fatigue, vibration, hot oil, gasoline, aeromatic (aviation, jet) fuels, organic solvents and prolonged exposure to temperatures up to 300°F and down to —60°F. It is a cold process, liquid synthetic, rubber based solvent type contact bonding adhesive and is all that is required for any application.

STANDARD CONTAINERS:

WASCOFLEX® Adhesive is available in 5 gal. pails and 1 gal. cans.

WASCOFLEX® SPECIFICATION

GENERAL SPECIFICATION

Material for flashing (insert areas or conditions) as shown on drawings shall be WASCOFLEX® Elastic Roof Flashing as manufactured by WASCO PRODUCTS, INC. Joining of material to itself shall be accomplished only with WASCOFLEX® Adhesive in accordance with manufacturer's recommendations. Hot mopping of asphalt or pitch or WASCOFLEX® Adhesive shall be used in securing to other materials depending upon the specific application.

If impractical to apply WASCOFLEX® Adhesive at joints at same time WASCOFLEX® is being mopped in, butt joints and adhere later a strip of WASCOFLEX® to cover joint as follows:

PREPARATION:

1. Clean dust and dirt from surface of WASCOFLEX® approximately 8″ both sides of joint with a dry cloth.
2. Cut 12″ wide stock into as many pieces as there are laps to be covered in a length equal to the height or width of the flashing.

APPLICATION:

Apply a thin coat of WASCOFLEX® Adhesive using a short nap 4″ solvent resistant roller to installed flashing, stripe at top and bottom joining a vertical strip both ends to form a box. The vertical stripes inner edge to be approximately 2½″ from the joint thereby allowing a space of approximately 5″ between the inner edges of stripe. Apply a 4″ stripping to one surface of the precut WASCOFLEX® pieces overlapping slightly at all four edges. Follow direction outlined for use of WASCOFLEX® Adhesive. Press adhesive surfaces together applying pressure with hand roller forming firm, watertight strip over each joint. Edge strip to complete installation.

Material for roof expansion joint as shown on drawings shall be WASCOFLEX® Roof Expansion Joint, factory manufactured by WASCO PRODUCTS, INC. and installed in accordance with manufacturer's recommendations.

APPLICATION

General Roof Flashing / Roof Expansion Joint / Expansion Joint Waterproofing / Fascia / Roof Drain / Shower Pan / Chimneys / Parapets Roof Risers / Conduits & Pipes / Columns / Skylights and Ventilator Bases / Valleys / Vent Stacks

 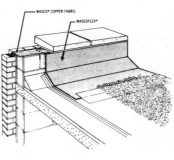 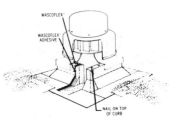

CHAPTER 14 REVIEW QUESTIONS

1. Define the following roof terms:
   (a) ridge board
   (b) back of rafter
   (c) center line of rafter
   (d) rise
   (e) birdsmouth
   (f) run
   (g) span
   (h) wall plate
   (i) cap plate
   (j) rafter tail
   (k) plumb and level cuts
   (l) pitch

2. Using the framing square tables given, complete the following table. The first item has been completed as a guide.

Framing Square Unit of Span	Cancel Down	Rise in Inches per Foot Run	Read on First Line of Rafter Tables
Rise over Span	Pitch	Pitch Times Span	Length of Rafter per Foot Run
$\frac{2}{24}$	$\frac{1}{12}$	2"	12.37
$\frac{3}{24}$	$\frac{1}{8}$		
$\frac{4}{24}$	$\frac{1}{6}$	4"	
$\frac{5}{24}$		5"	
$\frac{6}{24}$			
$\frac{7}{24}$	$\frac{7}{24}$		
$\frac{8}{24}$	$\frac{1}{3}$	8"	
$\frac{10}{24}$			
$\frac{12}{24}$	$\frac{1}{2}$		

3. By how much should a common rafter for a gable roof be shortened to fit against a ridgeboard that is $1\frac{1}{2}$" thick?

4. By how much should a jack rafter be shortened to fit against a hip rafter?

5. List four advantages of making on-the-job (or purchasing) trussed rafters.

6. Explain in writing the difference between certigrade shingles and certigrade shakes.

7. Sketch a section of the foot of a cavity masonry wall showing the application of flashing.

8. Explain and draw a neat freehand sketch of the structural detail for termite control.

9. Explain the difference between a cavity wall and a curtain wall.

10. Draw a neat sketch of a built-up roof as shown in this chapter.

Wall Cladding, Fiber Boards, Insulation, and Vapor Barriers

In this chapter we shall discuss wall cladding and some of the materials and methods used in wood frame wall construction to control the passage of heat, cold, and sound, and also used in the decorative finish of both inside and outside surfaces. Builders should keep abreast of all new developments in these fields, and be alert to the cost of materials, their applications, and the efficacy of the end results.

If just the correct amount of materials required by the workmen is placed in each residence, it will then be easy to identify careless and wasteful workmen and dispense with their services. Remember to have the surroundings for workmen clear of all debris at all times, so that nothing extraneous will impede their progress with the work on hand.

15.1 WALL CLADDING AND FIBER BOARDS

All materials attached to and forming an integral part of a wood frame wall are known as cladding. These include:

1. Fiber boards for both internal and external finish, which are usually manufactured in 4'-0" x 8'-0" sheets, (the height of most residential rooms) which may be of wood or simulated wood. For finished good class work, finishing panels

are secured to the studs with glue, or with nails of similar color to the panel; or the nail heads may be countersunk and stopped with matching color filling.

2. Plaster which is trowelled onto the wall and having expansive metal backing to reinforce over wall openings for doors and windows; external corners are reinforced with metal beads. Outside wall may be treated with stucco applied to a base of stucco wire with the knurls of the wire placed toward the wall surface. *This is important.*

3. Drywall which is manufactured in 4'-0" x 8'-0" standard sized sheets (or longer to order) and secured with manufacturers' specified nails at stated spacings. It is important that there be no joints over doorways; see Fig. 15.6, page 281.

4. Hard-pressed fiber board which may be purchased with a plain or decorative finish that will take all kinds of paint or wallpaper. Care should be taken to drill all holes for nailing to avoid bent nails and hammer face marks on the finished surface.

5. Plastic sheets in tile patterns used especially in bathrooms.

6. Asbestos sheets which are mandatory in some areas behind solid fuel stoves and heaters. They are also used as outside wall finish.

7. T & G (tongue and groove) boards which may be finished for interior use or used as an outside finish in varying patterns.

8. Aspenite sheets which are made from thin wafers of aspen wood bonded under heat and pressure with waterproof glue; unsanded or sanded. These panels, $\frac{1}{4}$" thick, may be used for a pleasing natural finish or for applying coloring stains and varnish. They are also used for outside finished walls. Panels are made in all thicknesses up to $\frac{3}{4}$" as used in floor sheathing.

9. Weldwood sheets which lend themselves to rich internal finishes.

10. Metallic sheets for outside siding.

11. Rough plywood sheathing that is available in standard sized sheets and in varying thicknesses to meet any specifications. The minimum thickness of plywood applied to outside wall sheathing such as shiplap is $\frac{1}{4}$". Unsheathed wall framing should have $\frac{3}{8}$" plywood sheets for studs 16" OC, and $\frac{1}{2}$" for studs placed 24" OC. The edges of plywood panels should be secured with corrosive resistant nails spaced 6" apart along the edges.

12. Shiplap or square-edged boards that may be used as sheathing both for walls and subfloors.

13. All wall sheathing which should be covered with building paper before applying the outside finish.

14. Outside finish that may be wood siding or other man-made fabrics simulating wood, rock, brick and so on. Rough-sawn cedar siding may be horizontally applied with the unmachined barked edge of the board exposed to give the lower edge a rustic appearance.

15. Masonry walls of brick or brick veneer, and concrete blocks in all colors (textured specially for their insulating qualities and having their cores filled with granulated inert insulation).

It is imperative that the starting panel of all the foregoing materials be placed correctly (for plumb and horizontals) since all other panels will be aligned against the first one placed. See the following five pages and Fig. 15.1.

BASIC PATTERNS

From these two basic types, manufacturers have developed a number of panelling profiles by varying the surface design. As a result, Western red cedar panelling is available with smooth or saw-textured surfaces; in clear or with knots; with vertical or flat grain; in different widths and thicknesses; in flush-joint; v-joint or channel patterns; with bevelled or bullnosed edges; with moulded or flat surfaces; and many other combinations of these variables. Three typical types of panelling are shown here:

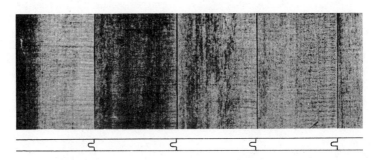

**FLUSH SURFACE
TONGUE AND GROOVE JOINT**
(saw-textured finish)

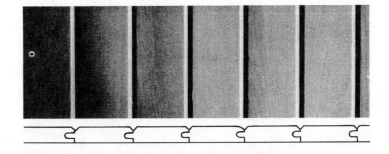

**V-JOINT SURFACE
TONGUE AND GROOVE JOINT**
(milled finish)

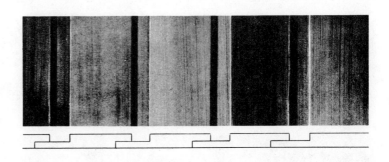

**CHANNEL SURFACE
LAPPED JOINT**
(milled finish)

Courtesy of the Council of the Forest Industries of British Columbia

For further information on available patterns and panelling manufacturers contact
Council of the Forest Industries of British Columbia

NAILING TECHNIQUES

Western red cedar panelling should be applied with standard finishing nails long enough to penetrate at least an inch into the framing members. For all panelling up to ¾-inch thick (1-inch nominal), 2-inch (6d) nails are recommended for both face nailing and blind nailing.

TONGUE AND GROOVE

Tongue and groove panelling up to six inches in width should be blind nailed through the base of the tongue into the cross-blocking or nailing strips of the supporting wall. The nails should be countersunk to allow for flush application of the next panel.

Tongue and groove panelling more than six inches in width requires surface nailing as illustrated below. Countersink all surface nails. Since wood tones vary with the type of finish selected, nail holes are best concealed with a matching putty *after the final surface finish has been applied.*

LAPPED

Lapped panelling less than six inches wide should be surface nailed using the standard countersink-and-fill method.

For lapped panelling more than six inches wide, use an additional surface nail taking care to avoid nailing through the underlying lap. Both nails should be countersunk below the surface of the panelling. Since wood tones vary with the type of finish selected, nail holes are best concealed with a matching putty *after the final surface finish has been applied.*

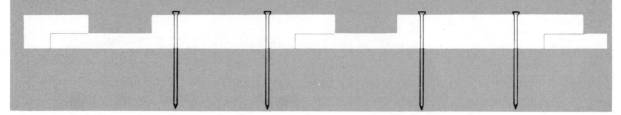

BASIC WALL CONSTRUCTION

Like all siding materials, the performance and permanence of Western red cedar will depend to a large extent on the quality and suitability of the frame to which it is applied. Correct installation will ensure excellent service and handsome appearance.

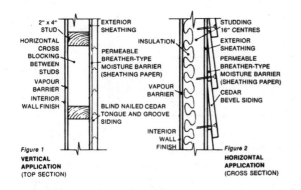

Figures 1 and 2 illustrate the recommended construction of walls for the application of cedar siding.

Structurally, the wall must have the ability to hold fastenings firmly in place, with standard studding and wood sheathing of either shiplap boards or plywood. The interior side of the wall studs should be overlaid with a non-permeable *vapour barrier*, while the exterior sheathing should be covered with a permeable breather-type *moisture barrier* sheathing paper before applying the siding.

Vertical application of siding *(figure 1)* calls for the installation of 2″ x 4″ horizontal cross blocking between the studs to provide a continuous base for nailing.

For siding up to ½-inch in thickness, cross blocking should be installed on 24-inch centres, floor to ceiling.

For ¾-inch material or thicker, blocking may be on 48-inch centres.

Horizontal application of siding *(figure 2)* does not require cross blocking. Nails are driven through the sheathing and directly into the studs.

TYPES OF NAILS

The type of nail selected for the application of siding will have a substantial bearing on the appearance and performance of the finished work.

The following recommendations are the result of a long-term research program conducted by the Western Red Cedar Lumber Association in laboratory tests and under conditions of actual use.

(a) QUALITIES

Ideally, siding nails
- will not rust or cause discolouring;
- will have good holding power with no tendency to pop out or pull through;
- will minimize splitting;
- will be easy to install yet strong enough to resist bending;
- will be relatively unobtrusive in the finished wall.

(b) DESIGNS

The four designs recommended for the installation of siding are illustrated below *(figure 3):*

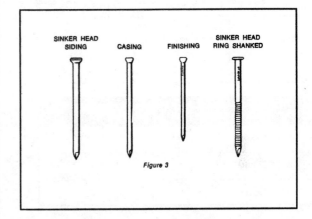

The two sinker head varieties and the casing nail are commonly used for face nailing. Of these, the ring shanked nail has superior holding properties. Sinker head nails are normally tapped flush with the surface of the siding; casing nails may be countersunk and puttied.

(c) FINISHES

The three corrosion-resistant nails recommended for siding are:
> high tensile strength aluminum
> stainless steel
> galvanized – hot-dipped

Other finishes may cause discolouration or staining.

Courtesy of the Council of the Forest Industries of British Columbia

(d) NAIL POINTS

Nails are available with a variety of points for a variety of purposes. For good holding power in the application of siding, with little tendency to cause splitting, the blunt or medium diamond point and the blunt or medium needle point with a ring-threaded shank can be recommended.

(e) SIZES

The correct size for siding nails will depend upon the type and thickness of the material being applied.

As a general rule, siding nails should be long enough to penetrate *at least 1½ inches* into studs and wood sheathing combined.

The following tables will be useful for estimating sizes and quantities.

SIZES			QUANTITIES			
Nail Length (in.)		U.S. Size	Siding Nails (no. per lb)		Pounds Required per 1,000 bd ft of Siding	
Aluminum	Hot-Dipped Galvanized		Aluminum	Hot-Dipped Galvanized	Aluminum	Hot-Dipped Galvanized
1⅞	2	6d	566	194	2	6
2⅛	2¼	7d	468	172	2½	6½
2⅜	2½	8d	319	123	4	9
2⅞	3	10d	215	103	5½	11

NUMBER OF NAILS REQUIRED PER THOUSAND SQUARE FEET OF SIDING	
6" Siding	1560
8" Siding	1180
10" Siding	960

BASIC PATTERNS AND APPLICATION

The four basic patterns of Western red cedar siding are: bevel, tongue and grove, channel, and board and batten.

Based on these patterns, siding manufacturers have developed a wide range of variations to meet personal preferences and the dictates of design: smooth surfaced or saw textured; clear or with knots; vertical or flat grain; different widths and thicknesses of material; v-joint or flush joint; rabbetted or plain bevel; straight-edged, wavy-edged, simulated log cabin style, and many others.

The application methods described here may be adapted to all variations of the four basic siding patterns.

For further information on available patterns and siding manufacturers contact Council of the Forest Industries of B.C.

(a) BEVEL SIDING *(Horizontal application)*

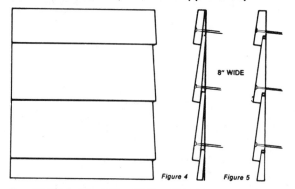

Figure 4   Figure 5

8" WIDE

For 4", 6" or 8" widths, ½" or ⅝" thick, use 2" (6d) nails; for 10" and 12" widths ¾" thick, use 2½" (8d) nails.

Use one nail per stud *(figure 4)* taking care not to nail through both courses of the siding. The nail should miss the feather edge of the underlying piece by ⅛". This allows for expansion and contraction. For rabbetted bevel siding *(figure 5)* the nail should be driven 1" above the thick edge of the piece.

Nailing should be snug but not tight, with nail heads either tapped flush with the surface of the siding or countersunk and filled.

(b) TONGUE AND GROOVE SIDING *(Vertical application)*

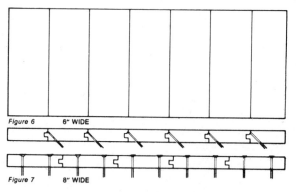

Figure 6   6" WIDE

Figure 7   8" WIDE

Narrower widths (4", 5" and 6") of tongue and groove cedar siding are normally blind nailed as shown *(figure 6)*. Use one 2" (6d) finishing nail per bearing, toenailing into the base of the tongue.

For widths over 6" *(figure 7)* face nail the siding using two 2½" (8d) siding nails per bearing. If a smooth finish is desired, countersink all nails slightly and fill with wood putty or other filler.

APPLICATION

In addition to proper nails and nailing techniques, a sound nailing surface is important for both new construction and remodelling.

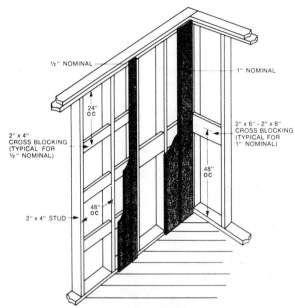

½" NOMINAL
1" NOMINAL
24" OC
2" x 4" CROSS BLOCKING (TYPICAL FOR ½" NOMINAL)
2" x 6" - 2" x 8" CROSS BLOCKING (TYPICAL FOR 1" NOMINAL)
48" OC
48" OC
2" x 4" STUD

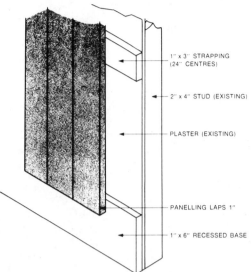

1" x 3" STRAPPING (24" CENTRES)
2" x 4" STUD (EXISTING)
PLASTER (EXISTING)
PANELLING LAPS 1"
1" x 6" RECESSED BASE

PANELLING APPLIED TO WALL FRAME

When applying panelling to an unfinished frame wall, 2" x 4" cross blocking between the wall studs provides a solid nailing surface. It is a good technique to turn the cross blocking so that the 4-inch surface is face-outwards. This provides a larger nailing surface and easy passage for electrical wiring. Spacing of cross blocking varies with the thickness of the panelling. For 1-inch panelling, 48-inch intervals are recommended. For ½-inch panelling, blocking should be set at 24-inch intervals. Extra care should be taken in aligning the first panel: do not rely on the adjacent wall to be plumb at the corner. Once the first panel has been nailed accurately in place, the other panels will align themselves.

PANELLING APPLIED OVER PLASTER

When applying panelling over an existing plaster wall, the area to be panelled should be strapped with a series of 1" x 3" furring strips set horizontally at 24-inch centres. As illustrated, a 1" x 6" strip at the floor level creates a recessed baseboard. In applying the strapping, use nails of sufficient length to penetrate the wall plaster and anchor firmly in the wall studs. The top length of strapping should be set at ceiling height. Accurate alignment of the first piece of panelling will ensure correct alignment of the panels to follow.

Plaster ceilings can be panelled using a similar technique. When applying ½-inch thick panelling, use furring strips set at 16-inch centres; for 1-inch panelling, set furring strips at 24-inch centres.

ESTIMATING

The table opposite will assist in estimating the quantity of panelling required to cover a given wall area using either the lapped or the v-joint tongue and groove pattern.

Nominal Size (in.)		Quantity Required to Cover 1,000 sq ft of Wall Area (bd ft)*	
V-Joint	Channel	V-Joint	Channel
1 x 4	—	1,333	—
1 x 6	1 x 6	1,200	1,263
1 x 8	1 x 8	1,185	1,231
1 x 10	1 x 10	1,143	1,176
—	1 x 12	—	1,143

* Allow small additional footage for cutting and fitting.

 COUNCIL OF THE FOREST INDUSTRIES OF BRITISH COLUMBIA
1500 Guinness Tower, 1055 West Hastings Street, Vancouver 1, British Columbia

Canadian Regional Offices: Edmonton, Toronto and Montreal.
United Kingdom Office: Templar House, 81/87 High Holborn, London WC1V 6LS, England.
Overseas Regional Offices: Liverpool, England; Paris, France; Karlsruhe, West Germany;
Utrecht, The Netherlands; Sydney, Australia.

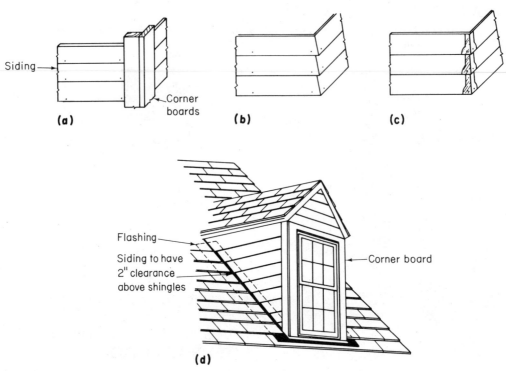

Fig. 15.1. Corner treatment of siding. *(Courtesy of the Forestry Products Laboratory, U.S. Department of Agriculture.)*

15.2 REQUIREMENTS FOR SOUND CONTROL

Sound insulation or acoustic separation is a material's ability to reduce or resist the transmission of sound; the ASTM *Sound Transmission Class (STC)* rating system, based on the decibel, measures the sound insulating performance of construction assemblies in buildings. The National Building Code of Canada (NBCC) uses this *STC* rating system for measuring sound transmission loss. (ASTM *E-90-66T, Laboratory Measurement of Airborne Sound Transmission Loss of Building Partitions.*)

The NBCC sets a minimum *STC* rating of not less than 45 for construction in some areas of multiple dwelling structures as follows:

- Between dwelling units in the same building.
- Between a dwelling unit and any space common to two or more dwelling units.
- Between dwelling units and service rooms or space serving more than one dwelling unit (e.g., storage room, laundry, workshop, building maintenance room or garage).

The NBCC does not cover situations other than those listed above, so the responsibility of selecting constructions with suitable *STC* ratings rests with the designer. He should use common sense and ensure that sound is properly controlled for the purpose for which the building will be used. It is difficult to satisfy all tenants all the time; a successful building is one that can accommodate the wide variety of tenants and tenant activities amicably.

Some things that should be considered are:

- In *apartment buildings* and other multiple dwelling structures, consider transmission of noise from one dwelling unit to another, as well as local individual noise.

- In *office buildings,* noise insulation requirements are less stringent than for apartment buildings because they are usually occupied for only about 8 out of 24 hours. A moderate amount of business noise is acceptable and interference with sleep is rarely a concern; the main requirement is speech privacy. In cases where absolute privacy is needed the insulation requirement might be about *STC* 35.

- In *classrooms,* concert halls and auditoriums, the primary consideration is the communication process. The occupants must be able to hear and understand the message being delivered. Room surfaces should be shaped to control and reflect sound and absorptive surfaces should be used to damp excessive reverberation.

Sound Control

Acoustical problems can, in most cases, be solved in advance by good design. The importance of acoustics varies greatly, depending on the purpose for which a building is designed, but the problems should be considered early in the design stage.

Site Selection

Building sites should be located as far as possible from potential sources of noise such as airfields, industrial plants, railroads and traffic arteries. Buildings should be placed to take advantage of rolling terrain, stands of trees and other buildings which are natural noise barriers. Final site selection may be governed by the intended use of a building and degree of noise control required.

Building Layouts

One of the most important things in sound control is segregating noisy areas from quiet areas; this simplifies the problem of achieving adequate noise insulation. A few rules are:

- In *manufacturing plants,* locate the offices as far as possible from sources of intense noise.

- In *multiple dwelling structures,* arrange layouts so that the most critical rooms (bedrooms, living rooms) are protected from adjoining apartments by a buffer zone of non-critical sound areas such as bathrooms, kitchens, closets and hallways. The next best arrangement is to place quiet rooms such as bedrooms on the two sides of a party wall.

- Keep windows and doors away from the noisy side of a building if possible.

Selecting Equipment and Services

In industry, vibration in the factory area is most important. Vibrations can be minimized by providing special foundations or vibration-isolating mountings such as rubber coil spring shock mountings.

Office equipment such as card-punching, sorting machines and reproduction equipment should be placed in a separate room. Air conditioners, furnace equipment and pumps for any building should be located in the basement, on roofs or on specially designed equipment floors.

Choosing Construction Assemblies

In sound conditioning a room, both sound absorption and sound insulation of floor, wall and ceiling materials should be considered. For sound emitted within a room, the objective is for the walls, floor and ceiling to absorb some of the sound energy. Absorptive requirements may be met in most residential rooms by ordinary furnishings such as drapes and carpets. For control of sounds emitted from a source outside a room, the sound insulation properties of the walls, floor and ceiling are important. The *STC* ratings for some typical window, door, wall and floor constructions are given in the text.

Windows are the weakest part of exterior walls for sound control; their effectiveness depends on the airtightness of the installation. Fixed windows provide more sound insulation than windows that can be opened, and double-glazed windows give more sound insulation than single-glazed windows.

Doors are the weakest part of interior walls for sound control; they determine the amount of sound control that can be provided. Solid-core doors provide better sound insulation than hollow-core doors. Gaskets can be installed between the door and jamb to improve sound insulation; this also softens the impact when the door is slammed.

Insulation of *walls* against air-borne sound depends on weight, stiffness, breaks in sound travel paths and sound absorption within the wall space.

Air-borne sound may be controlled with a heavy wall or one composed of several independent layers. To obtain high sound insulation without excessive weight, component members must be structurally isolated from each other. Generally speaking, a wall should be flexible to minimize effects of panel vibrations. Dividing a wall into two independent barriers will accomplish this, because the two relatively flexible layers replace a rigid structure. One highly efficient wall construction uses two sets of wood studs, staggered or slightly offset from each other, so that the two surfaces are not linked. Thus there can be no direct sound transmission except through the top and bottom plates.

Doubling the weight of the surfaces improves performance by about 3 db; adding insulation or a porous material between wall surfaces also helps considerably. Another structure equally efficient utilizes resilient clips or channels to fasten facing materials to wood framing. Again, absorbing material in the wall space is of great value. The walls should extend from floor to ceiling and doors, windows and service outlets should be well sealed.

For air-borne sound, the principles for *floors* are the same as for walls. In addition, a problem of special importance in floors is impact sound (e.g., footsteps) originating as a vibration in the separating structure itself. Impact sound, in many cases, can lead to more complaints than air-borne sound.

Impact noise transmission can be reduced by soft floor coverings, by suspended ceilings, by sound absorbing materials between ceiling and floor and by separating floor surfaces from structural parts of the floor with strips of wood furring or other material.

Controlling Sound Efficiently

The key to efficient sound control is good layout of building interiors; sound problems in many cases can be eliminated, or at the very least, minimized. After the building layout is fixed, the designer then can select types of construction and materials that will satisfactorily control the sounds liable to occur. Designers should remember that no material will solve all problems and that sound control systems are no more efficient than their weakest link. Windows, doors, service outlets and the perimeter of partitions, floors and ceilings should be well sealed to minimize sound leakage. Failure to observe these basic principles will result in annoying problems for tenants, workers and users. If alterations are necessary after a structure is complete, they will be expensive and probably ineffective.

Wood panel products and framing members provide effective, economical sound insulation and absorption when properly utilized. Careful attention to the details of design will result in performance as intended.

(Courtesy of the Canadian Wood Council.)

Decibels

Good insulation reduces the transmission of heat, cold, and sound through floors, walls and ceilings. Sound is measured in decibels using the symbol db. The range of sound runs from that of a whisper, say 0–10 db, to the average radio, say 70 db, and to the noise of a jack-hammer which ranges between 110–120 db and is literally deafening.

15.3 INSULATION AND VAPOR BARRIERS

Of the many different types of insulation and sound proofing materials used in building construction, the following is a partial list; others described in this book may be seen by consulting the index:

1. Batt-types mineral, also obtainable in rolls, see Figs. 15.2 and 15.3.
2. Batt-type fiber, also obtainable in rolls.
3. Lightweight cellular concrete blocks.
4. Lightweight chemically treated wood shavings and cement.
5. Loose fill—inert or fiber.
6. Lightweight cellular plastic types.
7. Straw (chemically treated) boards.
8. Foam glass boards in 2'-0" x 4'-0" sheets of varying thicknesses.
9. Foamed-in-place insulation.
10. Vermiculite, polystyrene, urethane, and perlite.
11. Gypsum boards, mineral fiber boards, wood fiber boards.
12. Rigid urethane laminated roof insulation.
13. Reflective insulation in sheets of aluminum, copper foil, or sheet metal.
14. Polyethylene vapor barriers from .002 to .010.

Further information may be obtained from: Superintendent of Documents, U.S. Government Printing Office, Washington, D.C. 20402; also from Publication Section, Division of Building Research, National Research Council, Ottawa, Ontario; and from Sweet's "Construction Catalog File," which may be seen at your library.

Make a close study of the drawings shown in Figs. 15.2 through 15.6.

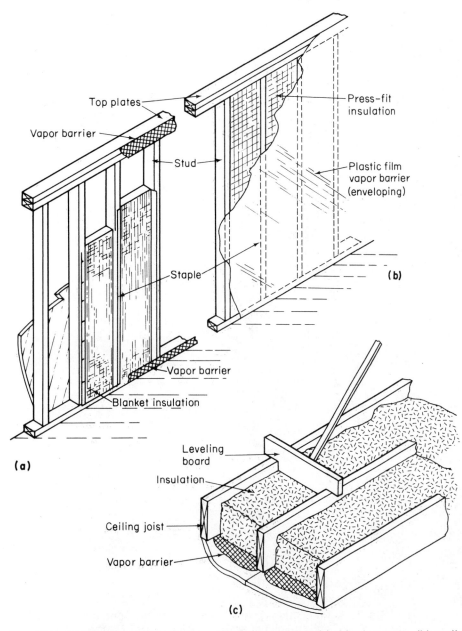

Fig. 15.2. Application of insulation: (a) wall section with blanket type, (b) wall section with "press-fit" insulation, (c) ceiling with full insulation. *(Courtesy of the Forestry Products Laboratory, U.S. Department of Agriculture.)*

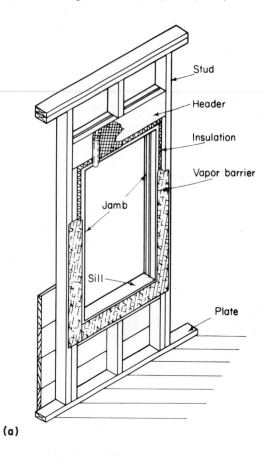

(a)

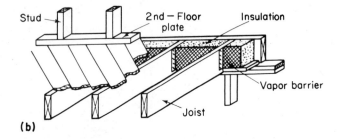

(b)

Fig. 15.3. Precautions in insulating: (a) around openings, and (b) joist space in outside walls. *(Courtesy of the Forestry Products Laboratory, U.S. Department of Agriculture.)*

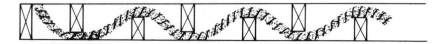

Fig. 15.4. Staggered stud partition with blanket roll insulation.

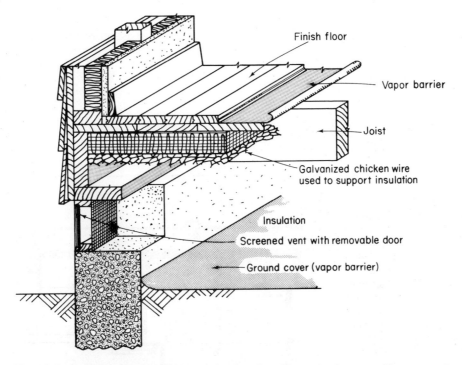

Fig. 15.5. Crawl space ventilator and vapor barrier ground cover. *(Courtesy of the Department of Forestry and Rural Development and the Central Mortgage and Housing Corporation, Ottawa.)*

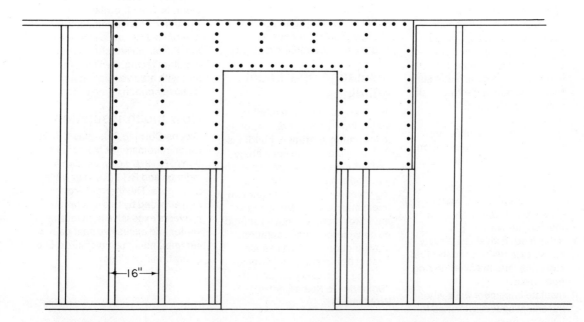

Fig. 15.6. To prevent checking, apply drywall without jointing over doorways.

The material on the following pages on Fiberglas building insulation has been excerpted from one of the publications of Owens-Corning Fiberglas Corporation, and is reproduced here with permission.

Fiberglas building insulation

Where to insulate

To achieve full thermal benefits from building insulation, the insulation must form a complete thermal blanket around the living portion of the house, with no gaps. Install insulation in all walls, ceilings, and floors that separate heated from unheated space. The drawing represents a composite house, showing the places in a building where insulation is required.

- Insulate all exterior walls completely.
- Insulate all ceilings having cold space above.
- Insulate dormer ceilings and walls.
- Where attic is finished as living space or in 1½-story houses, insulate knee walls, sloping roof section and ceiling, leaving open space above for ventilation. Caution: make sure that insulation at the knee wall butts snugly to the ceiling insulation.
- Insulate floor over unheated spaces such as garages, open porches, or vented crawl spaces.
- When a crawl space is enclosed as a plenum, insulate crawl space walls instead of the floor above.
- If the second level of a 2-story house extends beyond the first story, insulate the overhanging floor area.
- Insulate basement walls when basement space is finished for living purposes.
- Insulate between floor joists at the top of foundation walls in

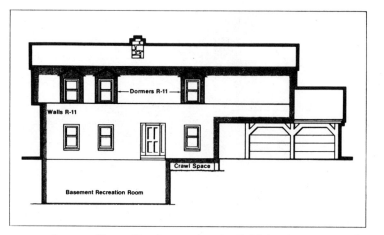

every house, with or without basement.

- Insulate the perimeter of a slab on grade.
- For protection against dirt and air, use sill sealer between sill plate and foundation masonry.

Insulating apartment buildings

The general rule for insulating single-family houses also applies to insulating apartment buildings: install insulation between heated and unheated spaces. When each apartment has a separate heating/cooling system, insulation in floors and walls between apartments for thermal reasons should be considered. This insulation may function as a part of sound control treatment. (See Owens-Corning Fiberglas publication, "Solutions to Noise Control Problems.")

Insulation in remodeling

A house being remodeled ideally should be insulated to the same extent as a new house. Blanket insulation can be applied where interior finish has been removed in the course of renovation. For upgrading existing insulation in ceilings, use Fiberglas* Unfaced Friction Fit Insulation. Simply lay it in place from above. Since Friction Fit has no attached vapor-resistant facing, it will not interfere with attic ventilation nor will it cause condensation problems.

How much insulation

Owens-Corning Fiberglas Corporation recommends the installation of R-19 (6″) of insulation in ceilings and R-11 (3½″) in walls and floors. The thermal protection afforded by this insulation prevents excessive heat loss in winter and excessive heat gain in summer, maximizing comfort and operational savings.

*Trademark Registered Owens-Corning Fiberglas Corporation

Installing insulation in walls

General procedure

When applying insulation in walls, start at the top of each stud space and work down. Be sure to butt insulation firmly against both top and bottom plates.

Cutting insulation. All types of Fiberglas Insulation cut easily

with a sharp knife. Place the insulation on a piece of scrap plywood or wallboard, compress the material with one hand, and cut with the other. When cutting faced insulation, keep the facing up.

Installing Friction Fit

Friction Fit batts come in 15″ and 23″ widths, to fit 16″ and 24″ o.c. framing. Friction Fit usually requires a separate vapor-resistant film. Wedge Friction Fit batts into place between studs, butting them snugly against top and bottom plates. Fill spaces around doors and windows with insulation. Unroll polyethylene film across entire wall area, including window and door openings. Staple securely to top and bottom plates, studs, and door and window framing. When polyethylene has been securely fastened, cut out window and door openings. (It is sometimes left in place until after painting and finishing to act as masking, thereby reducing cleanup.) Duplex-laminated kraft vapor barrier may be used instead of polyethylene and applied in the same way.
Foil-backed gypsum board may also serve as a vapor-resistant material. It is applied in the same manner as ordinary drywall, with

the foil side facing the insulation. To seal window and door framing, staple a 6″ wide strip of polyethylene connecting that framing and the structural framing of the building, prior to the drywall application.

Installing faced insulation

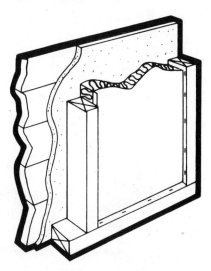

Inset stapling. Fiberglas Kraft-Faced and Foil-Faced Insulations may be inset stapled. Insulation comes in 15″ and 23″ widths to fit 16″ and 24″ o.c. framing.
Place insulation between studs. Staple both flanges snugly to the sides of the studs.
Where the insulation meets the top and bottom plates, peel back about 1″ of the facing, butt the blanket snugly against the plate, and staple the facing to the framing.

Installing insulation in ceilings

General procedure

Fiberglas Kraft-Faced and Foil-Faced Insulations may be inset stapled or face stapled to exposed ceiling joists. Friction Fit insulation is held in place between joists by its own resiliency until finish surface is applied.

If ceiling finish is already in place, faced insulation may be laid in from above with the facing down.

Owens-Corning recommends the installation of 6″ of insulation (R-19) in ceilings for maximum comfort and economy.

Installing Friction Fit

Unfaced Friction Fit insulation is applied in a ceiling by placing it between joists. The insulation should overlap the wall plate slightly, but not enough to block eave vents. The insulation should touch the top of the plate along its full width to reduce air infiltration and consequent heat loss. If the attic space is ventilated to meet applicable FHA requirements, a separate vapor barrier is not needed to prevent condensation damage. A separate vapor-resistant film will, however, help maintain controlled humidity levels within the house.

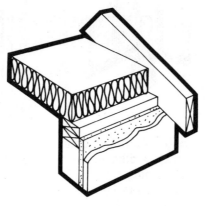

While insulation in a ceiling should always extend over the top wall plate, it is important that it is not allowed to block ventilation if eave vents exist.

Installing faced insulation

Inset stapling. In order to close the gap between inset stapled insulation and the plate at the end of a joist space and reduce air infiltration, pull the insulation down slightly to lie snugly against the top of the top plate. Peel the facing back and staple to the edge of the plate as shown. Provide excess facing to accommodate finish application without rupturing facing.

Use sufficient staples to hold facing snug to the framing member. If the interior wall finish is to be drywall, caution the drywall contractor not to tear the flaps that are stapled to the top plate.

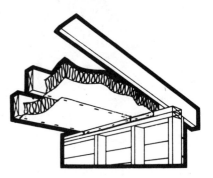

Face stapling. Insulation that has been face stapled to joists should be carried over the top of the top plate, as shown, to reduce air infiltration.

Installing insulation in floors

Reverse flange insulation

For insulating between floor joists from below, Owens-Corning provides a special Kraft-Faced insulation with the nailing flange attached to a special breather paper layer on the side opposite the kraft facing. This allows standard inset stapling techniques to be used, while keeping the vapor-resistant facing toward the warm-in-winter side.

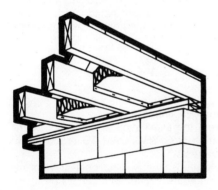

Other methods

Ordinary Kraft-Faced or Foil-Faced insulation may be installed in floors with the facing up and supported by one of the following techniques:

1. Heavy-gauge wire pointed at both ends is bowed and wedged in place under insulation.

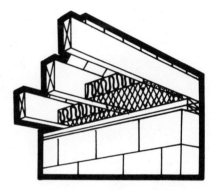

3. Chicken wire nailed to the bottom of floor joists will support the insulation.

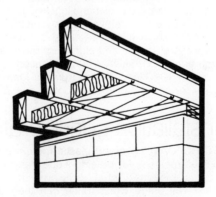

2. Nails are located at intervals along the joists and insulation is supported by wire laced back and forth on the nails.

**15.4 STYROFOAM IN WOOD FRAMING:
      FOUNDATIONS AND ROOFING**

The following material is published by courtesy
of Amspec, Inc.

STYROFOAM TG brand plastic foam replaces three steps in wood frame construction.

STYROFOAM TG brand plastic foam is the only insulation that effectively insulates 100% of the wall space in stud-frame construction.

STYROFOAM TG (tongue-and-groove) is installed just as conventional sheathing, but eliminates the need for vapor barrier and batt insulation. It was developed especially as an insulation / sheathing material, and has much greater thermal resistance than commonly used sheathing. STYROFOAM TG is its own vapor barrier, its strong plastic skin will not delaminate, and it has unusual resistance to moisture, rot and mildew. It is lightweight and easily handled on the job, and can be cut to size with knife or saw to fit windows and doors.

STYROFOAM TG is applied on the outside of the studs, just as conventional sheathing, using common carpentry tools and techniques. Headers, sills, plates, as well as the studs, are covered on the outside, providing complete foundation-to-roof insulation for all of the wall area. This increases the effectively insulated wall area as much as 25%. Because STYROFOAM TG has such extraordinary insulating qualities, the need for batt insulation between studs is eliminated. And because STYROFOAM TG never loses its insulating effectiveness, the home sidewalls remain fully insulated for a lifetime.

Advantages

This application of STYROFOAM TG brand plastic foam is accepted by regional and national building codes, and qualifies for FHA financing.

The substitution of one material for three, the elimination of one labor operation, plus the ease of handling and installation of STYROFOAM TG brand plastic foam results in a net savings in construction costs.

With no insulation required between the studs, there is no conflict with mechanical, electrical and plumbing trades or materials. Inside work can go on at any time after STYROFOAM TG is installed.

STYROFOAM TG brand plastic foam is a versatile performer, insulating effectively for both heating and cooling seasons.

Installation considerations

STYROFOAM TG brand plastic foam is non-structural. There are a number of approved methods for bracing wood frames that incorporate non-structural sheathing boards. Diagonal "let-in" bracing with 1 x 4's is the preferred method. (see photo "A") The entire wood frame can then be covered with STYROFOAM TG brand plastic foam insulation / sheathing for optimum insulating effectiveness. Another approved method of bracing includes steel tension straps. Siding materials and brick wall ties can easily be nailed through the foam into the studs.

**STYROFOAM TG brand plastic
foam is tongue and groove on
all 4 sides for tight joint seal.**

CORNER POST DETAIL

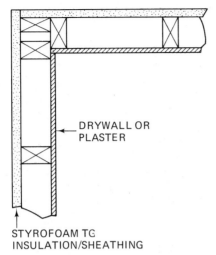

DRYWALL OR
PLASTER

STYROFOAM TG
INSULATION/SHEATHING

SOFFIT AND KNEE WALL DETAILS

PLENUM
AREA

OPEN
VENTED
SOFFIT

STYROFOAM TG
INSULATION/
SHEATHING

WINDOW/DOOR TRIM DETAILS

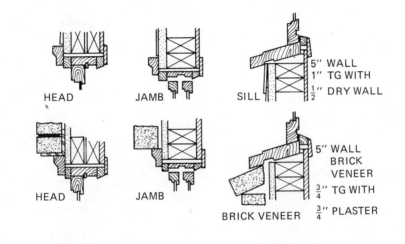

HEAD

JAMB

SILL

5" WALL
1" TG WITH
$\frac{1}{2}$" DRY WALL

HEAD

JAMB

BRICK VENEER

5" WALL
BRICK
VENEER
$\frac{3}{4}$" TG WITH
$\frac{3}{4}$" PLASTER

SILL PLATE DETAILS

STYROFOAM TG
INSULATION/
SHEATHING

STYROFOAM TG
INSULATION/
SHEATHING

PERIMETER
INSULATION

PERIMETER
INSULATION

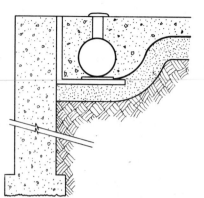

Perimeter Heating
Observe General Design Requirements
for Perimeter Insulation, Adding Additional
$\frac{3}{4}$ Inch to Design Thickness. Isolate
Heating Duct from Subsoil as Shown.

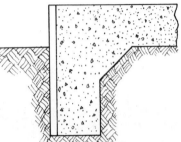

Monolithic Foundation-Slab
Place STYROFOAM SM Brand Plastic
Foam Against Exterior of Foundation
During Backfilling. Insulation Above
Grade Should be Covered With Sheet
Metal, Cement Asbestos Board or a
Similar Finish Adhered to Foam With
DOW General Purpose Mastic No. 11.

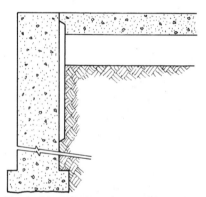

Vertical Installation
Place STYROFOAM SM Brand Plastic
Foam Against Interior or Exterior of
Foundation During Backfilling. Boards can
be Laid Dry Against the Foundation as
Backfill is Placed, or May be Adhered
With a DOW Mastic Prior to Backfilling.

Crawl Space
Cover Soil in Crawl Space With 4 Mil
Polyethylene and Seal at Foundation Edge
With Asphaltic Compound. Lap and Seal
Joints. Insulate Foundation Wall with
STYROFOAM SM Brand Plastic Foam.
Fasten Insulation to Sill Plate with Large
Headed Nails or Bond Insulation Directly
to the Wall With a DOW Mastic Adhesive.

THE IRMA system lends itself to fast, trouble-free installation.

The IRMA roof can be used on all types of structural roof decks — nailable, non-nailable and metal — with a maximum slope of up to 2 inches per foot. A minimum slope of ¼″ per foot is recommended. Positive drainage is required. Decks must be rigid, dry, clean and free of projections.

The IRMA system's conventional 3-ply, 15-pound asphalt-saturated membrane is put down in a single operation directly onto the roof deck. Bare felts are never left exposed at night or in inclement weather. Steep asphalt is used throughout application, with mopping continuous and uniform, and each ply is embedded by brooming or other mechanical means. Installed as specified in the IRMA system this built-up membrane gives improved performance.

The completed membrane is mopped with 45 pounds per square steep asphalt and STYROFOAM RM is embedded while the asphalt is fluid. Boards are walked in so the insulation is as nearly 100% bonded as possible. All joints are tightly butted.

Drains are set at the membrane level. All flashing — including drain flashing — is installed the same as on a non-insulated roof.

Bundles of STYROFOAM RM are lightweight and easy to handle. No problem to store, handy for use during installation.

STYROFOAM RM is easy to work with, easy to fabricate right on the job for fitting around vents and projections.

Final step in IRMA roof installation is spreading of crushed stone, with an average size of ¾", at the rate of 1,000 pounds per square. The stone gives a flame resistant finish, anchors the STYROFOAM RM in the adhering asphalt and protects it from ultraviolet degradation.

The finished IRMA roof. This roof was installed in November, 1968 on the Herbert H. Dow High School in Midland, Michigan.

Faster closing in — better scheduling.

As you can see, the IRMA system uses nothing new or untried. All materials and application techniques are standard, conventional, proven—accepted throughout the roofing industry. Minimum training is required. No new equipment. No additional personnel.

Another significant advantage of the IRMA system is that it allows more efficient use of roofing and other subcontractor crews during construction.

Because the membrane does not have to be laid in conjunction with the insulation — as in conventional built-up roofs — the roofer can utilize fair weather to lay it down in a single operation; putting the building under a dependable permanent roof sooner. He can put down the STYROFOAM RM brand plastic foam as much as three weeks later, even in questionable weather, and the stone can be applied as long as three weeks after *that* — in *any* kind of weather. The more parts of the system completed, the more the membrane is protected from physical abuse.

In the meantime, with this faster closing in, the general contractor can schedule his inside work better: get plumbers, electricians and other sub-contracted work crews into action without slowing down the progress of the overall job.

15.5 FOAMED-IN-PLACE INSULATION

The following material has been excerpted from one of the publications of the U. F. Chemical Corporation, and is here published with permission.

typical installations   7.14/Uf

foamed-in-place insulation

thermal and acoustical

what UFC foam is

UFC FOAM is a superior insulation which on an installed-cost-per-unit effectiveness basis is less expensive than poured or matted insulation material.

its many attributes:

- thermal insulation
- sound absorption
- low cost
- application ease
- dimensional stability
- moisture resistance
- pest control
- non-flammability

how UFC foam is applied

UFC-Foam is applied from a patented gun within which the foaming action takes place. There is no further expansion after the foam leaves the gun. Voids can be completely filled without fear of subsequent pressure build-up. It can be applied in any temperature as easily as spreading shaving cream.

UFC-Foam can be used to fill existing voids through holes as small as one inch, can be applied between open frames—floor, wall or ceiling—or can be foamed through metal lath. Once in place it can be smoothed with a trowel and sheathed over immediately. A typical between-studs void is completely insulated in less than 2 minutes.

Where U.F.C. foam has been used

Columbia University
St. Clair Place, N.Y.C., N.Y.
Architect: Brown, Guenther, Battaglia, Galvin

Walston Building
77 Water Street, N.Y.C., N.Y.
Architect: Emery Roth & Sons
Gen'l contractor: Diesel Construction Co., Inc.

Chanin Building
1411 Broadway, N.Y.C., N.Y.
Architect: Irwin S. Chanin
Gen'l contractor: Chanin Construction Corp.

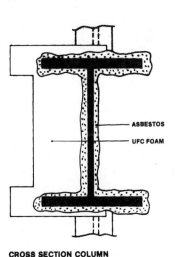

CROSS SECTION CORNER COLUMN

ASBESTOS
UFC FOAM

CROSS SECTION COLUMN

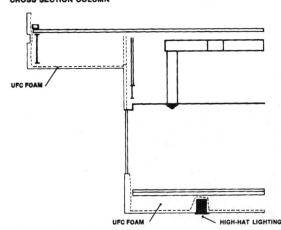

UFC FOAM

UFC FOAM   HIGH-HAT LIGHTING

UFC insulating and soundproofing foam is readily applied to vertical and overhead surfaces with use of metal lath, hardware cloth or lightweight non-burning netting (above) to create a cavity to contain the foam.

air and sound insulation

The unique thermal and acoustical properties of UFC foam, and its ability to completely fill odd-shaped crevices containing pipes, wires, ducts and fixtures, make it an ideal insulating material for pipe chase areas and other cavities adjoining lightweight walls or partitions.

Preventing the transmission of annoying or embarrassing sounds is effectively accomplished with UFC foam while providing efficient insulation for hot and cold conduits.

U. F. CHEMICAL CORP.

CHAPTER 15 REVIEW QUESTIONS

1. Define the term *cladding* as used in wood frame construction.

2. List five inside and five outside types of wall cladding.

3. Define the term *decibel,* and give an extract from your local building code concerning it.

4. List five different building materials used for sound control.

5. What considerations regarding noise should be taken into account before purchasing a building site for the erection of houses?

6. What is the function of a vapor barrier?

7. Make a freehand sketch of the plan of a staggered stud wall partition with blanket roll insulation.

8. Give three qualities of styrofoam T & G (tongue and groove) insulation.

9. Make a freehand sketch of a perimeter insulation detail for a heating duct and arrow and name the parts.

10. List eight characteristics of foamed-in-place insulation.

11. Make three neat freehand sketches of outside corner treatments of wood-wall siding.

12. Define *Styrofoam* and state two advantages in using it.

Stairs, Fireplaces, and Interior Trim

In effect, a stairway may be considered to be an inclined hall leading to other areas of a building. Remember that a good house is designed around the stairway. Stairs are provided for the purpose of passage of *persons of all ages in varying degrees of health,* and should be easy of ascent, and constructed strong enough to carry the weight of any predetermined load, such as persons carrying heavy pieces of furniture or equipment.

16.1 STAIR DEFINITIONS

Stairwell opening — The opening in a floor through which ascent is gained to an upper floor.

Fascia board — The finished trim around the inside of a stairwell which covers the rough flooring assembly.

Total rise of a stair — The perpendicular height from the *finished* lower floor to the top of the *finished* floor above.

Riser — The perpendicular height from the top of one tread to the top of the next tread above. Note that there is one more riser than tread to every staircase because the top riser leads to a floor, not a tread.

Run of stairs — The horizontal distance from the front of the first riser at the foot of a stairway to the front of the top riser.

Run of stair tread — The horizontal distance from the front of one riser to the front of the next riser above. The nosing is additional to this. See Fig. 16.1.

Angle of flight — The incline (slope) of a stair.

Headroom — The minimum allowable perpendicular height from a point on the angle of flight to the finished underside of the floor, or stair soffit above. See Fig. 16.5.

Stair soffit — The finished underside of a stair.

Story rod — A straight wooden measuring rod which is placed perpendicular from the finish of one floor to the finish of the floor above on which is marked the total height of the stairs. It is used by stair builders. See Fig. 16.2.

Stringers — (a) The sides of stairs into which the treads and risers are secured.

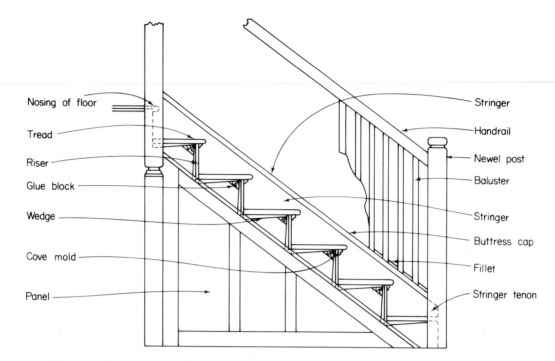

Fig. 16.1. Wooden stair with open stringer.

(b) A mitered stringer is one into which the risers are mitered to an open stringer, (i.e.) a stringer that is not covered by a wall. See Fig. 16.1.

(c) A notched stringer is one from which (on the upper side) the profile of the tread and riser is cut away. It is similar in some respects to a stair carriage. See Fig. 16.4.

(d) A housed stringer has grooves (trenches) cut into the stringer to receive the risers and treads. They are called housings in wood stringers and are usually half an inch deep. See Fig. 16.3.

Glue blocks —Triangular pieces of wood about three inches long and two inches wide on each side of a 90° angle. They are glued to the underside of wooden stairs. See Fig. 16.3.

Handrail —This is what its name implies. It should be placed thirty-two inches perpendicular above the face of the risers. At landings it should be placed thirty-six inches above the level of the floor.

Landings —Horizontal platforms between flights of stairs (usually to change the direction of the flight). They should be provided with a handrail.

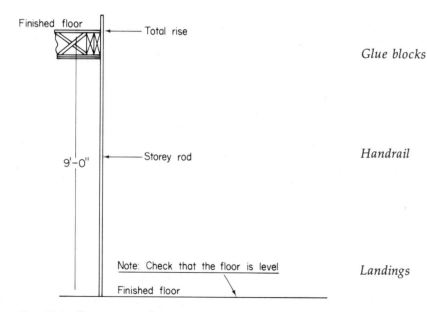

Fig. 16.2. The story rod in position for measuring the total rise from finished floor to finished floor.

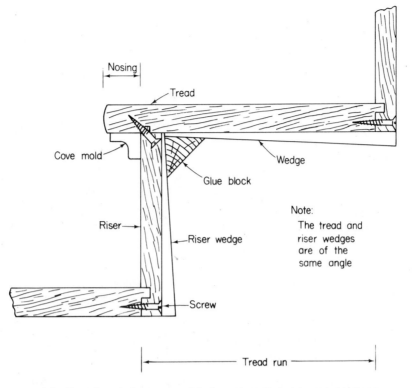

Fig. 16.3. Tread and riser assembly for a housed stringer staircase.

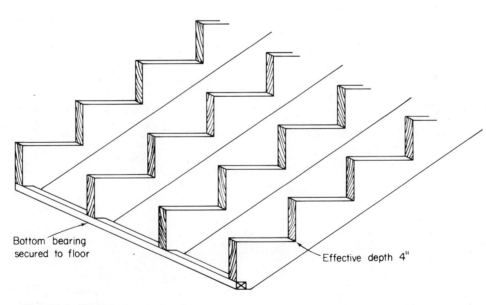

Fig. 16.4. Stair horse or carriage.

Newel post — A post on a stair to which is secured the stringer. (See Fig. 16.1.)

Balusters — Vertical members placed between the tops of stringers (or treads) and the underneath side of the handrail. They must be close enough together to prevent children from falling between them. (See Fig. 16.1.)

Stairhorse — A rough lumber carriage supporting a staircase between stringers. (See Fig. 16.4.)

It is important that the local building code be studied and complied with, especially in such areas as:

(a) Maximum allowable (unvarying) heights of risers and widths of treads in any one flight.

(b) Minimum widths of stairs.

(c) Minimum allowable perpendicular headroom clearance between a point on the angle of flight to the underneath side of the floor, (or soffit of another stair) above.

(d) All stairs in residential construction must be supplied with handrails. Stairs with less than three risers are a hazard and should be avoided if possible.

16.2 RISERS AND TREADS

There are several rules for determining the ratio-measurements between risers and treads. Some authorities assert that the sum of any two reasonable measurements for a riser and a tread, which together equals 17″ will give an acceptable ratio, and provide a stair that is easy of ascent. A classic example may be seen at the Heliopolis at Baalbek in Lebanon built about 2,000 years ago.

Example:

Rise	Tread	Inches
6	11	17
$6\frac{1}{4}$	$10\frac{3}{4}$	17
$7\frac{3}{16}$	$9\frac{13}{16}$	17
8	9	17 and so on.

The stairways for apartment buildings and new housing developments may be prefabricated either on or off the job, and they may be of wood, metal, or concrete construction, and all measurements are taken direct from the drawings and specifications.

16.3 FINDING THE LINE OF FLIGHT FOR A STAIRWAY

Assume that a stair builder has to design and build a staircase from the following references:

(a) rough stairwell opening   3′-3″ x 10′-4″

(b) height from finished floor to finished floor   9′-0″

(c) total thickness of floor assembly   12″

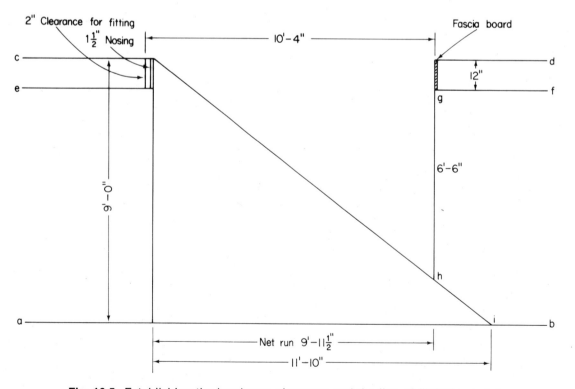

Fig. 16.5. Establishing the headroom clearance and the line of flight for a stair.

(d) headroom clearance (check local code) .... 6'-6"

(e) clearance allowance for fitting the stair to the upper level .... 2"

(f) allowance at upper level for tread nosing .... 1¼"

(g) allowance for fascia board to rough floor opening .... 1"

Solution:

Make a single-line scaled drawing as follows:

Step 1: Draw two parallel construction lines 9'-0" apart representing the two finished floors as at a-b and c-d, Fig. 16.5.

Step 2: Draw in the total thickness of the floor assembly. (12") as at e-f.

Step 3: Above line a-b lay off the length of the rough opening 10'-4".

Step 4: At the right hand end of the rough opening make an allowance for the thickness of the fascia board.

Step 5: Allow 2" at the upper left hand level for fitting the stair, and allow 1" for the thickness of the riser.

Step 6: Allow a further 1½" at the upper level for the stair nosing.

Step 7: From the bottom of the outside edge of the fascia board (right side of opening) project a 6'-6" perpendicular line as at g-h. This line represents the headroom clearance.

Step 8: From the top edge of the top riser at the upper floor level, draw a line of flight to intersect with -h- (for headroom), and continue the line to the lower floor as at -i-.

Step 9: Examine the drawing; this line of flight gives the best proportioned stairway from all the references given in this exercise.

16.4 RATIO OF RISE TO TREAD

To find an acceptable ratio for the rise and tread for the stairway shown in the foregoing exercise and assuming that the local building code states that no riser for this stairway may exceed 8" in depth, take the following steps:

Step 1: Arithmetically divide the total rise of 108" by 8, which equals 13½.

Step 2: Since 13½ risers are not permitted, divide 108 by 14. Using the rule of 17 for the ratio of riser to tread, this gives a rise of 7.71" and a tread of 9.29"; the sum of these figures equals 17".

Geometrical Division of a Parallelogram into Equal Width Parts

Now let us have a bit of fun. To find the center line of a piece of 2 x 4 stock, (note that 2" x 4" is the sawn size of the lumber, and the finished size is 1⅝" x 3⅝") place a tape, square across the 3⅝" face of the stock. Traverse the tape until the 4" mark may be read on the top edge of the stock. Spot the 2" mark which is the center. Set an adjustable square to the center spot and draw in the center line ₵. See Fig. 16.6.

To divide a blank piece of 8½" x 11" typing paper into seven equal width columns; place a rule square across the paper as shown in Fig. 16.7 and traverse it until the 10½" mark may be read on the opposite edge of the paper as shown. Spot mark every 1½" unit onto the paper; note that 1½" goes into 10½" seven times. Now draw parallel lines to the lengthwise edge of the paper intersecting each spot mark. **Now ask yourself the question, what units would you use on a diagonally placed rule to divide the same paper into nine equal width columns?**

Now let us revert back to the stairs with step three:

Step 3: Figure 16.8 is drawn to the same scale as that of Fig. 16.5. To divide geometrically the total scaled rise of 108" into 14 equal units, use as a focal point -b-, and traverse the scale rule to any convenient measure of 14 equal spaces and spot mark them (with a pin prick) as shown by the dotted line a-d.

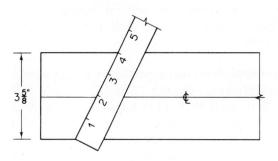

Fig. 16.6. Finding the center line ₵ of a piece of wood 3⅝" wide.

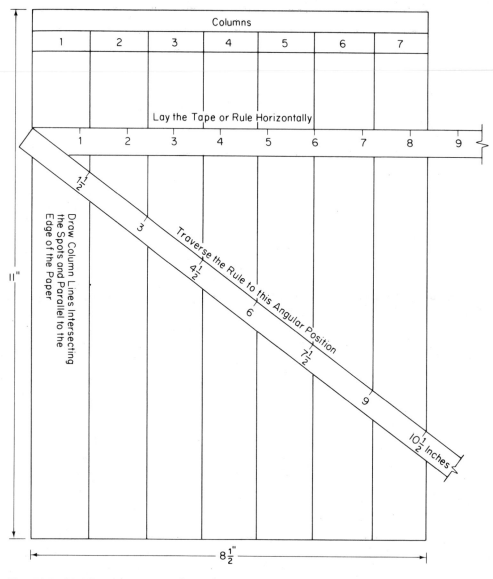

Fig. 16.7. Dividing a piece of 8½" x 11" paper into seven equal width columns.

Step 4: Draw horizontal lines parallel to the base line and intersecting the 108" perpendicular, the pricked spots, and the line of flight. These lines at the intersection with the line of flight represent the treads. Drop perpendiculars to show the rise to each successive tread.

Problem: Redo this exercise for a stair with similar specifications but with 15 risers.

Remember that stairs have one more riser than treads; this is because the top riser leads to a floor and not to another tread.

Handrails should be placed 32" perpendicular *above the front* of the risers; and they should be placed 36" above landings.

16.5 CHIMNEYS AND FIREPLACES

In the whole field of residential construction, there are few things that will discredit a builder more than selling a house with an inefficient chimney which causes smoke in the room where the fireplace for solid fuel is provided. Apart from the nuisance, there is the risk of fire. For this reason, so that a simple construction of a safe and efficient chimney and fireplace may be built, several authoritative sources of information are here given:

(a) The Structural Clay Products Institute of America.

(b) "Clay Masonry Manual," published by the Brick and Tile Institute of Ontario, Canada.

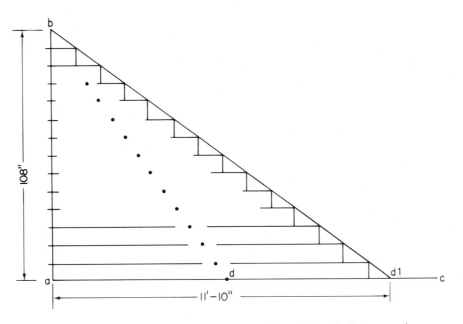

Fig. 16.8. The geometrical method of establishing the ratio between risers and treads.

(c) "Wood Frame House Construction," obtainable from the U.S. Government Printing Office.

(d) "Wood Frame House Construction," obtainable from any office of the Central Mortgage and Housing Corporation.

(e) "The Canadian Code for Residential Construction," obtainable from the National Research Council of Canada.

(f) Donley's, "Book of Successful Fireplaces—How to Build Them!"

(g) The National Code as recommended by the National Board of Fire Underwriters.

(h) Your local building code. Study it!

While there is nothing intricate nor difficult in the construction of a chimney and fireplace, it is important that the contractor have knowledge sufficient to supervise his own men (or specializing subcontractors) in this field.

The following features are especially important:

(a) The foundations for a chimney should be specially designed to support the heavy load.

(b) All chimney flues serving solid fuel fires for any one stack shall be kept separate.

(c) A metal cleanout opening with a tight fitting metal door shall be provided near the base of the chimney flue.

(d) Chimney flues shall not be inclined more than 45° to the vertical.

(e) Wood framing or other combustible material shall not be placed less than 2" from the walls of a chimney, and 4" shall be allowed at the backs of fireplaces.

(f) The space between the masonry of a chimney and any combustible material shall be fire-stopped with metal supported incombustible material.

(g) Chimney flashing shall be provided at the uppermost point of the roofing at its intersection with the chimney.

(h) The chimney shall be built at least 3'-0" above its intersection with the roofing, and it shall be not less than 2'-0" above the ridge or any other obstruction within 10'-0" of the chimney.

(i) Chimney caps shall be designed to shed water away from the flue; they shall extend at least $1\frac{1}{2}$" from all faces of the chimney; and be provided with a dripcap to prevent water from running down the face of the chimney walls.

16.6 A GLOSSARY OF CHIMNEY AND FIREPLACE TERMS

Fireplaces

ash dump; (see Fig. 16.9);
ash pit with metal door;
brick arch under finished hearth;
cleanout;

concrete slab under finished hearth;

damper;

flat concrete arch supporting chimney breast;

flue lining;

hearth, tiled;

mantle shelf;

metal arch supporting chimney breast;

mortar, cement above roof level and fireclay for enclosed brickwork;

smoke dome;

smoke shelf.

flashing at chimney level;

flue lining;

height of chimney over roof line; See local by-laws;

independent concrete reinforced footings;

insulation between masonry and woodwork;

parging; lime, cement and sand mortar;

pots;

revolving metal ventilators over the chimney cap.

Chimneys

bricks, common, decorative, fireclay;

concrete cap over brick chimney;

cowl over chimney top;

cricket, small roof structure between the back of the chimney and the sloping roof;

16.7. CHIMNEY ILLUSTRATIONS

The following drawings are presented as graphic illustrations of chimney and fireplace construction; *but, study your local building code, and some of the publications mentioned in this section.*

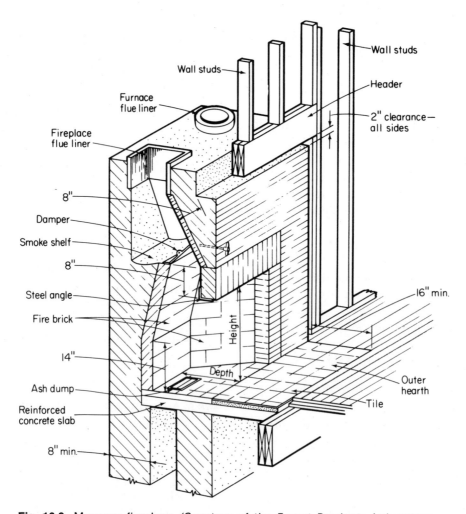

Fig. 16.9. Masonry fireplace *(Courtesy of the Forest Products Laboratory, U.S. Department of Agriculture.)*

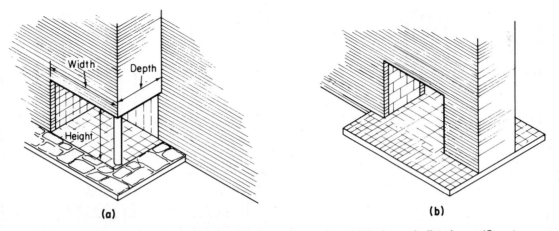

Fig. 16.10. Dual opening fireplace: (a) adjacent opening, and (b) through fireplace. *(Courtesy of the Forest Products Laboratory, U.S. Department of Agriculture.)*

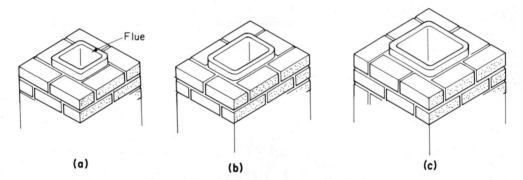

Fig. 16.11. Brick and flue combination: (a) 8 x 8 inch flue lining, (b) 8 x 12 inch flue lining, and (c) 12 x 12 inch flue lining. *(Courtesy of the Forest Products Laboratory, U.S. Department of Agriculture.)*

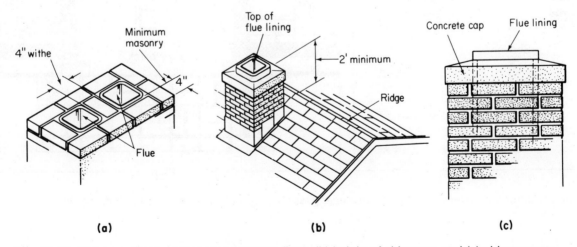

Fig. 16.12. Chimney details: (a) spacer between flues, (b) height of chimneys, and (c) chimney cap. *(Courtesy of the Forest Products Laboratory, U.S. Department of Agriculture.)*

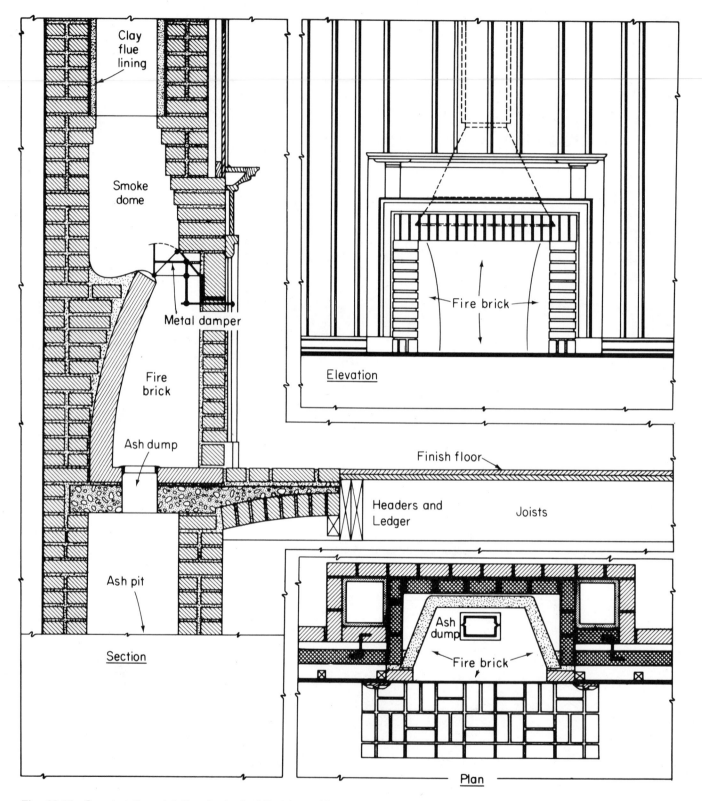

Clay
flue
lining

Smoke
dome

Metal damper

Fire
brick

Ash dump

Ash pit

Section

Fire brick

Elevation

Finish floor

Headers and
Ledger

Joists

Ash
dump

Fire brick

Plan

Fig. 16.13. Construction details of a typical fireplace. *(Courtesy of the Brick and Tile Institute of Ontario.)*

16.8. INTERIOR TRIM

The following material has been excerpted from a booklet published by Western Mouldings and Millwork Producers and is here reproduced by permission.

western wood moulding and millwork

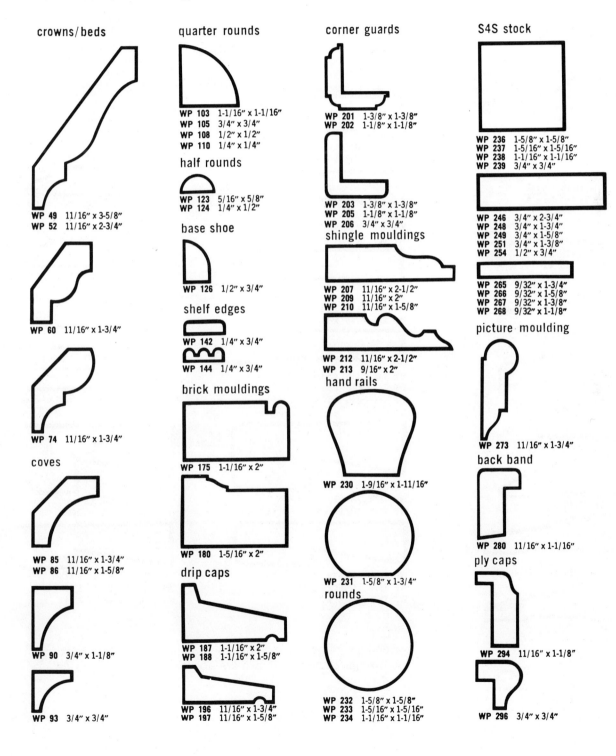

crowns/beds

WP 49   11/16" x 3-5/8"
WP 52   11/16" x 2-3/4"

WP 60   11/16" x 1-3/4"

WP 74   11/16" x 1-3/4"

coves

WP 85   11/16" x 1-3/4"
WP 86   11/16" x 1-5/8"

WP 90   3/4" x 1-1/8"

WP 93   3/4" x 3/4"

quarter rounds

WP 103   1-1/16" x 1-1/16"
WP 105   3/4" x 3/4"
WP 108   1/2" x 1/2"
WP 110   1/4" x 1/4"

half rounds

WP 123   5/16" x 5/8"
WP 124   1/4" x 1/2"

base shoe

WP 126   1/2" x 3/4"

shelf edges

WP 142   1/4" x 3/4"

WP 144   1/4" x 3/4"

brick mouldings

WP 175   1-1/16" x 2"

WP 180   1-5/16" x 2"

drip caps

WP 187   1-1/16" x 2"
WP 188   1-1/16" x 1-5/8"

WP 196   11/16" x 1-3/4"
WP 197   11/16" x 1-5/8"

corner guards

WP 201   1-3/8" x 1-3/8"
WP 202   1-1/8" x 1-1/8"

WP 203   1-3/8" x 1-3/8"
WP 205   1-1/8" x 1-1/8"
WP 206   3/4" x 3/4"

shingle mouldings

WP 207   11/16" x 2-1/2"
WP 209   11/16" x 2"
WP 210   11/16" x 1-5/8"

WP 212   11/16" x 2-1/2"
WP 213   9/16" x 2"

hand rails

WP 230   1-9/16" x 1-11/16"

WP 231   1-5/8" x 1-3/4"

rounds

WP 232   1-5/8" x 1-5/8"
WP 233   1-5/16" x 1-5/16"
WP 234   1-1/16" x 1-1/16"

S4S stock

WP 236   1-5/8" x 1-5/8"
WP 237   1-5/16" x 1-5/16"
WP 238   1-1/16" x 1-1/16"
WP 239   3/4" x 3/4"

WP 246   3/4" x 2-3/4"
WP 248   3/4" x 1-3/4"
WP 249   3/4" x 1-5/8"
WP 251   3/4" x 1-3/8"
WP 254   1/2" x 3/4"

WP 265   9/32" x 1-3/4"
WP 266   9/32" x 1-5/8"
WP 267   9/32" x 1-3/8"
WP 268   9/32" x 1-1/8"

picture moulding

WP 273   11/16" x 1-3/4"

back band

WP 280   11/16" x 1-1/16"

ply caps

WP 294   11/16" x 1-1/8"

WP 296   3/4" x 3/4"

Western Wood Moulding & Millwork ωσm 6.9/Wes
mouldings

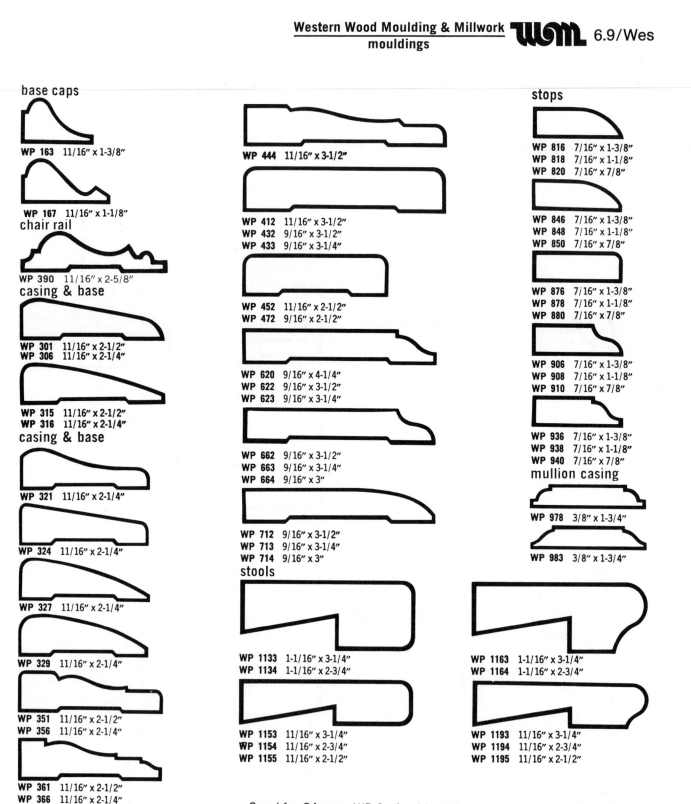

base caps

WP 163   11/16″ x 1-3/8″

WP 167   11/16″ x 1-1/8″

chair rail

WP 390   11/16″ x 2-5/8″

casing & base

WP 301   11/16″ x 2-1/2″
WP 306   11/16″ x 2-1/4″

WP 315   11/16″ x 2-1/2″
WP 316   11/16″ x 2-1/4″

casing & base

WP 321   11/16″ x 2-1/4″

WP 324   11/16″ x 2-1/4″

WP 327   11/16″ x 2-1/4″

WP 329   11/16″ x 2-1/4″

WP 351   11/16″ x 2-1/2″
WP 356   11/16″ x 2-1/4″

WP 361   11/16″ x 2-1/2″
WP 366   11/16″ x 2-1/4″

WP 376   11/16″ x 2-1/4″

WP 444   11/16″ x 3-1/2″

WP 412   11/16″ x 3-1/2″
WP 432   9/16″ x 3-1/2″
WP 433   9/16″ x 3-1/4″

WP 452   11/16″ x 2-1/2″
WP 472   9/16″ x 2-1/2″

WP 620   9/16″ x 4-1/4″
WP 622   9/16″ x 3-1/2″
WP 623   9/16″ x 3-1/4″

WP 662   9/16″ x 3-1/2″
WP 663   9/16″ x 3-1/4″
WP 664   9/16″ x 3″

WP 712   9/16″ x 3-1/2″
WP 713   9/16″ x 3-1/4″
WP 714   9/16″ x 3″

stools

WP 1133   1-1/16″ x 3-1/4″
WP 1134   1-1/16″ x 2-3/4″

WP 1153   11/16″ x 3-1/4″
WP 1154   11/16″ x 2-3/4″
WP 1155   11/16″ x 2-1/2″

stops

WP 816   7/16″ x 1-3/8″
WP 818   7/16″ x 1-1/8″
WP 820   7/16″ x 7/8″

WP 846   7/16″ x 1-3/8″
WP 848   7/16″ x 1-1/8″
WP 850   7/16″ x 7/8″

WP 876   7/16″ x 1-3/8″
WP 878   7/16″ x 1-1/8″
WP 880   7/16″ x 7/8″

WP 906   7/16″ x 1-3/8″
WP 908   7/16″ x 1-1/8″
WP 910   7/16″ x 7/8″

WP 936   7/16″ x 1-3/8″
WP 938   7/16″ x 1-1/8″
WP 940   7/16″ x 7/8″

mullion casing

WP 978   3/8″ x 1-3/4″

WP 983   3/8″ x 1-3/4″

WP 1163   1-1/16″ x 3-1/4″
WP 1164   1-1/16″ x 2-3/4″

WP 1193   11/16″ x 3-1/4″
WP 1194   11/16″ x 2-3/4″
WP 1195   11/16″ x 2-1/2″

Send for 24-page WP-Series Moulding Patterns booklet for full scale drawings. Drawings shown are approximately ⅔ full scale.

CHAPTER 16 REVIEW QUESTIONS

1. Define the following stair building terms:
   (a) stairwell opening (f) newel post
   (b) fascia board   (g) landing
   (c) run of stairs   (h) stair horse or carriage
   (d) angle of flight   (i) baluster
   (e) story rod   (j) stringer

2. State any rule for determining the relationship of the height of a riser to the width of a tread in stair building.

3. According to your local building code, what is the maximum height for a riser and the minimum width of a tread for basement stairs?

4. How many risers has a stair with twelve treads?

5. List three authorities from whom information may be obtained on how to erect a simple, safe and efficient chimney for a solid fuel fireplace.

6. List eight important features to observe when constructing a chimney for a solid fuel fireplace.

7. How would you estimate the total number of bricks required for any given residential chimney? Answer in not more than 120 words.

8. What is the minimum allowable height for a residential chimney above the highest point of the roof-line in your area?

9. Take a piece of blank writing paper and divide it into 12 equal width columns.

10. Make a neat freehand sketch of each of the following traditional wood moulds: (a) a quarter round; (b) a shelf mould; (c) a brick mould; (d) a window stool; (e) a mullion casing.

17

Painting, Decorating, Floor Finishing, and Extruded Metal

In this chapter we shall deal with some aspects of the above-mentioned subjects. It is important that speculatively built residences, both houses and apartment blocks, be painted, decorated and finished in the decor recommended by specialists. The builder cannot be all-knowing in every field of residential construction, and many suppliers give an inclusive service of decor advice for any materials that they supply.

17.1 PAINTING

Before any painting is commenced, the floors should be cleaned of all plaster droppings and other gritty materials, the ceilings and walls should be swept clear of dust and the floor swept clean.

When a builder is using his own work force for painting, it is advantageous to paint (from the bare subfloor) all the interior walls before the trim around doors, windows, other openings, and baseboards is applied. This method affords a great saving in time; the subfloor suffers no harm if it becomes splattered with paint. In addition, this method saves time in brush cutting-in around such places as door and baseboard trim (cutting-in is to paint up to, but without spotting paint onto, other fixed members of different color, such as baseboards). Some builders will

spread a large piece of polyethylene in the middle of the subfloor, pour a gallon of paint onto it, propel the paint roller into it, and roll-paint the ceilings and the walls from a standing position. Without any impediment in the usage of the paint roller by fixed trim or different color, rooms are speedily and efficiently painted. To seal small cracks on surfaces to be painted, a paste filler may be made from plaster of Paris mixed with the actual paint with which the surface will be covered.

Wood trim may be stained or painted in another area of the house, but this presupposes that the finishing carpenters are clean, competent and careful workmen; otherwise, they may scuff and scratch-mark the painted surfaces. When all the trim has been fixed, all countersunk nail holes should be stopped with matching colored putty, or with color matched nails. The rooms are then ready for the application of the finished flooring. After the flooring material is laid, all necessary touching up of paint, stain, and varnish work is completed.

There is a constant demand for paints of different quality for use in various conditions, and as a consequence, there is great competition among manufacturers for the market. It is a truism that one gets that which one pays for—this is particularly true of paint. It is recommended that purchases of paint be made from well-known and established dealers who have their reputations to maintain by the quality and price of their wares. Constant laboratory tests are made by manufacturers, and it is imperative that their instructions for application be complied with rigidly. Some workmen may have their own pet ideas about painting, without realizing there are frequent changes in the composition of the paint they are to use. Where a customer adamantly wants a special color of paint, have her select the paint herself at the supplier's shop. Most people are amazed to find that the color of a painted wall is very different from the color they thought they had selected.

17.2 WALLPAPER, MIRRORS, AND CHANDELIERS

As with paint, there is a wide variety and quality in the types of wallpaper and chandeliers available. It requires an experienced person to hang wallpaper; the supplier will usually supply and fix the materials for an inclusive price.

The selection of the color and design of wallpapers should be left to experts, and the quality of the paper must be made against the funds allotted for the purchase. It is easy and cheap to change the color of a painted wall, but it is more expensive to change a wallpaper.

Large Mirrors and Chandeliers are now almost standard units supplied by speculation builders. Mirrors lend an added feeling of spaciousness, and they save the house purchaser the problem of buying them separately and of having to pay someone to fix them. The chandeliers contribute a feeling of elegance to a room. However, they may present a problem for the electrician to fix after the purchase of the house; when supplied by the builder, the electrical outlets are designed to receive them.

Some builders also include wrought ironwork in the halls and stairways. Remember that the framing for residential construction varies little per square foot from one project to another. It is the difference in the quality of the finish and in the supplying of units that makes the difference in the cost of the final product. Remember, a residence should be built in any given area to meet the needs of the class of wage earners for whom it is designed. Not one penny more than is necessary to meet these conditions should be expended. *It costs just as much to build a good house on a poor lot as a similar house on a good lot.*

17.3 RESILIENT FLOOR COVERINGS

Before any finished flooring is applied over a subfloor, it is imperative to check that all nail heads are sunk below the surface of the subfloor; that loose boards are securely nailed; and that all loose and gritty material such as knobs of dried plaster droppings, plumbers solder drippings, pieces of dry putty, clippings of electric wire be removed, and the floor swept clean. Unless the subfloor is clear and clean, resilient flooring (and carpeting) will reveal the underlying matter by sight, by foot, or both. The debris will cause the floor coverings to wear through more quickly and unevenly.

Resilient flooring includes: linoleum in either tiles or rolls; asphalt tile; vinyl asbestos; cushioned and self adhesive all purpose vinyl tiles, and some with rubber facings; cork; non spark and others. It is important to remember that no resilient flooring should be laid except on a subfloor underlay of fiber boards such as sheet plywood and so on.

17.4 CARPETED FLOORS

Most residential units are marketed with fitted wall to wall carpets in all areas except kitchens, bathrooms and work areas, and since the introduction of indoor-outdoor carpeting in rolls or tile form, there is no place in a modern home that need be without carpeted flooring. Types of carpeting range from exotic Indian and Arabian carpeting to those of natural fibers. Among man-made materials are nylon, viscose, polypropylene, acrilan, polyester/nylon; self adhesive carpet in tile form; and indoor-outdoor carpeting.

The speculative builder must be aware of the latest trends in all kinds of floor coverings, and study carefully the initial cost of the materials, *plus the cost to install them.* He must select the materials for quality, color, design, and appeal to the class of wage earners for whom he is building.

17.5 WOOD FLOORING

There is a wide selection of both hard and soft wood for finish flooring; the most commonly used T & G flooring is oak, birch, Douglas fir, hemlock, and southern pine. See Fig. 17.1 for side and end matched T & G flooring. Narrow boards are the most expensive because they require more handling, cut proportionately more to waste in the manufacturing process, and require more time to lay. As an example, there is just as much wood cut to waste in milling the tongues and grooves for $1\frac{1}{2}''$ boards, as there is for $2''$ boards; and the floor layer can place a piece of $2''$ board just as quickly as he can place a $1\frac{1}{2}''$ board. When estimating the number of square feet of floor to be covered with hardwood, first determine the face width of the boards required, and then consult the table in Chapter 4, page 77. Add the percentage allowance for waste that is shown.

It is important that a good quality building paper be laid between the subfloor and the hardwood. This will help to eliminate a squeaky floor, and it will act as a dust barrier to unfinished floors below.

Study Figs. 17.1 through 17.5. Extra nailing of flooring should be made in hallways and doorways

and anywhere that will have a heavy traffic flow. For minor bedrooms a cheaper softwood finish flooring will be adequate. This will make a saving in initial purchase of the material, especially in applying it.

Carefully study Fig. 17.3b which shows a spacing to be left at each longitudinal sidewall of the boards; this is to allow for the floor boards to expand if exposed to moisture. I have seen a hardwood floor raised three feet down the center of a room because provision was not made for expansion; such would be the consequence or else the force would displace one or both of the side walls.

The ancients used to quarry enormous stones by dressing them (on four sides) on their natural bed in the quarry, then applying dry wood between the back dressed side and the quarry face, and then ap-

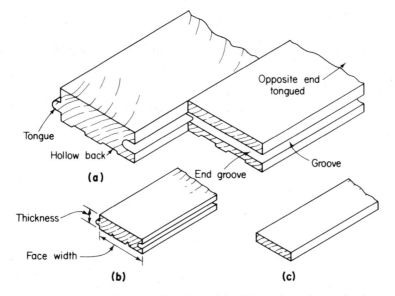

Fig. 17.1. Types of strip flooring: (a) side-and end-matched, (b) thin flooring strips, matched, and (c) thin flooring strips, square-edged.

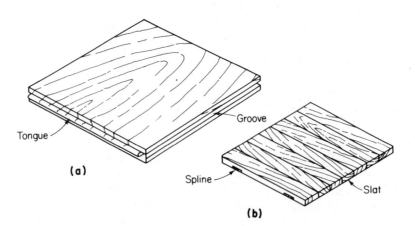

Fig. 17.2. Wood block flooring: (a) tongued and grooved, and (b) square edged, splined. *(Courtesy of the Forestry Service Products Laboratory, U.S. Department of Agriculture.)*

plying a water treatment causing the wood to expand and cracking the stone at its base. Near the site of the Heliopolis at Baalbeck, in Lebanon, is the largest stone that was ever quarried by man. It is 14'-0" in section and 64'-0" long, and was quarried by this method two thousand years ago. It is still there for all to see. Be careful!

Another type of wood flooring is parquetry wood block flooring. These blocks are accurately machined for size, and some are made up (under quality controlled shop conditions) into panels which facilitate the final floor laying, especially for large areas.

Other types of wood flooring include 4'-0" x 8'-0" sheets of birch plywood; also aspenite (see Chapter 15.1). Such floors are usually finished with varethane which is a plastic (hard wearing) varnish. Other types of fiber boards are also used for flooring, in little used areas, such as basement storerooms; the fiber boards are then finished with quality enamel paint.

When planning the laying of a wood finished floor over a concrete base, it is imperative that the concrete surface be waterproofed, and that the sleepers (1 x 4 or 2 x 4) be treated and anchored to the slab with concrete nails, and that provision be made for air to circulate between the top of the concrete slab and the underside of the flooring, see Fig. 17.5.

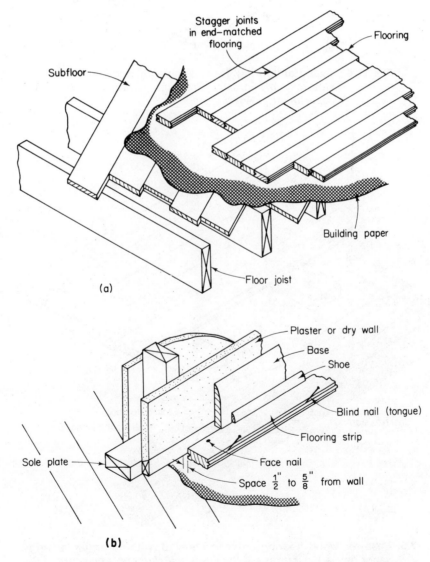

Fig. 17.3. Application of strip flooring: (a) general application, (b) starting strip. *(Courtesy of the Forestry Products Laboratory, U.S. Department of Agriculture.)*

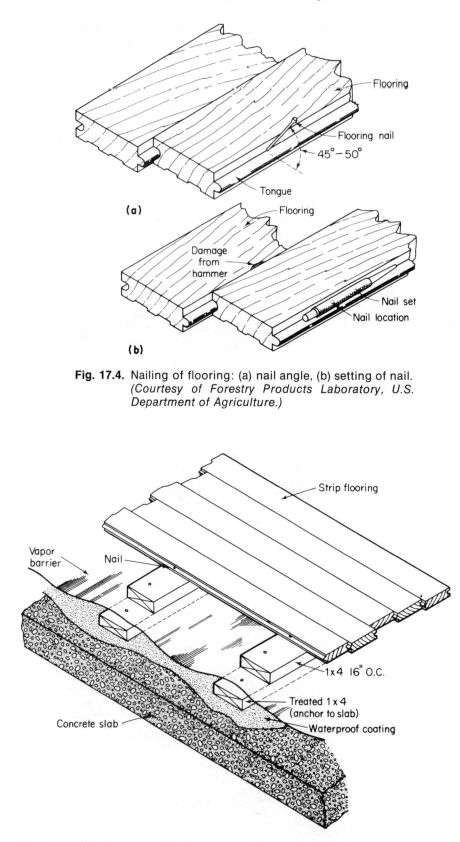

Fig. 17.4. Nailing of flooring: (a) nail angle, (b) setting of nail. *(Courtesy of Forestry Products Laboratory, U.S. Department of Agriculture.)*

Fig. 17.5. Base for wood flooring on concrete slab (without an underlying vapor barrier). *(Courtesy of the Forestry Products Laboratory, U.S. Department of Agriculture.)*

17.6 ABRASIVE SAFETY STAIR AND WALKING PRODUCTS

The following material has been excerpted from a Wooster Products Inc. catalog, and is here reproduced by permission.

STAIRMASTER SAFETY TREADS

Ideal for repairing worn and slippery stairs

Stairmaster Safety Treads Type No. 500 are specially designed for the repair and modernization of all kinds of stairs. Made in one style and dimension with lengths up to 12' 0", they provide an economical and fast method of renewing worn and slippery stairs. Built to dependable Wooster quality standards they will outwear any other material under the heaviest traffic.

Landing treatment shown at left is made up of special sections shown on page 8. Coverage can be full as shown, or in a depth of 12", 15", 18", etc.

Write for stair repair literature and sample.

aluminum SUPER-GRIT LADDER TREADS INDUSTRIAL MARINE
TYPE 850

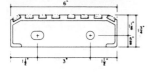

Treads are a complete unit, ready for bolting to stringers. The extruded aluminum base, alloy 6061-T6, is light weight with exceptional strength. Eight abrasive-filled ribs offer a non-slip walking surface even under wet or oily conditions. Standard 6" width. Other widths available in quantities sufficient for special set-up. Lengths to order to 2' 6". Standard hole spacing shown below for 3/8" bolts makes the tread reversible. Fasteners not included.

aluminum SUPER-GRIT
Expansion Joint Cover

Uniform width, thickness, and smooth square edges makes it readily adaptable for use with various types of floor covering material.

Flat abrasive surface eliminates smooth slippery surfaces in walking areas, providing a positive non-slip walk surface.

Expansion joint covers available only in the widths shown — lengths up to 12'-0".

ANGLES BY OTHERS

TYPE 645   TYPE 610

Holes for fastening on approximate 12" centers factory drilled and ¼" machine screws for attaching to metal angles included.

CHAPTER 17 REVIEW QUESTIONS

1. List four preliminary operations that should be complete before painting the inside of a newly built residence.

2. Why is it advisable that customers select their own paint colors from the suppliers?

3. What is meant by the term "cutting-in" as used by painters?

4. Define five types of resilient floor covering.

5. List five types of carpeting as used in modern residential construction.

6. Why should hardwood floors have a building paper between them and the subfloor?

7. Draw a freehand sketch of the method to drive the nails of a hardwood floor above the tongue of the boards.

8. Why should the first and last strip of hardwood flooring be kept $\frac{1}{2}''$ to $\frac{5}{8}''$ from the walls?

9. Why is the wastage in the manufacture of hardwood flooring so great? See the tables on page 77.

10. Give four places in residential construction where extruded thresholds may be used to advantage?

18

Subtrades

In this chapter we shall discuss the persons involved in residential construction, and particularly the relationship between the GC (general contractor) and his subcontractors.

The number of individual entrepreneurs compared to large organizations in the construction industry is decreasing, similar to that of the corner store and the supermarket. The pattern is set for this drift to continue. At this time, more and more individuals are specializing in fewer and fewer specific areas, but are becoming more skillful in their own specific fields. This demands ever better organizing ability by the GC, who must become efficient at having different subtrades and suppliers of materials of service units move onto, and from, the buildings at predetermined times. It also requires a wider, and up to date, knowledge of construction techniques on the part of the GC who also must be capable in the handling of men, finance, machine usage, and in anticipating future trends in the industry.

See CPM (Critical Path Method) Chapter 3, art. 3.6.

18.1 THE PROPOSAL TO BUILD

The concept of the erection of every building originates in the mind of one individual who, on his own, or in company with others, may exploit the idea.

Immediate considerations involve such questions as: where to build residences, what design and price range of completed units, what financial requirements will be necessary, what types of service units (such as refrigerators and so on) should be installed; what building codes must be complied with. The GC should also keep in mind the prospect of building to meet the requirements of the "leisure industry" in such areas as lake or seaside homes or making one time logging camps into modern summer (or permanent) homes and so on. He should also take a close look at the prefabrication industry for home building. This is a fast developing area in residential construction.

As a rule of thumb, a family can afford to buy a house at roughly 2.5 times its annual income. Assume that a family income is $20,000.00, the family can afford to buy a house in the $50,000.00 range. *It is important for the GC to keep in mind that he build suitable residences, in appealing areas, to meet the demands of people earning specific salaries.*

18.2 THE APPRAISER

Before finalizing a design for a residence, it is useful to consider how an appraiser might value the proposed residence. Among other things, he uses a

Measure-Master. This is an odometer wheel attached to the bottom of a long handle, which clicks off the footage as it is wheeled along the floor; it saves using a tape measure. This instrument is also used to check the traffic flow pattern, say, from the center of the kitchen to the front and back doors, to the laundry room, the garage, and so on. Assuming that it is 37'-0" from the central point in the kitchen to the laundry room, then it is 74'-0" for the round trip.

It is said that in a well planned house one ought not to have to walk into the living room unless one wishes to do so; it should be a self-contained room to be used only for its specific role in the home and should not afford means of passage to any other area.

It is also considered by many people that the number of outside corners to a residence is one indication of its appeal. Four makes a house rectangular, six adds a recessed area, and anything over ten outside corners in a home may be considered as excellent. But the more corners, the greater the expense in concrete formwork, framing, roofing, and so on. From the underpinning to the chimney cap, the proposed residence should be assessed as to construction costs and sales appeal.

To remain competitive, a contractor must be adaptive to modern techniques and innovations, and be prepared to install many services such as:

(a) Air conditioning.

(b) Automatic push button self cleaning oven.

(c) Gas ranges and refrigerators.

(d) Automatic rotisserie and electrically operated thermometer.

(e) Double stainless steel kitchen sink with government approved waste disposal unit (sometimes called a garburetor).

(f) Automatic electric dishwasher.

(g) Fully formed formica counter tops and breakfast bars.

(h) Quick recovery large capacity water heater.

(i) Laundry room with installed washer and dryer, and an ironing area with ample electric outlets including a 220 volt outlet for dryer; outlets in this area for telephone and television; and ample storage for linen.

(j) Total electric or other heating.

(k) Door chimes.

(l) Chandeliers.

(m) The whole residence to be prewired for cable television, telephone in strategic areas, door chimes, and thermostatic controls.

(n) All rooms and areas to be provided with vacuum cleaner outlets to carry dust direct to one large bag located in the garage; the bag may only require emptying every six months.

(o) Silent switches.

(p) Good plumbing fixtures, architectural metalware.

(q) Ample mirrors and cabinets in the bathroom.

(r) The latest and most attractive in metal ironwork and carpets.

(s) Burglar alarm.

Try drafting a newspaper advertisement for your houses before you build them. Then compare it for value, appeal, and location with current advertising for similar houses. This will fix in your mind the type of competition you are to face.

18.3 RESPONSIBILITIES OF THE GENERAL CONTRACTOR TO THE SUBCONTRACTOR

It is usual in residential construction for the GC to employ his own men for erecting and removing concrete forms, for all wall and roof framing, and in some cases to complete the outside and inside trim.

The GC should provide each subcontractor with a pertinent extract of the progress schedule and should endeavor to have the dates for starting and completing each portion of work complied with, (see articles 3.6 and 3.11). It requires great organizing ability on the part of the GC to arrange to have his own work crew advanced in their work, at each stage of production, so that the subtrades may commence their operations. As an example, once a residence is closed, i.e., the walls roughed in and the the roof framing and roofing completed, the building may then be fitted with temporary doors. The plumbers, electrician, air conditioning contractors, telephone and cable vision technicians, and the tinsmiths may all run in their lines.

Remember that all these lines must be inspected by the local authority having jurisdiction before they may be covered with insulation, wall boards or any other covering. Failure to have such work inspected may result in the walls having to be opened up for inspection.

Remember that if any of the subtrades is late in completing his work, every other operation may be equally delayed. The longer it takes to complete a unit, the greater the delay in the sale of the project, and the longer it will have to be financed by the GC.

In addition, the prime months for selling units may have passed.

It is important that the GC check that all his subcontractors are bondable because the GC may be held responsible for completing the work and financing the work of a defaulting subcontractor.

All subcontractors should sign a subcontractor's form of agreement, and the GC should make a practice of consulting with his attorney before, *not after,* signing any such legal document.

Some subcontractors work with the same GC, year in and year out, with amicable relations. It once came to my notice that a GC had placed a successful proposal bid for a project and had forgotten to include one page of an estimate sheet which amounted to more than $18,000.00. He called a meeting of all his subcontractors, explained the situation, and they mutually agreed to cut their prices proportionately to the total contract price. They all made a profit, and they all respected each other. The memory is refreshing.

The GC should see that the area of operations for the subcontractors is always clear of debris and ready for them to start their work; also he must arrange that they each have an allotted place to store their materials.

18.4 THE EXCAVATOR

On small projects, the general contractor may use his own men, equipment, and knowledge of local ground conditions to clear a building site, grade the land to correct level, excavate, and take care of the water table, if necessary, by pumping, shoring or other means. But the GC must be aware that if he employs a subcontractor to do this work for him, such contractor must cover himself and his work crew by workmen's compensation and any other mandatory insurances. If this is not done, the responsibility for body or property damage is that of the GC.

Heavy construction equipment is expensive and requires continuous expert maintenance, accounting procedures, and a knowledge of allowable depreciation for income tax purposes. The GC must decide how much work to do himself, and how much to subcontract to others in their specialized fields. Although it is undesirable that the GC become merely the purveyor of other men's work, nevertheless it is usual for the GC to do no more than, say, 15% to 20% of all the work involved; this, in turn, means that he is only financing about that proportion of the total cost, but he may be held responsible for the safety of men, the public, adjoining property, and so on in case of default of any of his subcontractors. See articles 1.2 and 1.3.

Irrespective of who does the earthwork, the GC or his supervisor should be on the job from the turning of the sod to completion of the project. *A new building project that is correctly located on its legal location, with the first floor at its correct elevation in space is off to a good start.* Be careful with all preliminary building operations. See 6.7 for depreciation.

18.5 CONCRETE SUBCONTRACTORS

On most small residential projects, it is usual for the GC to use his own work crew to erect the concrete forms, place the reinforcing steel and the concrete, and remove the forms after use. These operations require the services of a supervisor who can interpret drawings, understand concrete form construction and the reason why reinforcing steel must be placed as specified. Also the supervisor must understand the design and control of concrete mixes, see Chapter 8.

While you are reading this, somewhere at this moment of time, there has been a failure in the formwork for concrete, resulting in a loss of life, injuries to workmen, and presenting a very difficult job of cleaning up the mess and starting all over again. This delays the work and is an avoidable error. *Be doubly sure about the strength of the formwork.* Many companies specialize in this field, and for an inclusive price will erect and dismantle their own concrete forms, place the reinforcing steel and the concrete.

In rural districts of North America, and especially in developing countries, it is often necessary to mix concrete in the field, in which case the supervisor must be very competent and obtain the specified designed strength of concrete for every individual batch. See Chapter 8.

18.6 A GUIDELIST OF SUBTRADES

The following list is presented as a guide to the different specializing subtrades; the list should be kept up to date and may be referred to when designing residential units. Remember that firm legal bids must be obtained from every subtrade before work is commenced. *Under no circumstances, should word-of-mouth agreements be accepted.*

1. land levelling and grading
2. water drainage and control

3. excavating

4. ditching and trenching

5. concrete contractors and finishers

6. concrete: prestressed, and post tensioned

7. concrete forms

8. concrete block

9. reinforced steel erectors

10. septic tank and field specialists

11. plumbing

12. electric and power

13. heating

14. air conditioning

15. telephone and television linesmen

16. tinsmiths

17. masons

18. swimming pool installers

19. roofing specialists

20. metalwork

21. drywallers

22. plastering and stucco wire

23. acoustic ceiling

24. chimney builder

25. doors and windows

26. glass block, glass and glazing

27. mirrors

28. chandeliers

29. tiles; earthenware, ceramic, plastic

30. insulators

31. beams "I"; wood solid, wood built up

32. flooring: hardwood, linoleum, tiles, broadloom, indoor-outdoor carpeting

33. hardware suppliers

34. painting and decorating

35. black top, gravel, and other drives

36. janitorial services

37. landscapers

38. sale and closing procedures

In addition to the foregoing, builders, supply houses and suppliers of units such as refrigerators should be regarded similarly to the subtrades, in that it is imperative that they make their deliveries at predetermined times, so that the progress schedule may be complied with. It is only in this manner that display homes may be opened for public inspection, and the advertising campaign be instituted to take advantage of the best selling seasons of the year.

CHAPTER 18 REVIEW QUESTIONS

1. List five pertinent considerations that should be taken into account before proposing to build residences for sale.

2. As a rule of thumb, what percentage of the family income can be expended on living accommodation?

3. List fourteen sales appealing units (such as washing machines and so on) that a speculative builder should consider installing in speculatively built residences.

4. Explain how an overall building schedule is related to each subcontractor. See pages 51 and 52.

5. What is meant by, "The general contractor should check that all his subtradesmen are bondable"?

6. List twenty-five subtrades.

7. Once a general contractor has contracted with an owner to erect an apartment block for a stated price, who is responsible to complete the work of a subcontractor who absconds or goes bankrupt?

8. Assuming that the plumbing subcontractor is three weeks late in roughing in the plumbing lines for an apartment block, how does this affect other subtrades and the final completing date of the project?

9. Define the duties of an appraiser.

10. List 10 units and/or fixtures that a builder should install in a new residence.

19

Landscaping, Final Inspection, Notarization of Documents, and Property Sale

In this chapter we shall discuss driveways, landscaping, the final inspection of the property, land titles, notarization of documents, advertising, and, finally, the sale of the property.

19.1 DRIVEWAYS, PATHS, AND OUTSIDE CONCRETE STEPS

As stated earlier, once a proposed building has' been correctly located for its location and elevation on its legal lot, the builder is off to a good start; this also evidences itself in the final grading, drainage, landscaping, and in the building of driveways, paths, and outside concrete steps on the property. Any series of steps with three risers or more should be provided with a handrail.

The surfaces of all driveways, paths, and concrete steps should be fashioned to be non-skid, and to shed water quickly without leaving any low spots to form puddles of water or icy patches, either from rain, snow, or from a water sprinkler system. There are several methods to produce a non-skid surface to concrete such as:

(a) Brooming, which will give a pleasing non-skid surface, see Chapter 8.

(b) Finishing the surface with either a wood or cork float.

(c) Sprinkling a fine particle abrasive over the surface such as emery, flint, or aluminum oxide, and finishing it lightly with a wood or cork float.

Where frigid winter conditions prevail, consideration should be given to providing thermostatically controlled electric heating units under areas subject to snow and icing conditions.

Some aspects of the preparation of driveways and paths is explained in the following seven paragraphs and the ensuing two pages which have been excerpted from publications titled, "Site Planning Book," and "Principles of Small House Grouping," respectively, published by the Central Mortgage & Housing Corporation, Ottawa. Publications may be obtained free upon request.

Driveways-General. It is recommended that driveways serving more than four units lead onto minor streets at least 100 feet from an intersection, where possible. Developers proposing to use exterior ramps to internal garages, with gradients in excess of 10%, should use some form of slab heating for the winter months and should provide a transition slope.

The alignment and gradient of driveways should be coordinated with the grading plan to prevent the passage of large flows of water on or across the driveways.

Size of Driveways. The minimum width of a one way driveway or driveway to serve a maximum of 4 dwelling units shall be 8 feet. If the driveway is also to serve as a walkway, the combined width shall be not less than 10 feet clear of all projections.

The minimum width of a two way driveway to serve over 4 dwelling units shall be 18 feet. If a walkway is to be combined with a driveway, it shall be additional to the minimum driveway width and shall be clearly demarcated.

Gradients of Driveways and Ramps. Maximum cross slope for driveways and ramps shall be 7 inches in 10 feet.

Where the gradient is less than 2 inches in 10 feet, the minimum cross slope shall be not less than 2 inches in 10 feet.

Walkways-General. Walkways or combined walkway-driveways shall be provided from all required entrances and exits of residential buildings to parking areas and adjacent public streets.

The alignment and gradient of walkways should be coordinated with the grading plan to prevent the passage of large flows of surface water on or across the walkways.

Sizes of Walkways. The minimum width of walkways shall conform to the following:

Minimum Width Main Walkways	Secondary Walkways	Maximum No. of Dwelling Units Served
2'-6"	2'-0"	4
3'-0"	2'-6"	8
4'-0"	3'-0"	16
5'-0"	4'-0"	more than 16

Where walkways adjoin a parking lot where cars may overhang, there shall be a minimum 3 feet width of walkway, clear of all cars.

Gradients of Walkways. The maximum gradients for walkways shall be 12 inches in 10 feet.

Maximum cross slope for walkways shall be 7 inches in 10 feet.

Where the gradient is less than 2 inches in 10 feet, the minimum cross slope shall be not less than 2 inches in 10 feet.

Steps in Walkways. Where steps are required in a walkway, there shall be not less than 2 risers in a flight of steps. The steps shall be as wide as the walkway and shall be provided with a handrail when there are more than 3 risers in a flight.

19.2 TYPES OF DRIVEWAYS

It is important that all driveways be provided with a level turnaround; this obviates the hazard of reversing a car onto a thoroughfare. The turnaround is also advantageous for the temporary parking of the cars of guests. Study all the Figures 19.1 through 19.4 in this article.

Concrete Driveways in natural color or tinted should be placed on a well-tamped gravel base. To minimize cracking they should be stabilized with reinforcing rods, steel mesh, and/or temperature rods and be provided with control joints every eight or ten feet.

Flagstone Driveways (also ribbon type) should be set on a sand or gravel base topped with $2\frac{1}{2}''$ of concrete into which the flagstones are maneuvered into adjoining lineable positions. The concrete stabilizes

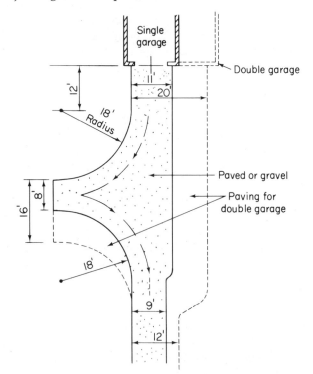

Fig. 19.1. Driveway turnaround. *(Courtesy of the Forestry Products Laboratory, U.S. Department of Agriculture.)*

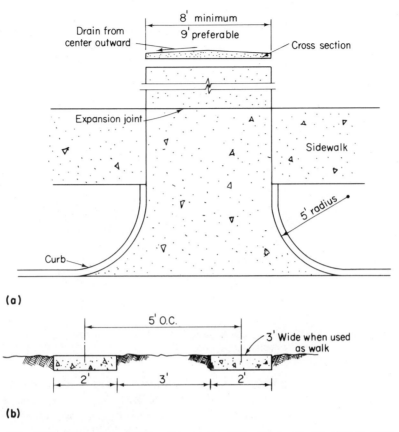

Fig. 19.2. Driveway details: (a) single-slab driveway, (b) ribbon-type driveway. *(Courtesy of Forestry Products Laboratory, U.S. Department of Agriculture.)*

the rocks and prevents weed growth between the joints. In some areas this type of driveway is called "crazy paving."

Bituminous Pavement should be placed on well drained ground with a well-tamped gravel base.

Masonry Paving should be placed on a well-drained and well-tamped gravel base topped with 2″ of concrete, into which the units are embedded true to line.

Concrete Slab Unit Driveways embedded on sand are very satisfactory in some areas. Such driveways may be easily re-leveled in the early spring of each year.

Gravel Driveways are acceptable in some areas where the gravel is of select color and ranges from $\frac{3}{8}$″ to 1″ in sizing: such gravel should be contained within the confines of wood or masonry edging. A disadvantage is the probability of gravel being thrown from car wheels onto the lawns.

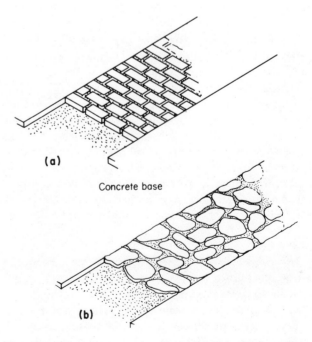

Fig. 19.3. Masonry paved walks: (a) brick, (b) flagstone.

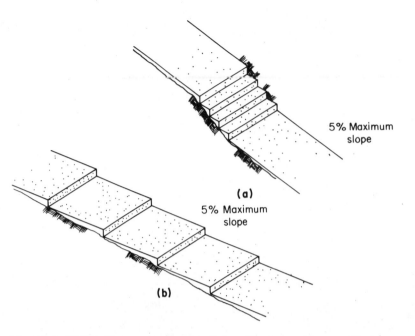

Fig. 19.4. Sidewalks on slopes: (a) stairs, (b) stepped ramp. *(Courtesy of Forestry Products Laboratory, U.S. Department of Agriculture.)*

19.3 LANDSCAPING

It will be recalled that in article 6.3, it was suggested that among other preliminary building operations the building site should first have the top soil stripped and stockpiled on the site, the surface brought to grade, and ultimately the soil dispersed over the lot after the driveways and paths have been built. Particular care should be taken wherever deep trenches have been dug, as these areas will subside unless the ground has been well compacted after the laying of service lines. Only then may the total area be landscaped.

Landscape gardening applies to the leaving in position (whenever possible) some native trees, then making a pleasing arrangement of grassed (seeded or sodded) areas, plants, shrubs, other trees, rocks, water, hedges, decorative fences, wrought iron work, and so on, in such a manner that they will complement each other and their immediate surroundings. But, all this must be accomplished while having regard for the drainage of surface water from all sides of the building.

In some arid regions, the layout may be made attractive by the placing of several different colored gravels to form designs on either side of the driveway and paths. Monotony can be relieved by having several cactus plants set in complementary array.

The nature of any landscaping should lend charm, a feeling of restfulness, and sales appeal to the property. The builder must give careful thought to this subject. Some purchasers love gardening, others abhor it. In this age of leisure many people want to be off to their games, lakesides, or travels, and leave the chores of gardening to others. It is suggested that a minimum area be planted. Plants such as hedges which require continual trimming should be avoided. The choice to plant or not to plant should be left to the purchasers. Even the watering of lawns, unless done automatically by underground feeder lines to sprinklers, may be a chore to some people. For apartments, condominiums, and rented places, the care and upkeep of surroundings may be taken care of by private contract between owners and landscape gardeners.

The following two pages have been excerpted from a booklet which was published by the Central Mortgage and Housing Corporation, Ottawa, and is here reproduced with permission.

Flat Landscape

The site which is flat or almost level is comparatively easy to develop. The grouping picture is governed by the street pattern rather than the topography. The landscaping should be used to assist in developing the street scene. Trees and hedges may be used to help identify spaces that are formed by the buildings. Visual barriers, such as front gardens and shrubs, which tend to confuse rather than define areas should be avoided.

If the building site is well treed, certain trees should

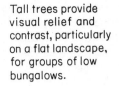

Tall trees provide visual relief and contrast, particularly on a flat landscape, for groups of low bungalows.

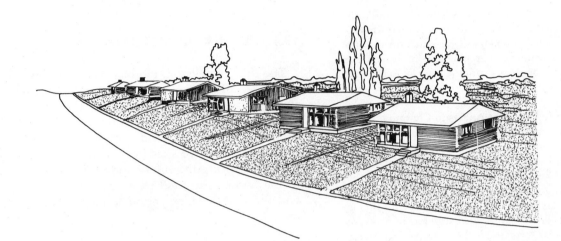

Houses on roads which cross contour lines at right angles should be chosen with similar pitched roofs and sited so that these roofs create an even-stepped rhythm, either up or down the hill.

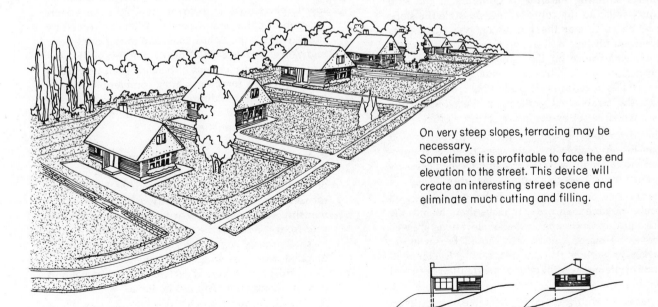

On very steep slopes, terracing may be necessary.
Sometimes it is profitable to face the end elevation to the street. This device will create an interesting street scene and eliminate much cutting and filling.

Fig. 19.5.

be selected for retention before building operations commence. Careful consideration should be given to the locations and the future shapes and sizes of these trees. Trees that seed profusely should be avoided. Coniferous trees can sometimes be retained as windbreaks. In absolutely flat areas it may be desirable to employ tall trees as a relief and contrast to a row of low bungalows.

Sloping Landscape

There are two situations that occur when building sites are on steeply sloping terrain. Both require different landscape treatments. One situation is a result of streets running at right angles to the contour lines, the other of streets running parallel to the contour lines.

Streets Running at Right Angles to Contours. When streets are at right angles to contour lines it is necessary to resolve the conflict in appearance between the slope of the land and the horizontal lines of the building elevations. The houses should be sited in such a way that an even rhythm is obtained by the stepped roofs. One way of achieving this is to have roofs of the same pitch. The lots in this case may have to be terraced. If the slope is very steep, set the houses endwise to the road. This will create a pleasant grouping, well related to the landscape, while eliminating much of the cutting and terracing required if the houses were placed lengthwise down the slope. When lots are terraced care must be taken to ensure drainage of surface water away from the side yards.

Streets Running Parallel to Contours. If the road runs parallel to the contours it is possible the slope may be such that the houses on one side of the street are on lots well above the established grade level while those on the opposite side are on lots lower than the established grade. It is important that all lots have finished grades that are determined from the established grade from the crown of the road which is set by the city engineer's office.

The lower lots present the greatest problems as it may be impossible in some cases to effect connections to the main sewer. Extra fill may be necessary to raise the houses to a suitable elevation.

The houses on the higher side of the street need careful treatment in order to make them fit into the surrounding landscape. If the lot elevations are well above the street, the houses should be set as far back as possible so that flat areas can be developed in front of the houses, thereby reducing their apparent height.

A gentle ascending slope is preferable to steps as an approach to a house that is higher than the street. A path not exceeding a 10 percent slope is satisfactory. When steps are necessary the banks should parallel the stairway and should not be so steep that turfing is impracticable.

If there are good views from the site, houses should be placed to take advantage of them. This involves a study of each lot to determine a satisfactory placing of the house in relation to the slope of the land, the neighboring houses and to any houses, trees or other obstacles that may come between the lot and the view.

19.4 THE FINAL INSPECTION

The final inspection, the maintenance of the completed building for a stated time, and the method of final payments to the contractor are all part of the general conditions of the contract set forth in the original specifications and contract documents. The following extract of specifications are typical of the clauses covering the final handover of the building to the owner, the final payments to the contractor, and the maintenance period of liability of the contractor.

Extract of Specifications

Payments. On the twentieth of each month, the contractor shall submit to the architect an estimate on all work done and material supplied onto the job. Contractor shall attach to his progress estimate receipted accounts and wage sheets covering the items entering into his estimate. Upon this total the Board will allow 10 percent (10%) to cover the Contractor's overhead and operational expenses. When approved by the Architect, the Board shall pay to the Contractor eighty-five percent (85%) of the amount of his estimate. The final payment of balance shall be paid to the Contractor forty days after completion and acceptance of the building, provided satisfactory evidence is shown that all accounts for material and labor have been satisfied.

Correction of Work After Final Payment. (a) Neither the final certificate of payment nor any provisions of the Contract documents shall relieve the Contractor from responsibility for faulty materials or workmanship which shall appear within a period of one year from the date of the completion of the work, and he shall remedy any defects, and pay for any damage to other work resulting therefrom which shall appear within such period of two years, but beyond that the Contractor shall not be liable.

(b) No certificate, payment, partial or entire use of the building or its equipment by the Owner shall be construed as an acceptance of defective work or material.

Completion Date. The completion date shall be established in writing by the Architect and Contractor, after notice is presented to the Architect that work has been completed.

Progress Schedule. During the construction of the building the general contractor must keep very close watch on the actual work accomplished against the amount estimated on the progress schedule, so that the building may be handed over at the contracted time. If for any reason the building has been falling behind schedule, the contractor must consider the wisdom of paying premium rates of pay for overtime work to his men. In many contracts there is a penalty clause for each day of delay in handing over the completed project. (See the progress schedule on p. 51.)

Sidewalk and Landscaping

When concrete sidewalks are not placed in the late fall because of extremely cold weather, the architect may retain a holdback of cash sufficient to do that work in the spring. Landscaping is usually an entirely separate contract outside the scope of the general contractor's work.

General Contractor's Inspection

A short time before the final inspection it is usual for the general contractor, in company with the superintendent and the foreman, to make a careful inspection of the whole work, at which time notations are made of all details to be completed and a time set for these to be done. It is most likely that some broken glass will have to be replaced, door locks adjusted, paint work retouched, and so on.

Window-Cleaning

On a large job this is a subtrade; for smaller jobs the contractor's own men may do the work, but in either case it must be allowed for in the estimates. Upper story windows should be reversibly hinged so that outside surfaces may be cleaned inside the room.

Janitor Services

Some jobs will require the services of janitors to clean thoroughly the whole building. Many specifications state that all fingerprints shall be removed from all paint and polished work, and that the building shall be left ready for immediate occupation. This may also be a subtrade.

Architect's Final Inspection

When the general contractor is quite sure that the building is completed, he will notify the architect in writing and request a date and time for the final inspection.

The final inspection is made by the architect with the clerk of the works, the owner, and the general contractor and his superintendent.

Any adjustments are noted and corrected, and the building is then ready for handing over, provided there are no liens or other legal encumbrances against the property.

Maintenance

Many contracts state that after a period of six months the contractor will examine the building and ease all doors and windows and make good any settlement cracks in any of the walls. It is usual for buildings to settle; and the better the class of workmanship in the fitting of doors, the more likely it is that such doors will require a little easing after a few months. The actual maintenance period is shown in the contract. It is your duty as a builder to make a close study of the general conditions of the contract so that you will be in a position to allow for your company to make provision for the expense of maintenance.

Finally, use your schedules and remember that if any item is forgotten on the estimate, such item must still be purchased and installed.

19.5 LAND TITLES

The following material has been excerpted from the book *Real Estate, Principles and Practices*, 7th edition, by Alfred A. Ring, published by Prentice-Hall, Inc., and is here reproduced with the permission of the copyright holders.

Protection of Title. As landed property became more valuable and ownership of real estate more diverse, it became increasingly important to provide safeguards to protect the true owner from loss of title by claim, error, or fraud. With cooperative aid of public and private agencies, transfer of title by the rightful and true owner and continued, though absentee, ownership were safeguarded by one or more of the following methods of title assurance: (1) recording acts and examination of title, (2) title insurance by private companies, (3) land title registration by public agencies in accordance with legislative enactments.

Recording of Conveyances. Possession of property is notice to the world that the possessor claims or has some interest in the property. An owner in possession under a valid deed may be discovered to the actual knowledge of anyone who goes to the property. However, it is not always practicable for the owner actually to be in possession. He may own many buildings, or the structure may be an office or factory building or vacant land. One might go to the premises many times and not find the owner. Some method of *constructive* notice of ownership had to be devised as a substitute for *actual* knowledge, to protect the owner by relieving him of the necessity of remaining constantly in possession and to protect persons who, desiring to deal with the property, would wish to ascertain the real owner. Otherwise A, an owner, might sell his land to B, giving him a deed; and if B did not take possession, A might turn about and sell it to C. Or he might give a mortgage to D to secure a loan after he had sold to B. To prevent such frauds, recording acts have been enacted in all states. These provide that all instruments affecting real property may, when properly proved, be recorded in a certain public office in the county where the property is located. All such instruments are copied on the records and indexed. When so recorded, they are notice to the world with exactly the same effect as if the owner were actually in possession.

Constructive notice is just as good as actual notice. Consequently, one dealing with real estate is bound by all recorded instruments. Suppose A sells a piece of land to B and B fails to record his deed; A then sells it again to C, who knows nothing of the prior sale to B, and records his deed before B's deed is recorded. Under the theory of notice, C's right to the property is ahead of B's because B was not in possession and the records at the time C bought the property showed title in A. B should have protected himself by recording his deed as soon as he received it. Constructive notice is, however, no better than actual knowledge. If C, in the case above, had known of B's purchase, he could not obtain any right superior to B by recording his deed first.

Proof of Execution. No instrument may be recorded unless proved as required by the law of the state. Proof varies in the several states. Some require a subscribing witness; others an acknowledgment; others both. The following are the officials who are authorized to take acknowledgments: notaries public, commissioners of deeds, justices of the peace, judges of courts of record, mayors of cities, ambassadors and ministers residing abroad, consular agents, and commissioners of deeds appointed by the governors of states to take acknowledgments in other states. Each of these officials has definite limits of authority. He cannot act outside the area of his authority. Within his area of authority he may take

an acknowledgment of an instrument to be recorded elsewhere. When he does this, the instrument cannot be recorded elsewhere without a certificate attached from the clerk of the court of the county or city in which the official is qualified to act, stating that the official is qualified to take acknowledgments of instruments intended to be recorded in that state, that the signature of the official is known to the clerk, and that the signature affixed to the certificate of acknowledgment is genuine.

Examination of Records. It is readily seen that one who contemplates a real estate transaction not only must inspect the realty involved but also must procure a thorough examination of the records to ascertain the owner and the condition of the title and all instruments concerning which the law presumes everyone to have notice. The examination reveals the entire history of the title from the earliest record to the present time and shows the chain of deeds, wills, and actions by which the property has passed from owner to owner, as well as mortgages, leases, restrictive and other agreements, and instruments encumbering or affecting the title or use of the property. The examiner first *abstracts* all the instruments conveying the title; that is, he makes a separate digest of each. This gives him what is known as a chain of title. He may find his chain very simple, consisting perhaps of a grant from the state to A and successive deeds from A to B, B to C, C to D, D to E, and E to F, F being the present owner. Usually, however, someone in the chain has died owning the property. In that event he may find deeds from A to B and B to C and no deed from C, although F claims ownership. The probability is that C died owning the property. In that case his will (if he left one) was probated and is on record in the court. If he left no will, it is usually found that an administrator of his personal property had been appointed, and the papers on file for that purpose state the names of his heirs. The examiner accordingly turns to the records of deaths and wills to fill the gap and finds the will or record of death of C. This supplies him with the names of C's devisees or heirs, and he then resumes his search by locating the deed from them to D, and so continues his chain. The chain is often broken by some legal action, such as foreclosure. Some person in the chain may have mortgaged the property so that the chain of title stops in D. A search of the records of legal actions shows that D was cut off in a foreclosure suit. Examination of the judgment in the action reveals the name of the official who sold and gave a deed of the property. Search against him will show his deed, and the chain is resumed. After the chain of title is completed, separate search is made against each owner for the period he owned the property to ascertain what encumbrances he may have placed upon the property.

The examiner's completed works is an *abstract of*

title. In many states the abstract passes with each sale of the property, being kept up to date by the addition of a memorandum of each new transfer. It is deemed so valuable that in some states it is customary to provide in the contract of sale that the seller shall deliver the abstract of title at or before the delivery of the deed.

The Title Examiner. The law of real property is complicated and technical. The average person dealing in real estate has no knowledge of these rules, nor has he time to examine the title. He usually employs counsel or a conveyancer to do this work for him—someone who is familiar with the records, their location, indexes, and, more important, the law applicable to the various situations in the title which the examination might reveal. The responsibility of the examiner to his employer should be noted. He does not guarantee the result of his search. He simply asserts (1) that he has sufficient knowledge and experience to be a competent examiner of titles and (2) that he will use his knowledge honestly and diligently in accordance with the appropriate rules of law. His report of title is only his opinion, backed, to be sure, by his legal training and a careful scrutiny of the records. The records are copies of instruments; he is not responsible if the signature on some deed in the chain later proves to be a forgery. C may have died intestate owning the property. X and Y thereafter conveyed the property by deed, reciting that they were the only heirs of C. Z may thereafter claim to have been an heir as well. The examiner is not to blame. He may pass upon some situation in the title in accordance with the law as then in force. Later a court may reverse the decision upon which the examiner based his opinion. For none of these things is the examiner liable, yet his employer may lose large sums as a result.

As the records in the county and other offices grow in size and complexity, it is safe to have searches made only by someone familiar with them. Specialists in this field of work are at every county seat. Customs and the volume of work will regulate their type and methods of conducting business. To a large extent in the rural counties and even in the cities, a great deal of title searching is done by lawyers or conveyancers. In some places the searching is customarily done by men who make a specialty of making up abstracts and who supply them to lawyers on order. The lawyer then reads the abstract and certifies the title.

In more active counties, abstract companies do this work. Every abstract company has a force of employees, who duplicate and offer additional records, maps, and surveys, and supply complete and accurate abstracts.

Title Insurance. No system of title searching is perfect. As previously indicated, errors may creep in, or forgeries and other things that cannot be guarded against may cause loss. To remedy this situation, title insurance has come into use in the larger cities. It is a direct growth of the abstract company. Many of these companies, years ago, devised the idea of insuring not only their abstracts but going a step beyond, and for an additional fee, reading and insuring the title as well. Like all other insurance, title insurance is a distribution of loss among all insured. Title companies are organizations authorized by law to examine and insure titles. They charge a fee or premium for their service. The amount of the premium is usually based upon the value of the property and covers not only the expense of the examination and abstract, but also an additional amount that is placed in a general fund to cover the losses insured against. The company makes a careful examination of the title. If it is satisfied that there are no apparent defects in the title, it insures against any loss. Should there later be a loss, by reason of forgery or any other defect arising prior to the insurance, the title company pays the loss. This, in brief, is the theory of title insurance.

In seeking title insurance, the person about to acquire the title or some interest in the real property first applies to the title company. He agrees to pay a certain fee for examination of the title. The title company on its part obligates itself to make an examination of the title and to insure against undiscovered defects. It does not, however, agree to insure against defects and encumbrances that may appear from the examination.

After the examination is completed, the applicant should therefore insist on being given a *report of title*, which is a statement setting forth a description of the property, the name of the record owner, and a detailed list of all objections to the title, that is, encumbrances and defects found upon the records. The reason for having this report is simple: it enables the applicant to know the exact condition of the title. If he is a purchaser, his contract stipulates that he shall take title subject to certain encumbrances. The report sets forth all the encumbrances found on the records. The purchaser demands that the seller dispose of all those not agreed upon in the contract before delivering a deed. If the applicant has agreed to make a mortgage loan, he insists that the owner render his title free and clear before the loan is made.

After the objections not agreed upon have been removed, the title is closed and the instruments passing title are delivered and recorded. The title company now prepares to issue its policy of the title insurance. There may, of course, still be encumbrances on the property which have been agreed upon. For example, the transaction may be a sale of the property subject to one or more mortgages. The policy should be carefully examined to see that the property is properly insured without any exceptions other than those agreed upon.

Title Insurance Policy. The usual form of title insurance policy contains four parts:

1. Agreement of insurance
2. A schedule describing the subject matter of insurance
3. A schedule of exceptions
4. Conditions of the policy

The agreement of insurance usually reads as follows:

[The company] in consideration of the payment of its charges for the examination of this title to it paid doth hereby insure and covenant that it will keep harmless and indemnify . . . . (hereafter termed the assured) and all other persons to whom this policy may be transferred with the assent of this company, testified by the signature of the proper officer of this company, endorsed on this policy, against all loss or damage not exceeding . . . . . . dollars which the said assured shall sustain by reason of defects, or unmarketability of the title of the assured to the estate, mortgage, or interest described in Schedule A hereto annexed, or because of liens or encumbrances charging the same at the date of this policy. *Excepting* judgments against the assured and estates, defects, objections, liens, or encumbrances created by the act or with the privity of the assured, or mentioned in Schedule B or excepted by the conditions of this policy hereto annexed and hereby incorporated into this contract, *the loss* and the amount to be ascertained in the manner provided in the annexed conditions and to be payable upon complieance by the assured with the stipulations of said conditions and not otherwise.

This agreement is dated and executed by the proper officers of the company under its corporate seal.

The company's charge is a fixed rate based usually on the amount of insurance named. Unlike other insurance, it is a flat fee, paid but once. Customarily the company insists that the property be insured for at least its full value. The insured also should want the property insured for its full value, since the company is in no case obligated to pay more than the amount set forth in the policy. The insured may, if he contemplates improving the property, have his title insured for a sum greater than its value at the time of transfer. The date of the policy is very important. The company insures only against loss to the insured arising from some defect at or prior to the date of the policy. The insured should insist that the policy be dated at or after the time title is closed. Because the policy is issued under seal, the time to sue upon it does not begin to run until a loss is sustained. The statute of limitations may be twenty years. The loss might not occur until fifteen years after the policy was issued. In such cases the right to sue on the policy would not expire until thirty-five years after the date of the policy.

The schedule describing the subject matter of insurance usually follows the agreement of insurance. It is divided into three parts. First it states the estate or title of the insured. Next comes a brief description of the premises covered by the policy. This description should be sufficiently detailed so that the property may be easily identified. The policy covers not only the land but all buildings and fixtures thereon. It does not cover personalty. The insured should see to it that the description is clear.

The schedule of exceptions is virtually the most important part of the policy. It sets forth a detailed list of all encumbrances and defects against which the company does not insure. No loss arising from any of these exceptions is covered by the policy. The insured should insist that only such encumbrances as he has agreed to shall be inserted in the schedule. Much trouble has arisen on this point, and many companies insist, before the closing of title, that the insured consent in writing to such objections to the title as have not been removed. Nearly all companies refuse to insure against the rights of tenants and persons in possession of the property; therefore those rights usually appear in the schedule. All encumbering facts shown by a survey are excepted, or if there be no survey, the policy will except "any state of facts an accurate survey may show."

The last part of the policy is a statement of the conditions of the policy. These conditions are seldom read but are very important. They specify the terms of the company's liability and the relations between the company and the insured. First it is stipulated that the company will, at its own cost, defend the insured in all actions founded on a claim of title or encumbrance prior to the date of the policy and thereby insured against. This stipulation not only assures the insured against loss but saves him the inconvenience and expense of litigation.

Should the insured contract to sell the property and the purchaser reject the title for some defect not excepted in the policy, the company reserves the option of either paying the loss or maintaining at its own expense an action to test the validity of the defect. In such a case the company is not liable under the policy until the termination of the litigation.

If the policy is issued to a mortgagee, the company's responsibility arises only in the event that the mortgage, upon foreclosure, is adjudged to be a lien upon the property of an inferior quality to that described in the policy, or in the event that the purchaser at the foreclosure sale is relieved by the court of completing his purchase by reason of some defect not excepted in the policy.

The conditions of the policy also provide for arbitration, in certain cases, of disputes as to the validity of objections to the title insured. The policy covers the insured even after he has sold the property, should he be sued upon the convenants in his deed. The policy is not transferable, except that if it insures a mortgagee and he sells the mortgage, his rights under the policy may be passed to his assignee. But even the company's consent must be obtained.

Should there be a loss under the policy, the company, having settled the claim, acquires all the rights and claims of the insured against any other person who is responsible for the loss. This right is based upon the legal doctrine of subrogation. The title company may be able to collect all or part of the loss from the person who caused the loss.

In any case, if the company has paid a loss totaling the amount of the policy, it reserves the right to take over the property from the insured at a fair valuation. There is a very good reason for this provision. In some instances the title is defective but can, with time and effort, be cured. The company in such cases pays the fair value to the insured, receiving a deed from him. It then owns the property and at its pleasure can take such action as may be necessary to remove the defects in the title.

Encroachments on Others. The building on the lot may cover more land than is within the lot lines. The municipality either owns the street or has an easement to use the street for public purposes. In either event no one has a right to encroach upon it, except by legal permission. Such an encroachment by a permanent structure may render the title unmarketable. Likewise, if the building encroaches upon a neighbor's land without his consent, the neighbor may be able either to recover damages for the encroachment or to compel the removal of so much of the building as encroaches on his land. A purchaser could not be compelled to accept a title under such conditions. The survey should indicate party walls, and it should also be examined with reference to any restriction upon the property and as to whether or not such restrictions are violated by the building. The effect of such conditions could be determined only by one familiar with the law applicable in each case.

Encroachments by Others. The entire building may stand upon the proper lot, and a nieghbor's building may encroach. The title to the portion of the lot that is not encroached upon is marketable, but it may be doubtful whether or not the encroachment affects the marketability of the lot as a whole. This is a question not so much of law as of commercial utility, and the courts consider it to be the latter. If the court finds the encroachment does not substantially lessen the value and extent of the property, it will compel a purchaser to accept it. If the encroachment does substantially lessen the value and extent of the land, the purchaser may refuse to take it, or he may take it and

be given an allowance for the land lost by the encroachment.

Use of Title Policy. Of course, the insured seldom needs to resort to his policy to recover a loss, but he should always refer to it in subsequent transactions with reference to the property. It tells him at once just what property he owns and what are the encumbrances on it. If he later enters into a contract to sell the property, he should use the description in the policy and undertake to give a title subject to just those encumbrances stated as exceptions in the policy.

Land Title and Registration. Throughout the history of title recording, means and ways have been sought to make transfer of real property as simple and safe as the transfer of other property, doing away, if possible, with the repeated search and examination of titles. The search for some permanent form of title registration is still prompted by the tedious and often difficult and cumbersome method by which title changes and, in addition, by the ever-present fear that ownership in farm or home, despite title insurance, may be invalidated by court order because of faulty or illegal property transfer somewhere in the chain of title grantors. Although the development and perfection of title insurance has speeded the search and transfer of title, particularly in urban areas, the necessity to reestablish and recheck the chain of ownership and title encumbrances every time a sale takes place still impedes the use of realty as a readily transferable, liquid asset or form of investment.

Under present operation of the recording acts, deeds and other instruments affecting rights or property in realty are placed on public record. These records are generally maintained in the office of the registrar by the clerk of the county in which the property is located. They are notice to the world as to constructive ownership in realty, and an examination of all recorded instruments affecting the property in question is necessary to determine the validity and condition of title. In early days, far fewer instruments were on record; it was therefore possible to examine a title with fair speed and consequent reasonable time and expense. Now, the more active counties have such voluminous records that the operation is slow, time consuming, and expensive, and, in addition, this condition is becoming worse as time passes and more and more instruments are recorded. Yet, theoretically, it is necessary for each title examination to go back to the date of conquest by war or revolution, or to the date of original gift or sale by a sovereign ruler; some states—Illinois, Indiana, Iowa, Massachusetts, Michigan, Wisconsin, and others—have established by statute a maximum period of time varying from thirty to fifty years beyond which an otherwise valid title need not be searched to prove ownership good and marketable. These laws do not estop an individual, however, who claims an interest in the property arising from events antidating the statute of title limitation from filing an affidavit of claim and from recording such claim for all to know and heed. Obvious interests that are evident by physical inspection, such as easements or party walls, need not be recorded to be protected in these states against title search statutes of limitations. In most states, however, there is still no legal date that can be fixed beyond which there is no need of going back to search a title. The lack of such a basic date, in brief, is the problem faced by title searchers.

In actual practice, title and abstract companies usually "assume title good at some fairly early date." There is nevertheless an irritating duplication of work on successive title examinations. And some meticulous attorneys are not satisfied unless the examination is carried back to the earliest date, as witness the following story:

In a legal transaction involving transfer of property in New Orleans, a firm of New York lawyers retained a New Orleans attorney to search the title and perform other related duties. The New Orleans attorney sent his findings, back to the year 1803. The New York lawyers examined his opinion and wrote again to the New Orleans lawyer, saying in effect that the opinion rendered by him was all very well, as far as it went, but that title to the property prior to 1803 had not been satisfactorily answered.

The New Orleans attorney replied to the New York firm as follows:

I acknowledge your letter inquiring as to the state of the title of the Canal Street property prior to the year 1803.

Please be advised that in the year 1803 the United States of America acquired the territory of Louisiana from the Republic of France by purchase. The Republic of France acquired title from the Spanish Crown by conquest. The Spanish Crown had originally acquired title by virtue of the discoveries of one Christopher Columbus, sailor, who had been duly authorized to embark upon the voyage of discovery by Isabella, Queen of Spain. Isabella, before granting such authority, had obtained the sanction of His Holiness, the Pope; the Pope is the Vicar on Earth of Jesus Christ; Jesus Christ is the Son and Heir Apparent of God. God made Louisiana.

A number of suggestions have been made to remedy this date situation. All of them hinge on some method of registering title and stem from the system first suggested by Sir Robert Torrens of Australia.

Origin of the Torrens System. Sir Robert Torrens was a businessman who had been a collector of cus-

toms in charge of shipping. In this position he became familiar with a law under which ships were registered. The registry showed the name or names of the owners of every vessel and all liens and encumbrances against it, and it revealed briefly and simply the condition of the title. Later Torrens became registrar-general of South Australia. His experience with shipping led him to believe that the principle of registration of titles could be applied to land as well.

19.6 FACTS TO ASCERTAIN BEFORE DRAWING CONTRACT OF SALE

1. Date of contract.
2. Name and address of seller.
3. Is seller a citizen, of full age, and competent?
4. Name of seller's wife and whether she is of full age.
5. Name and residence of purchaser.
6. Description of the property.
7. The purchase price:
   a. Amount to be paid on signing contract;
   b. Amount to be paid on delivery of deed;
   c. Existing mortgage or mortgages and details thereof;
   d. Purchase money mortgage, if any, and details thereof.
8. What kind of deed is to be delivered: full covenant, quit claim, or bargain and sale?
9. What agreement has been made with reference to any specific personal property, i.e., gas ranges, heaters, machinery, partitions, fixtures, coal, wood, window shades, screens, carpets, rugs, and hangings?
10. Is purchaser to assume the mortgage or take the property subject to it?
11. Are any exceptions or reservations to be inserted?
12. Are any special clauses to be inserted?
13. Stipulations and agreements with reference to tenancies and rights of persons in possession, including compliance with any governmental regulations in force.
14. Stipulations and agreements, if any, to be inserted with reference to the state of facts a survey would show: i.e., party walls, encroachments, easements, and so forth.
15. What items are to be adjusted on the closing of title?

16. Name of the broker who brought about the sale, his address, the amount of his commission and who is to pay it, and whether or not a clause covering the foregoing facts is to be inserted in the contract.

17. Are any alterations or changes being made, or have they been made, in street lines, name, or grade?

18. Are condemnations or assessment proceedings contemplated or pending, or has an award been made?

19. Who is to draw the purchase money mortgage and who is to pay the expense thereof?

20. Are there any covenants, restrictions, and consents affecting the title?

21. What stipulation or agreement is to be made with reference to Tenement Building Department and other violations?

22. The place and date on which the title is to be closed.

23. Is time to be of the essence in the contract?

24. Are any alterations to be made in the premises between the date of the contract and the date of the closing?

25. Amount of fire and hazard insurance, payment of premium, and rights and obligations of parties in case of fire or damage to premises from other causes during the contract period.

Upon the Closing of Title, the Seller Should Be Prepared with the Following

1. Seller's copy of the contract.

2. The latest tax, water, and assessment receipted bills.

3. Latest possible meter reading of water, gas, or electric utilities.

4. Receipts for last payment of interest on mortgages.

5. Originals and certificates of all fire, liability, and other insurance policies.

6. Estoppel certificates from the holder of any mortgage that has been reduced, showing the amount due and the date to which interest is paid.

7. Any subordination agreements that may be called for in the contract.

8. Satisfaction pieces of mechanics' liens, chattel mortgages, judgments, or mortgages that are to be paid at, or prior to the closing.

9. List of names of tenants, amounts of rents paid and unpaid, dates when rents are due, and assignment of unpaid rents.

10. Assignment of leases.

11. Letters to tenants to pay all subsequent rent to the purchaser.

12. Affidavit of title.

13. Authority to execute deed if the seller is acting through an agent.

14. Bill of Sale of personal property covered by the contract.

15. Seller's last deed.

16. Any unrecorded instruments that affect the title, including extension agreements.

17. Deed and other instruments that the seller is to deliver or prepare.

19.7 ADVERTISING AND SELLING THE FINAL PRODUCT

Part of the cost of a speculatively built home, warehouse, office block, or other structure is for advertising. Some of the methods of bringing and keeping the builder's name and product before the public, and by which most people find the home they purchase are as follows:

(a) Most people locate the home they want by driving around in the family car and by hearing of a reliable builder.

(b) Show homes (especially if furnished) attract many visitors who are potential buyers.

(c) Newspapers are a good source of advertising, especially if portraying a finished home ready for occupation.

(d) Friends often recommend a builder, a district, and type of home.

(e) People look for accessibility to schools, shopping, public utilities, transportation, churches, and an acceptable district.

(f) Company trucks, cars, and mobile field offices should prominently display the builder's name.

(g) Billboards on the building site, listing the names of the builder and subcontractors, attract attention.

(h) Radio and TV keep the public informed.

(i) Real estate companies are experienced in introducing prospective buyers. They can explain the different means of financing; draw up necessary papers and notarize them.

It is important that the builder realize all of his investment of cash, wages, and a minimum of 10% profit, in the sale of a house. Financiers advance money on first mortgages up to about 60% of the assessed value of the property; this leaves a balance of much more than many people can afford as a down payment. There is a temptation for some builders to hold a second mortgage in the sum of say 20 to 25%, making it easier for the prospective purchaser to buy the property.

A builder must be very circumspect before leaving part or all of his salary and profit in the property by way of a second mortgage. It must be remembered that he will not only have to pay income tax on the salary and profit that he leaves in the house, but he will become more pressed for working capital for his future operations. *The builder must remain in the construction business; not enter the financial business.*

Let us assume that a house is listed to sell for $70,000.00 which includes the real estate agents' fee of 5%. There is to be a first mortgage of 60% of the assessed value and the builder agrees to hold a second mortgage in the sum of 20% of the assessed value. The position would be as follows:

Listed price of the property		$70,000.00
Real Estate fee of 5%	$ 3,500.00	
First mortgage of 60%	42,000.00	
Second mortgage of 20% held by the builder	14,000.00	
Down payment of purchaser	10,500.00	
	$70,000.00	$70,000.00

It can be seen from the above record that a builder would only have to repeat this type of transaction five times, to have more than $70,000.00 of his own funds tied up in real estate mortgages. Mortgages can be discounted, but the percentage is consider-able, especially on second mortgages. It is common for second mortgages to be discounted from 20 to 40%. Assume that a second mortgage for $14,000.00 (as shown) was discounted at a rate of 30%; this would yield to the builder $9,800.00, the discount being $4,200.00 on $14,000.00.

Speculatively built homes must be of the right type, erected in a district compatible with price value, and aimed at the class of people that can afford the amount of down payment required to close each deal.

To maintain a good public image, the speculative builder must make good any legitimate complaints by purchasers for a short time after occupation.

CHAPTER 19 REVIEW QUESTIONS

1. List three methods of producing non-skid surfaces to concrete walks and driveways.

2. What is the minimum allowable width for main walkways to private property in your area for serving a maximum of sixteen dwelling units?

3. What is the minimum number of steps allowable (in one place) in concrete walkways?

4. List six types of materials used in building residential driveways. State your preference with the reasons.

5. List seven methods of advertising (for sale) speculatively built residences.

6. Assuming that a private party wanted to purchase directly from you your first speculatively built home, what steps would you take to close the deal? *This is an important question!*

7. Define the duties of a title (deed to property) examiner in your area.

8. What is the reason for title insurance in some states?

9. Briefly explain the implications of land title registration.

10. What is the address of the land title registration office in your area?

Index